Confluence of Cosmology, Massive Neutrinos, Elementary Particles, and Gravitation

Confluence of Cosmology, Massive Neutrinos, Elementary Particles, and Gravitation

Edited by

Behram N. Kursunoglu
Global Foundation, Inc.
Coral Gables, Florida

Stephan L. Mintz
Florida International University
Miami, Florida

and

Arnold Perlmutter
University of Miami
Coral Gables, Florida

Springer Science+Business Media, LLC

Library of Congress Cataloging-in-Publication Data

Confluence of cosmology, massive neutrinos, elementary particles, and gravitation/edited by Behram N. Kursunoglu, Stephan L. Mintz, and Arnold Perlmutter.
p. cm.
"Proceedings of an international conference on confluence of cosmology, massive neutrinos, elementary particles, and gravitation, held December 17–21, 1998, in Fort Lauderdale, Florida"—T.p. verso.
Includes bibliographical references and index.

1. Neutrinos Congresses. 2. String models Congresses. 3. Cosmology Congresses. I. Kurşunoğlu, Behram, 1922– II. Mintz, Stephan L. III. Perlmutter, Arnold, 1928–
QC793.5.N42C66 1999
539.7'215—dc21 99-39538
CIP

Proceedings of an International Conference on Confluence of Cosmology, Massive Neutrinos, Elementary Particles, and Gravitation, held December 17–21, 1998, in Fort Lauderdale, Florida

ISBN 978-1-4757-8637-8 ISBN 978-0-306-47094-3 (eBook)
DOI 10.1007/978-0-306-47094-3

Originally published by Kluwer Academic / Plenum Publishers, New York in 1999.
Softcover reprint of the hardcover 1st edition 1999
10 9 8 7 6 5 4 3 2 1

A C.I.P. record for this book is available from the Library of Congress.

PREFACE

Just before the preliminary program of Orbis Scientiae 1998 went to press the news in physics was suddenly dominated by the discovery that neutrinos are, after all, massive particles. This was predicted by some physicists including Dr. Behram Kusunoglu, who had a paper published on this subject in 1976 in the *Physical Review.* Massive neutrinos do not necessarily simplify the physics of elementary particles but they do give elementary particle physics a new direction.

If the dark matter content of the universe turns out to consist of neutrinos, the fact that they are massive should make an impact on cosmology. Some of the papers in this volume have attempted to provide answers to these questions. We have a long way to go before we find the real reasons for nature's creation of neutrinos. Another neutrino-related event was the passing of their discoverer, Fredrick Reines:

> *The trustees of the Global Foundation, members of the Orbis Scientiae 1998, dedicate this conference to Fredrick Reines of the University of California at Irvine. The late Professor Reines was a loyal and active member of these series of conferences on the frontiers of physics and cosmology since 1964. He also served as one of the trustees of the Global Foundation for the past three years. Professor Reines discovered the most elusive particle, the neutrino, in 1954. We are proud to say that we recognized the importance of this discovery by awarding him the J. Robert Oppenheimer memorial Prize 15 years before the Nobel Foundation's recognition of him in 1995. We shall all miss Fred. We extend our condolences to all the members of his family: his wife Sylvia Reines, his son and daughter, and four grandchildren.*

This conference was supported in part by a grant from the National Science Foundation. The Trustees and the Chairman of the Global Foundation wish to extend a special thanks to Edward Bacinich of Alpha Omega Research Foundation for his generous support of this conference.

Behram N. Kursunoglu
Stephan L. Mintz
Arnold Perlmutter

ABOUT THE GLOBAL FOUNDATION, INC.

The Global Foundation, Inc. was established in 1977 and utilizes the world's most important resource . . . people. The Foundation consists of senior men and women in science and learning; outstanding achievers and entrepreneurs from industry, governments, and international organizations; and promising, enthusiastic young people. These people form a unique and distinguished interdisciplinary entity, and the Foundation is dedicated to assembling all the resources necessary for them to work together. The distinguished senior members of the Foundation convey their expertise and accumulated experience, knowledge, and wisdom to the younger membership on important global issues and frontier problems in science.

Our work is a common effort, employing the ideas of creative thinkers with a wide range of experiences and viewpoints.

GLOBAL FOUNDATION BOARD OF TRUSTEES

GLOBAL FOUNDATION'S RECENT CONFERENCE PROCEEDINGS

Making the Market Right for the Efficient Use of Energy
Edited by: Behram N. Kursunoglu
Nova Science Publishers, Inc., New York, 1992

Unified Symmetry in the Small and in the Large
Edited by: Behram N. Kursunoglu and Arnold Perlmutter
Nova Science Publishers, Inc., New York, 1993

Unified Symmetry in the Small and in the Large - 1
Edited by: Behram N. Kursunoglu, Stephan Mintz and Arnold Perlmutter
Plenum Press, 1994.

Unified Symmetry in the Small and in the Large - 2
Edited by: Behram N. Kursunoglu, Stephan Mintz and Arnold Perlmutter
Plenum Press, 1995.

Global Energy Demand in Transition: The New Role of Electricity
Edited by: Behram N. Kursunoglu, Stephan Mintz and Arnold Perlmutter
Plenum Press, 1996.

Economics and Politics of Energy
Edited by: Behram N. Kursunoglu, Stephan Mintz and Arnold Perlmutter
Plenum Press, 1996.

Neutrino Mass, Dark Matter, Gravitational Waves, Condensation of Atoms and Monopoles, Light Cone Quantization
Edited by: Behram N. Kursunoglu, Stephan Mintz and Arnold Perlmutter
Plenum Press, 1996.

Technology for the Global Economic, Environmental Survival and Prosperity
Edited by: Behram N. Kursunoglu, Stephan Mintz and Arnold Perlmutter
Plenum Press, 1997.

25th Coral Gables Conference on High Energy Physics and Cosmology
Edited by: Behram N. Kursunoglu, Stephan Mintz and Arnold Perlmutter
Plenum Press, 1997.

Environment and Nuclear Energy
Edited by: Behram N. Kursunoglu, Stephan Mintz and Arnold Perlmutter
Plenum Press, 1998.

Preparing the Ground for Renewal of Nuclear Power
Edited by: Behram N. Kursunoglu, Stephan Mintz and Arnold Perlmutter
Plenum Press, 1999.

Confluence of Cosmology, Massive Neutrinos, Elementary Particles, and Gravitation
Edited by: Behram N. Kursunoglu, Stephan Mintz and Arnold Perlmutter
Plenum Press, 1999.

CONTRIBUTING CO-SPONSORS OF THE GLOBAL FOUNDATION CONFERENCES

National Science Foundation
Alpha Omega Research Foundation, Palm Beach, Florida

Glߋbal Foundation, Inc.

A Nonprofit Organization for Global Issues Requiring Global Solutions, and forProblems on the Frontiers of Science

Center for Theoretical Studies

P.O. Box 249055
Coral Gables, Florida 33124-9055
Phone: 305-669-9411
Fax: 305-669-9464
e-mail: kursungf@gate.net
Website: http://www.gate.net/~kursungf

INTERNATIONAL CONFERENCE
On
ORBIS SCIENTIAE 1998

CONFLUENCE OF COSMOLOGY, MASSIVE NEUTRINOS, ELEMENTARY PARTICLES, & GRAVITATION

(27th Conference on High Energy Physics and Cosmology Since 1964)

December 17 - 21, 1998
Lago Mar Resort
Fort Lauderdale, Florida

Dedication:

The trustees of the Global Foundation, members of the 27th Orbis Scientiae 1998, dedicate this conference to Fredrick Reines of the University of California at Irvine. The late Professor Fredrick Reines has been a loyal and active member of these series of conferences on the frontiers of physics since 1964. He has also served as one of the trustees of the Global Foundation for the past three years. Professor Reines discovered the most elusive particle the neutrino, in 1954. We are proud to say that we recognized the importance of this discovery very early by awarding him the J. Robert Oppenheimer Memorial prize 15 years before the Nobel Foundations' recognition of him in 1995. We shall all miss Fred. We extend our condolences to all the members of his family: his wife, Sylvia Reines, his son and daughter, and his four grandchildren.

Sponsored by:
GLOBAL FOUNDATION, INC.
P. O. Box 249055
Coral Gables, Florida 33124-9055
Phone: (305) 669-9411
Fax: (305) 669-9464
E-mail: kursungf@gate.net
Website: http://www.gate.net/~kursungf

Conference Hotel:
Lago Mar Resort
1700 South Ocean Lane
Fort Lauderdale, Florida 33316
Reservations: 1-800-255-5246
(954) 523-6511
Fax: (954) 524-6627
T he special group rate: $115/night
One Bedroom Suite

ORBIS SCIENTIAE 1998
PROGRAM

FRIDAY, December 18, 1998 LAKEVIEW ROOM

8:00 AM - Noon REGISTRATION

1:30 PM **SESSION I: MASSIVE NEUTRINOS AND SOLAR NEUTRINO PROBLEM**

Moderators: **HYWEL WHITE,** Los Alamos National Laboratory, New Mexico

Dissertators: **JOHN BAHCALL,** Institute for Advanced Study, Princeton
"Solar Model"
ALFRED K. MANN, University of Pennsylvania
"Evidence for Neutrino Mass and Mixing",
JORDAN GOODMAN, University of Maryland
"Search For Neutrino Mass with Super-KAMIOKAMDE"
KEVIN LESKO, Lawrence Berkeley National Laboratory
"Sudbury Neutrino Observatory"

Annotators: **ALAN KOSTELECKY,** Indiana University

Session Organizer: **HYWEL WHITE**

3:00 PM **COFFEE BREAK**

3:15 PM **SESSION I: CONTINUES**

3:45 PM **SESSION II : GRAVITATION**

Moderators: **BEHRAM N.KURSUNOGLU,** Global Foundation Inc.
SYDNEY MESHKOV, California Institute of Technology

Dissertators: **BEHRAM N.KURSUNOGLU**
"Gravitating Massive Neutrino Multiplicity"
EDWARD TELLER, University of California, Lawrence Livermore National Laboratory
"The Mass of the Neutrino" presented by Behram N. Kursunoglu
ROBIN STEBBINS, University of Colorado
"LISA--Space Based Interferometer"
DAVID REITZE, University of Florida, Gainesville
"Current Status of LIGO"

Annotators: **SYDNEY MESHKOV**

Session Organizer: **SYDNEY MESHKOV**

5:45 PM Orbis Scientiae adjourns for the day

6:00-7:00 PM - ***Welcoming Cocktails - Courtesy of Lago Mar Resort ,FOUNTAINVIEW LOBBY***

SATURDAY, December 19, 1998

8:30 AM **SESSION III: MASSIVE NEUTRINOS AND IMPACT ON COSMOLOGY**

Moderators: **SYDNEY MESHKOV**

Dissertators: **NIKOLAOS IRGES**
"Quark Hierarchies and Neutrino Mixing"
RICHARD WOODARD, University of Florida, Gainesville

"Quantum Gravitational Inflation"
KERRY WHISNANT, Iowa State University
"Three-neutrino vacuum oscillation solutions to the Solar and Atmospheric Anomalies"

Annotators: **STEPHAN MINTZ**, Florida International University

Session Organizer: **PIERRE RAMOND**

10:00AM **COFFEE BREAK**

10:15 AM **SESSION IV:** **RECENT PROGRESS ON OLD AND NEW IDEAS I**

Moderators: **PAUL FRAMPTON,** University of North Carolina at Chapel Hill

FREDRIK ZACHARIASEN, CALTECH

Dissertators: **SHELDON L.GLASHOW**, Harvard University
" More about neutrinos"
PAUL H. FRAMPTON,
"Orbifold Field Theory"
ROMAN JACKIW, Massachusetts Institute of Technology
"Lorentz Invariance Violation in Electromagnetism"
ALAN KOSTELECKY, Indiana University
"Tests of Lorentz and CPT Symmetries"

Annotators: **PASQUALE SODANO**, University of Perugia, Italy

Session Organizer: **ALAN KOSTELECKY**

12:30 PM Orbis Scientiae Adjourns for the Day

7:30-10:30 PM - ***Conference Banquet***, PALM GARDEN ROOM
Courtesy of Maria and Edward Bacinich

SUNDAY, December 20, 1998

8:30 AM **SESSION V:** **PROTON SPIN CONTENT**

Moderator: **WOLFGANG LORENZON**, University of Michigan

Dissertators: **XIANGDONG JI** , University of Maryland
"How Does QCD Build The Proton Spin"
BRAD FILIPPONE, California Institute of Technology
"Spin Structure Measurements With Lepton Beams"
JOEL MOSS, Los Alamos National Laboratory
"RHIC Spin Physics"

Annotators: **RICHARD ARNOWIT**

Session Organizer: **WOLFGANG LORENZON**

10:30 AM **COFFEE BREAK**

10:45 AM **SESSION VI:** **STRINGS**

Moderator: **LOUISE DOLAN**, University of North Carolina at Chapel Hill

Dissertators: **CHIARA NAPPI**, Institute for Advanced Study, Princeton, N.J
"Quantized Membranes"
FREYDOON MANSOURI, University of Cincinnati
"Super Non-Abelian Stokes Theorem and Superstrings"

LOUISE DOLAN
"Gauged Supergravities and Superstring Theory"
IGOR KLEBANOV, Princeton University
"From Threebranes to Large N Gauge Theories"

Annotators: **PRAN NATH**

Session Organizer: **LOUISE DOLAN**

12:45 PM **LUNCH BREAK**

1:30 PM **SESSION VII:** **SPIN AND STATISTICS**

Moderators: **O.W. GREENBERG**, University of Maryland

Dissertators: **MICHAEL V. BERRY**, University of Bristol, UK
"Quantum indistinguishability"
JON MAGNE LEINAAS, University of Oslo
"Spin and Statistics in 2 Dimensions"
O.W.GREENBERG
"Bounds on violations of statistics"

Annotators: **PAUL FRAMPTON**

Session Organizer: **O.W.GREENBERG**

3:00 PM **COFFEE BREAK**

3:15 PM **SESSION VIII:** **MASSIVE NEUTRINOS' IMPLICATIONS FOR ELEMENTARY PARTICLE PHYSICS**

Moderators: **STEPHAN MINTZ ,** Florida International University

Dissertators: **INA SARCEVIC**, University of Arizona
"Can Oscillations be Detected with A Neutrino Telescope ?"

STEPHAN MINTZ,
"High Energy Neutrino Reactions in Nuclei and Neutrino Backgrounds"
RICHARD ARNOWITT, Texas A&M University
"Neutrino Masses and Grand Unification"
TONY GABRIEL, ORNL, Knoxville, TN
"Neutrino Physics at the Spallation Neutron Source"

Annotators: **JORDAN GOODMAN**

Session Organizer: **STEPHAN MINTZ**

5:30 PM Conference Adjourns for the Day

MONDAY, December 21, 1998

8:30 AM **SESSION IX:** **RECENT PROGRESS ON OLD AND NEW IDEAS II**

Moderator: **DON LICHTENBERG**, Indiana University

MONDAY, December 21, 1998

Dissertators: **DON LICHTENBERG**
"Spin-Dependent Forces Between Quarks in Hadrons"
PRAN NATH, Northeastern University

"Super K Data and SUSY Grand Unification"
MARTIN SAVAGE, University of Washington, Seattle
"Perturbative Nuclear Physics"

Annotators: **TONY GABRIEL**

Session Organizer: **DON LICHTENBERG**

10:15 AM **COFFEE BREAK**

10:30 AM **SESSION X:** **RECENT PROGRESS ON OLD AND NEW IDEAS III**

Moderator: **THOMAS CURTRIGHT**, University of Miami

Dissertators **THOMAS CURTRIGHT,**
" Duality and Wigner Functionals"

DON COLLADAY, College of Wooster, Ohio

"Breaking Lorentz Symmetry in Quantum Field Theory"

KERRY WHISNANT

"Are There Four or More Neutrinos?"

Annotators: **BEHRAM N. KURSUNOGLU**

Session Organizer: **ALAN KOSTELECKY**

12:30 Noon **ORBIS SCIENTIAE 1998 ADJOURNS**

CONTENTS

SECTION I
Neutrino Physics

SECTION II
Recent Progress on New and Old Ideas

SECTION III
Spin and Statistics

SECTION IV
Strings

Section I
Neutrino Physics

ON THE MASS OF THE NEUTRINO

Edward Teller
Senior Research Fellow

Hoover Institution
Stanford University
Stanford, California

The original proposal of a neutrino and its role in the beta decay by Pauli, Fermi and others assumed that the energy distribution between the electron and the neutrino was simply proportional to the volume in phase space for the electron and the neutrino. Thus, the decay probability did not otherwise depend on the energy or spin of the electron and the neutrino.

A radical change in this beta decay theory was proposed by Lee and Yang. They proposed to compose the neutrino wave function from two states: one with the neutrino spin parallel and the other anti-parallel to the neutrino momentum. This, of course, is an entirely permissible description of the Pauli/Fermi proposal. The radical novelty consisted by adding that the spin and momentum vectors must be anti-parallel; the state with paralleled vectors does not exist.

This simplistic description of the theory of Lee and Yang is inconsistent with a statement that the neutrino has a non-zero mass.

Indeed, a neutrino with a finite mass, has a velocity less than the light velocity c. After an appropriate Lorentz transformation, the neutrino would appear to be at rest and performing the transformation with an even greater velocity, the neutrino would move in the opposite direction, i.e., its momentum would have changed its sign. At the same time, the angular momentum remains unchanged if the direction of the Lorentz transformation coincides with the direction of the spin. Thus, an assumption that spin and momentum are always anti-parallel and never parallel puts an unusual and I think, unacceptable limitation on the Lorentz transformation.

All of this does not lead to the obvious statement: "Lee and Yang had eliminated the possibility of a neutrino with a finite mass." The results of Lee and Yang can be made compatible with a finite mass, but only by introducing a considerable complication in their theory. One may assume, for instance, that contrary to Pauli and Fermi, the decay probability depends upon the scalar of product of the momentum and angular momentum of the neutrino. One might, for instance, assume that the Pauli-Fermi assumption holds for sufficiently large negative values of the scalar product; one may then further assume that the

Confluence of Cosmology, Massive Neutrinos, Elementary Particles, and Gravitation
Edited by Kursunoglu *et al.*, Kluwer Academic / Plenum Publishers, New York, 1999.

decay probability approaches zero when the scalar products approaches zero and that the decay probability is zero for positive values of the scalar product.

If this more involved suggestion is correct, then one might find reduced decay probabilities for low neutrino energies, i.e., for near maximum values of the energy of the electron.

The maximum energy of the electron is usually at least a few kilovolts. Experiments at Livermore[i] which have been done most carefully set an upper bound of seven electron volts for mc^2 of the neutrino. Therefore, even more precise measurements near the maximum energy of the electron would be needed to establish the mass of the neutrino.

Actually, the probability of emitting a near maximum energy electron within an energy interval ΔE close to the maximum energy is low on account of the low momentum space of the neutrino. A theory for a non-zero mass neutrino should give an even lower probability. The additional factor goes to zero as the maximum energy of the electron is approached. Contrarywise, in the Livermore experiment, there is an indication of an increase of probability.

Oscillations of states between electron neutrinos, muon neutrinos and tau neutrinos (*Physical Review Letters*, Volume 81, #8, pp. 1562) which can be related to a mass are compatible with the assumption that spin and momentum of the neutrinos are anti-parallel.

---

[i] *Anomalous Structure in the Beta Decay of Gaseous Molecular Tritium*, Wolfgang Stoeffl and Daniel J. Decman, *Physical Review Letters*, 30 October 1995, Vol. 75, No. 18, pp. 3237-3240.

THE TWO GRAVITATING MASSIVE NEUTRINO PAIRS

Behram N. Kursunoglu

Global Foundation Inc.
P. O. Box 249055
Coral Gables, Florida 33124-9055

INTRODUCTION

This paper is, in a belated sense and in view of the current experimental results on neutrino mass, a sequel to one that I published in 1976 in The Physical Review which will be referred to here as (I)[1]. The two important subjects discussed in (I) included: (i) a demonstration that particles with neutrino symmetries carry a mass and that there exist no massless neutrinos, (ii) the discovery of a new concept that was described as a *condensation of magnetic charges or monopoles*. Now, at this time, it happens that both of these subjects are of experimental and theoretical interest which provide me with an opportunity to write this paper and expand on the presentation of (I).

The reason for choosing the title of the paper related to gravitation comes from the likelihood that the universe contains a large amount of dark matter consisting mostly of massive neutrinos. It is an interesting approach to try to explain the mass of the neutrino in terms of its gravitational interaction only. There is a similarity with Einstein's general relativity in the presence of an electromagnetic field. Just as the electromagnetic field is the source of the gravitational field, massive neutrinos can also act as the source of a gravitational field, especially using the assumption that the distribution of matter in the universe contains at least 90% neutrino based dark matter. The force of gravity implies the multiplicity of neutrino masses. It is quite clear that just this much is not enough to describe a neutrino. We need to identify its spin and the nature of its interaction with other particles. Prediction of two massive neutrino pairs will place the τ-neutrino into the pair with properties differing from the pair containing the electron and muon neutrinos.

All these thoughts originate from the generalization of the concept of gravity to include the electromagnetic field. In 1975 I noted, while working on (I), some fundamental differences between my own 1951 version of the unified field theory and that of Einstein's 1949 non-symmetric theory as well as the alternative that was proposed by Erwin Schrödinger. These three different versions of the non-symmetric theory are discussed in the literature almost always together, the direct consequence of which is the unnoticed

---

[1] B. Kursunoglu, Phys.Rev.D6, Vol.13, Number 6, 1538 (1976)

Confluence of Cosmology, Massive Neutrinos, Elementary Particles, and Gravitation
Edited by Kursunoglu *et al.*, Kluwer Academic / Plenum Publishers, New York, 1999.

missing link in both Einstein's and Schrödinger's theories i.e., the absence of the fundamental length r_o. Information regarding these differences were not noted in Abraham Pais' book[2], page 348. Pais is essentially quoting from a paper[3] by Bruria Kaufman, Einstein's last assistant, which contains a summary of nonsymmetric theories at the Bern conference on the 50th Anniversary of Relativity in 1955. In my case I prefer to name the theory in more descriptive language as the *Generalized Theory of Gravitation : The Second Phase Of The General Theory Of Relativity.*

The aim of the paper (I) was to discover exact solutions to the field equations and compare all three versions of the non-symmetric theory. For this reason, my paper at that time had the most uninspiring title, "*Consequences of Non-Linearity in the Generalized Theory of Gravitation*" instead of the simple title of "*Masses of Neutrinos*" in which case a few physicists might have read that paper. The same, of course, applies to the fact that the same paper contained a brief presentation on the *condensation of monopoles.*

On this occasion I would like to point out the fundamental differences between my version of the non-symmetric theory and those of Einstein and Schrödinger. The discovery of a fundamental length, which appeared, *as a consequence of non-symmetric formulation,* for the first time in my December 1952 Physical Review paper[4], is a novel approach to the unification of gravitation and electromagnetism. It was a most stimulating and inspiring discovery to see that an invariant parameter plays a fundamental role in the evolution of the universe from its microcosmic to its macrocosmic states by assuming values of the order of Planck length and those values of the order of the size of the universe, respectively. The existence of this parameter that comes embedded with a structure of the equations makes it possible to conceive of the unification of elementary particle physics and cosmology. In fact without the fundamental length r_o we cannot define a mass. It has further been found that the laws of motion of an electric charge in a field cannot be deduced from the field equations without the presence of this fundamental parameter r_o. The fundamental length r_o is independent of the coordinates and all solutions of the field equations are functions of r_o. The physics of the theory is determined by the calculated values of r_o.

In the Big Bang creation of the universe its evolution is governed by starting with the small values of r_o which, as the expansion of the universe continues, is increasing proportionately. In fact the recent observations seem to indicate the expansion of the universe is taking place with increasing acceleration which eventually will reach a value of r_o of the order of a number greater than 10^{27}cm (observed or assumed size of the universe), namely all the solutions belong to a flat space-time from which we deduce that the universe is in fact *flat*. It is rather remarkable that the evolutionary behavior of the universe can be linked up with the fundamental parameter of the length r_o. Actually, for a more realistic confrontation of the theory with observation we need spherically symmetric and time dependent solutions of the field equations.

This same parameter, as a result of monopoles' condensation, plays the basic role in the creation of elementary particles with a composite structure consisting of the confined layers of magnetic charges with alternating signs and decreasing amounts. It, thus, results in the statement that there are no free monopoles; they are all confined to form elementary particles whose constituents consist of the magnetic charges g_n, ($n = 1,2,3...$), *quintessential matter*. Do the elementary particles really consist of this kind of a neutral structure, i.e., the sum of whose magnetic charge constituents is zero? This result poses a challenge both for the generally assumed quark structure versus this theory's prediction of

---

[2] Abraham Pais, *Subtle Is the Lord. . . The Science and the Life of Albert Einstein*, Oxford University Press, 1982.

[3] B. Kaufman, Helv. Phys. Acta Suppl.4, 227 (1956).

[4] B. Kursunoglu, Phys.Rev.88, 1369 (1952); Phys.RevD9, 2723 (1974).

infinite layers of magnetic charges as constituents of an elementary particle. Perhaps we could, at this point, say that elementary particles are still subject to an understanding and constitute the most important topic for 21st century physicists.

All of these considerations, be it for the innermost structure of matter or the outermost structure of the universe, are consequences of r_o's behavior as it appears in the field equations. It must be noted that the large values of r_o for the new cosmological parameter $\lambda = r_o^{-2}$, where now r_o differs from Einstein's concept of a cosmological constant in a fundamental way since in this case a single parameter is part of the non-symmetric structure and prevails over the entire evolution of the universe. For $\lambda = \infty$ one obtains the field equations of general relativity where the fields are decoupled from the electric and magnetic charges but they source the gravitational field itself. The case $\lambda = 0$ yields a flat space-time. We should, therefore, call the new λ *a running cosmological parameter*.

Why did Einstein not consider the necessity and highly visible existence of r_o? Was he mostly influenced by the disappointment with his own cosmological constant? Actually, Einstein answers this question, presumably based on our correspondence during 1950-1952 on the subject matter of constants in the field equations, in his famous book [5], page 146, "All such additional terms bring a heterogeneity into the system of equations, and can be disregarded, provided that no strong physical argument is found to support them." I must point out that without r_o we can't even obtain from the field equations the classical laws of motion of a charged particle in an electromagnetic field. I obtained additional results to refute entirely Einstein's claim quoted above[6]. The most striking fact can be found in the definition of mass - any mass: elementary particles, the sun, the Earth, a black hole, neutron star etc. obtained as

$$M=(c^2/2G)r_o=(1/2)(r_o/r_p)m_p \quad , \tag{1}$$

where c and G represent speed of light and gravitational constant, respectively, and where r_p and m_p represent Planck length and Planck mass, respectively. In the result (1) above, the parameter r_o is to be interpreted as the "gravitational size" of an object and is calculated from the solutions of the field equations. Thus, mass is measured in units of Planck mass which is the only mass scale in physics. The equation (1) yields the fundamental mass ratio,

$$(M / m_p) = (1/2)\,(r_o/r_p)\,, \tag{2}$$

where

$$r_p = \sqrt{(\hbar G/c^3)}, \quad m_p = \sqrt{(\hbar c/G)}\ . \tag{3}$$

From an approximation of the spherically symmetric field equations we obtain for r_o the result,

$$r_o^2 = (2G/c^4)\,(e^2 + g^2)\,N^2, \tag{4}$$

where e and g represent electric and magnetic charges, respectively, and where N^2 is, in some way, related to the ratio of the gravitational force and the sum of electromagnetic

[5] Albert Einstein, *The Meaning of Relativity*, Princeton University Press, 1953.
[6] Most physicists do not believe in the validity of Einstein's and Schrödinger's versions of the non-symmetric theories, and despite its success they have, so far, ignored my version of the theory, as well.

and strong forces and is of the order of 10^{-40}. It will be shown in the next section that for e=0 the field equations acquire neutrino symmetries violating the parity and charge conjugation. In this case the neutrino mass relation can be expressed in the form

$$M_\nu = (N_\nu g_\nu)/\sqrt{(2G)}, \qquad (5)$$

where the two free parameters N_ν and g_ν are to be determined. It must be observed that we need to know the spin angular momentum to complete the identification of the particle as a neutrino. From the equation (4) above we see that, because of the discrete distribution of the magnetic charge layers with decreasing thickness toward the edge with the alternating signs, the distribution of the electric charge will consist of layers of increasing thickness toward the edge with the same signs. In other words, if we were to portray the electric charge distribution in terms of discrete units e_n (n= 0,1,2,.....), all, of course, with the same signs, then the result would be more like an inverted form of the case for the magnetic charge. Thus, for the electric charge distribution the layers starting from a diverging configuration will converge and crowd in, around the origin, like the waves that result when a pebble is dropped into a tranquil lake.

Thus, most of the electric charge in an elementary particle (*orbiton*) "resides" on its "surface". This fact is revealed in Hofstadter's high energy electron and nuclei scattering experiments[7] to measure the nature of the electric charge distributions in protons and neutrons. Experiments by A. D. *Krisch*[8] *et al.*, pertaining to spin physics through the scattering of polarized beams from polarized targets, and very high energy scattering of electrons experiments by J.L. Friedman and H.W. Kendall are also relevant for the study of electric charge distributions. All of these distributions clearly imply that at high energy scattering Coulomb coupling increases (interaction near the surface) while strong coupling decreases (interaction away from the origin) and vice versa, when strong coupling increases (interaction near the origin) the Coulomb coupling decreases (small layers of electric charge). Hence we see that there is a *running coupling* parameter or constant like, for example, the ratios $e^2/(e^2 + g^2)$and $g^2/(e^2+ g^2)$.

BASIC SYMMETRIES OF THE FIELD EQUATIONS, OSCILLATIONS OF AN ORBITON

The special case of time-independent spherically symmetric field equations contain a wealth of information on the physics of the generalized theory of gravitation. In this paper I shall give a summary of the 1976 paper where massive neutrinos were predicted. The spherically symmetric form of the 16 component non-symmetric tensor $\hat{g}_{\mu\nu}$ can be expressed in terms of the four functions exp (ρ), exp (u), υ, and Φ which can be simplified into the form:

[7] Robert Hofstadter, *Annual Review of Nuclear Science*, Vol. 7, page 231, 1957, and the article in Vol. 22, page 203, 1972 by J. L. Friedman and H. W. Kendall. *Annual Reviews, Inc.*

[8] A. D. Krisch *et al.*, Phys.Rev.Letters 63, 1137 (1989).

$$[\hat{g}^s{}_{\mu\nu}] = \begin{bmatrix} -\frac{1}{\upsilon^2}\exp(-u) & 0 & 0 & \frac{1}{\upsilon}\tanh\Gamma \\ 0 & -\exp(\rho)\sin\Phi & \exp(\rho)\cos\Phi\sin\theta & 0 \\ 0 & -\exp(\rho)\cos\Phi\sin\theta & -\exp(\rho)\sin\Phi\sin^2\theta & 0 \\ -\frac{1}{\upsilon}\tanh\Gamma & 0 & 0 & \exp(u) \end{bmatrix}$$

The field equations can now be written in terms of the four functions $\exp(\rho)$, $\exp(u)$, υ, and Φ as

$$\frac{1}{2}\, r_o^2 f[fS\exp(\rho)\Phi']' = R^2\cos\Phi + (-1)^S \ell_o^2 \sin\Phi, \tag{6}$$

$$\frac{1}{2}\, r_o^2 f[fS\exp(\rho)\rho']' = -R^2\sin\Phi + (-1)^S \ell_o^2 \cos\Phi + \exp(\rho), \tag{7}$$

$$\frac{1}{2}\, r_o^2 f[f\exp(\rho)S']' = \left(1 - \frac{\sin\Phi}{\cosh\Gamma}\right)\exp(\rho), \tag{8}$$

$$\rho'' + \rho'\frac{f'}{f} + \frac{1}{2}(\rho'^2 + \Phi'^2) = 0, \tag{9}$$

where a prime indicates differentiation with respect to the coordinate r and where the functions S and R are defined by

$$f = \upsilon\cosh\Gamma, \qquad S = \frac{\exp(u)}{\cosh^2\Gamma}, \tag{10}$$

$$\cosh\Gamma = (R^2 + r_o^2)\exp(-\rho), \qquad [(\exp(2\rho) + \lambda_o^4)]^{1/2} = R_o^2 + r_o^2, \tag{11}$$

$$\ell_o^2 = q^{-1}|g|, \quad \lambda_o^2 = q^{-1}|e|, \tag{12}$$

with the constants of integration g and e representing magnetic and electric charges, respectively. The fundamental length r_o in (11) was calculated for Φ = constant, in terms of the constants[9] of integration λ_o and l_o as in the following:

$$r_o^2 = (\ell_o^4 + \lambda_o^4)^{1/2}, \text{ where } \ell_o^2 = N^2|g|(e^2 + g^2)^{1/2}, \tag{13}$$

$$\lambda_o^2 = \frac{2G_o}{c^4} N^2|e|(e^2 + g^2)^{1/2}, \tag{14}$$

[9] B. Kursunoglu, Phys. Rev. D 9, 2723 (1974); 12. 1850(E) (1975).

and where G_o is the gravitational constant, e and c represent the unit of electric charge and speed of light, respectively. Thus the existence of a correspondence principle (i.e., the $r_o = 0$ limit yields general relativity plus the electromagnetic fields) provides a powerful basis for the unique and unambiguous physical interpretation of the theory. Various solutions of these equations have been discussed in (I) . We shall now introduce a new variable β by

$$dr = f d\beta. \tag{15}$$

In terms of the new variable β, the spherically symmetric field equations can be written as

$$\frac{1}{2} r_o^2 \frac{d}{d\beta} (S \exp(\rho_{n\tau})\ \dot{\Phi}_{ns\tau}) = R^2 \cos\Phi_{ns\tau} + l_o^2 \sin\Phi_{ns\tau}\,, \tag{16}$$

$$\frac{1}{2} r_o^2 \frac{d}{d\beta} (S \exp(\rho_{n\tau})\ \dot{\rho}_{ns\tau}) = -R^2 \sin\Phi_{ns\tau} + l_o^2 \cos\Phi_{ns\tau} + \exp(\rho_{n\tau})\ , \tag{17}$$

$$\frac{1}{2} r_o^2 \frac{d}{d\beta} (\dot{S} \exp(\rho_{n\tau})) = \exp(\rho_{n\tau})\ \left(1 - \frac{\exp(\rho_{n\tau}) \sin\Phi_{ns\tau}}{R^2 + r_o^2} \right), \tag{18}$$

$$2\ddot{\rho}_{n\tau} + \dot{\rho}_{n\tau}^2 + \dot{\Phi}_{ns\tau}^2 = 0 \tag{19}$$

where

$$\rho = \frac{d\rho}{d\beta}\ ,$$

and where now S, ρ, and Φ can be regarded as functions of the new variable β and where the equation (19) is derivable from the three equations (16)- (18) which can be written as

$$\left(\frac{d^2}{d\beta^2} + \omega^2 \right) \exp(\tfrac{1}{2}\rho) = 0\ , \quad \omega = \tfrac{1}{2}\dot{\Phi}\ , \tag{20}$$

implies oscillations , when left free, of the orbiton magnetic charge layers. At distances large compared to r_o the function $\exp(\frac{1}{2}\rho) \rightarrow r$. Hence we see that the structure of an orbiton, in view of the variable frequency ω, oscillates like a pendulum whose length is changing. The frequency of the oscillations as a function of the electric and magnetic charges may be related to the mass of an orbiton. In view of the invariance of the field equations (6)-(9) under the transformation $f \rightarrow -f$, the definition (15) implies that the range of the new variable β extends from $-\infty$ to ∞. It is interesting to note that the equations resulting from (6)-(9) by taking $f = \pm 1$ are, formally, the same (except being functions of r whose range extends from 0 to ∞) as the equations (16)-(19).

The new forms (16)-(19) of the field equations involve only three unknown functions S, ρ, Φ satisfying four field equations. This kind of over-determination of the field variables S, ρ, Φ might imply a spurious character for the eliminated field variable f. Actually this interesting property of the field equations is, as will be demonstrated, a virtue since it established relations between the constants r_o, ℓ_o, and λ_o. The field equations are compatible only through definite relationships between these constants. The oscillations described by the equation (20) is a *new result* and needs to be related to experiment on elementary particles. Does, for example, a proton oscillate? What are the consequences of the proton oscillations induced, perhaps, by the short-range forces of the magnetic charge layers. In order to fully understand this phenomenon we need the exact solutions of the field equations (16)-(19).

Now, the field equations (6)-(9) are independent of the sign of the electric charge and therefore the solutions refer to both positive and negative electric charge irrespective of the positive and the negative energy states. There are thus two signs for electric charges for a particle as well as for an antiparticle. Furthermore, the energy density has a linear dependence on f and therefore it changes sign under the transformation

$$f \rightarrow -f, \tag{21}$$

under which the field equations are unchanged. Hence the field equations have both positive and negative energy solutions for particles with positive as well as with negative signs of electric charge. The role of the symmetry (21) for the field equations (6)-(9) is taken over in the new form (16)-(19) of the field equations by the invariance under $\beta \rightarrow -\beta$.

The spherically symmetric fields and the corresponding electric and magnetic charge densities are given by

$$\mathcal{E}_e = \frac{q}{\upsilon} \tanh\Gamma = \frac{q \sinh\Gamma}{f} = \frac{\pm e}{\upsilon(R^2 + r_o^2)} \tag{22}$$

for the charged electric field,

$$\mathcal{E}_0 = q\, r_o^2\, f \sinh\Gamma \rho' S' = q\, r_o^2 \frac{1}{f} \sinh\Gamma \dot{\rho}\, S = r_o^2\, \mathcal{E}_e\, \rho\, \dot{S} \tag{23}$$

for the neutral electric field [vacuum polarization induced by the electric field $\mathcal{E}_e$] and

$$\mathcal{B} = q \exp(\rho) \cos\Phi \sin\theta \tag{24}$$

for the neutral magnetic field and

$$\mathcal{H}_o = \frac{q}{\upsilon} \frac{\cos\Phi}{\cosh\Gamma} = \frac{q \cos\Phi}{f} \tag{25}$$

for the neutral vacuum magnetic field. The corresponding charge densities are

$$j^4_e = \frac{q}{4\pi} (\exp(\rho) \tanh\Gamma \sin\Phi)' \sin\theta = \frac{\pm e}{4\pi} \frac{1}{f} \frac{d}{d\beta} \left(\frac{\sin\Phi}{\tanh\Gamma}\right) \sin\theta \tag{26}$$

for the electric charge density,

$$j^4_o = \frac{\pm e r_o^2}{4\pi} \frac{1}{f} \frac{d}{d\beta} \left(\frac{\sin\Phi}{\tanh\Gamma} \dot{\rho}\, \dot{S}\right) \sin\theta\,, \tag{27}$$

for the neutral electric charge density, and

$$s^4 = \frac{q}{4\pi}\,(\exp(\rho)\cos\Phi)'\sin\theta = \frac{q}{4\pi}\,\frac{1}{f}\,\frac{d}{d\beta}\,(\exp(\rho)\cos\Phi)\sin\theta\ , \tag{28}$$

$$\zeta^4 = \frac{q}{4\pi}\,\left(\frac{\exp(\rho)\cos\Phi\,\sin\Phi}{\cosh\Gamma}\right)'\sin\theta = \frac{q}{4\pi}\,\frac{1}{f}\,\frac{d}{d\beta}\left(\frac{\exp(\rho)\cos\Phi\,\sin\Phi}{\cosh\Gamma}\right)\sin\theta \tag{29}$$

for the magnetic charge densities, where

$$\int j_e^4\, dr\, d\theta\, d\varphi = \pm e\ , \qquad \int j_o^4\, dr\, d\theta\, d\varphi = 0\ , \tag{30}$$

$$\int s^4\, dr\, d\theta\, d\varphi = 0\ , \qquad \int \zeta^4\, dr\, d\theta\, d\varphi = 0\ . \tag{31}$$

We note that the r integration in (31) are carried out over the interval $(0, r_c)$, where r_c represents the indeterminate distance of the magnetic horizon (at and beyond which g=0) from the origin. In this case, for the functions $\Phi(r)$ and $\rho\,(r)$ *we have the relations*

$$\Phi_{ns\tau}\,(r_c) = (\tfrac{1}{2})\pi\,(-1)^S \pm (2n + s + \tau)\,\pi \tag{32}$$

$$\rho_{n\tau}\,(r_c) = 1\mathrm{n}r_c^{\,2} \pm I\,(2n + \tau)\,\pi\ ,\ \ n = 0, 1, 2, \ldots, \tag{33}$$

where $\Phi(r_c) = (\tfrac{1}{2})\pi$ refers to the critical value of the angle function pertaining to the exterior solutions of the field equations. The invariance of the field equations under the transformation

$$\Gamma \rightarrow -\Gamma \tag{34}$$

implies electric *charge multiplicity* invariance (i.e., existence of $\pm\,e$, $-e$, 0 where the latter occurs for $\Gamma = 0$) which leaves $\mathcal{B}$, $\mathcal{H}_0$, s^4, and ζ^4 unchanged and reverses the signs of $\mathcal{E}_e$, $\mathcal{E}_0$, j_e^4, j_0^4. The sign change of energy under (21) leads to the change of signs of $\mathcal{E}_e$ and $\mathcal{E}_0$. Hence, because of $f \rightarrow -f$ invariance, it follows that particles and antiparticles carry equal and opposite signs of electric charge. However, under (21) the neutral magnetic field $\mathcal{B}$ and magnetic charge density ζ^4 remain unchanged, and therefore particles and antiparticles can have the same magnetic field and magnetic charge density.

Now, if the transformations (21) [i.e. $(-1)^\varepsilon$, ε=0,1] and (34) are followed by a change of magnetic charge sign by spin inversion [i.e., $(-1)^s$]or by parity inversion [i.e., $(-1)^\tau$], then we obtain the results

$$\mathcal{E}_e \rightarrow (-1)^{J+\varepsilon+\tau}\,\mathcal{E}_e\,,\ \mathcal{E}_0 \rightarrow (-1)^{J+\varepsilon+\tau}\,\mathcal{E}_0,\ \ \mathcal{B} \rightarrow (-1)^s\,\mathcal{B},\ \mathcal{H}_0 \rightarrow (-1)^s\,\mathcal{H}_0 \tag{35}$$

and

$$j_e^4 \rightarrow (-1)^{J+\varepsilon}\,j_e^4,\ \ j_o^4 \rightarrow (-1)^{J+\varepsilon}\,j_o^4,\ \ s^4 \rightarrow (-1)^S s^4,\ \zeta_o^{\,4} \rightarrow (-1)^S \zeta_o^{\,4}\ , \tag{36}$$

where J=0, 1 select positive and negative charge. The fields $\mathcal{E}_0$ and $\mathcal{H}_0$ represent the electric and magnetic fields of the *vacuum pairs*. The vacuum charge densities j_o^4 and $\zeta_o^{\,4}$

change signs under charge conjugation and spin inversion, respectively. The latter statement applies also to the charge densities j_e^4 and s^4. The relations of $(-1)^S$ and $(-1)^\tau$ to spin and parity inversions, respectively, are discussed in sections III and IV of this paper.

BROKEN SPACE-TIME SYMMETRIES

The direction of the spin and the signs of the magnetic charges g_n are correlated. This result implies that in the coupling of particles and antiparticles at high energies with parallel spins and opposite parities, because of the equality of their magnetic charges, the annihilation process must slow down. In the case of antiparallel spins, because of the opposite signs of their magnetic charges, the annihilation process is faster than in the previous case. Thus, the strongly bound magnetic layers or *magnetic levels* of an elementary particle with its antiparticle in a parallel spin state (i.e., same signs of their magnetic charges) results in bound states of a new particle of spin 1 and negative parity. The energy levels of the new particle are determined by electromagnetic, strong, weak (and even gravitational) interactions at short distances. For such systems (e.g., proton + antiproton) the slow annihilation could lead to a discrete spectrum of photons. In fact, the observed ψ_n (n=1, 2, 3 so far) or **J** particles[6] could well be due to the formation of such bound states of particles and antiparticles.

The quantity $\mathsf{g}_n^2/\hbar c$ represents the magnetic coupling between the *n*th layer of the particle and the field. Thus in the range of $(0, r_c)$ corresponding to each magnetic layer $(n=0,1,2, \ldots)$ there exists an infinite number of couplings between the field and the particle, the strength of which decreases as $n\to\infty$. Beyond $n\to\infty$ (i.e., beyond $r=r_c$) the coupling between the field and the particle is measured by the fine-structure constant alone.

We have thus established that the field equations (16)-(19) are invariant under the following symmetry operations.

(i) $\Gamma\to -\Gamma$ corresponds to electric charge multiplicity C_e [$\equiv (-1)^J$],
(ii) $\Phi\to \pi - \Phi$, $\ell_0^2 \to - \ell_0^2$ correspond to magnetic charge conjugation C_m,
(iii) $\Phi\to \pi + \Phi$, $\exp(\rho) \to - \exp(\rho)$ correspond to parity operation $\mathscr{P}$ [$\equiv (-1)^\tau$],
(iv) $\Gamma\to -\Gamma$, $q\to -q$ correspond to time reflection operation T [$\equiv (-1)^S$],
(v) $q\to -q$ corresponds to electric and magnetic charge conjugation C_eC_m,
(vi) $\upsilon\to -\upsilon$ corresponds to reversal of the sign of mass (and energy) [$\equiv (-1)^\epsilon$]

Some of these symmetries do, in a special situation, break down and are, therefore, not conserved. For example, if the constant of integration λ_0^2 vanishes (i.e., if the electric charge e=0), then the symmetries C_m and $\mathscr{P}$ are not conserved. This can be seen by noting that the field equations (6)-(9) in the limit λ_0^2=0 reduce to

$$(\tfrac{1}{2})\, \ell_0^2\, \upsilon[\upsilon \exp(\mathcal{U}+\rho)\Phi']' = [\exp(\rho)-\ell_0^2] \cos\Phi + (-1)^S \ell_0^2 \sin\Phi, \tag{37}$$

$$(\tfrac{1}{2})\, \ell_0^2\, \upsilon[\upsilon \exp(\mathcal{U}+\rho)\rho']' = - [\exp(\rho)-\ell_0^2] \sin\Phi + (-1)^S \ell_0^2 \cos\Phi + \exp(\rho) \tag{38}$$

$$(\tfrac{1}{2})\, \ell_0^2\, \upsilon[\upsilon \exp(\mathcal{U}+\rho)\mathcal{U}']' = \exp(\rho)\,(1-\sin\Phi) \tag{39}$$

$$\rho'' + \rho' \frac{\upsilon'}{\upsilon} + (½)(\rho'^2 + \Phi'^2) = 0 \tag{40}$$

In this case the solutions

$$\Phi_{ns\tau} = \pm (2n + s + \tau)\pi + (-1)^S\Phi, \ \rho \pm (2n + \tau)\, i\pi = \rho_{n\tau}, \ 0 \le \Phi(r) \le (½)\pi \tag{41}$$

involving both spin and parity do not satisfy the field equations (37)-(40). However, the solutions

$$\Phi_{ns}(r) = \pm (2n + s)\pi + (-1)^S\Phi(r) \tag{42}$$

without the parity quantum number do satisfy the field equations (37)-(40) and therefore they can be replaced by

$$(½)\, \ell_0^2\, \upsilon[\upsilon \exp(\mathcal{U}+\rho)\Phi'_{ns}]' = [\exp(\rho) - \ell_0^2] \cos\Phi_{ns} + \ell_0^2 \sin\Phi_{ns} \ , \tag{43}$$

$$(½)\, \ell_0^2\, \upsilon[\upsilon \exp(\mathcal{U}+\rho)\rho']' = -\, [\exp(\rho) - \ell_0^2] \sin\Phi_{ns} + \ell_0^2 \cos\Phi_{ns} + \exp(\rho), \tag{44}$$

$$(½)\, \ell_0^2\, \upsilon[\upsilon \exp(\mathcal{U}+\rho)\mathcal{U}']' = \exp(\rho)\,(1 - \sin\Phi_{ns}), \tag{45}$$

where, as before, the discrete indices n, s for the functions υ, $\mathcal{U}$, ρ, and for the ℓ_o have been suppressed and where

$$\Phi_{ns} = \pm (2n + s)\,\pi + (-1)^S\Phi, \ n = 0, 1, 2, \dots . \tag{46}$$

Thus the solutions of the equations (43)-(45) are not invariant under parity and magnetic charge conjugation operations. In this case we have only ∞^2 distinct solutions. Because of the invariance under $\upsilon \to -\upsilon$ we still have particle-antiparticle solutions, each with two spin states s=0 and 1 [i.e., $(-1)^S$]. These particles have no electromagnetic interactions and they couple through the magnetic charge alone. The absence of parity and charge conjugation symmetries for e=0 imply that these symmetries are of electromagnetic origin. Conversely, intrinsic parity and charge conjugation are space-time symmetries induced by electric charge. The chargeless particles have no continuum solutions occurring beyond the magnetic horizon since for $g_n = 0$ (ℓ_o=0) the equations (43)-(45), as a consequence of the relations $\lim_{\ell\to 0} \cos\Phi_{ns} = 0$, $\lim_{\ell\to 0}\sin\Phi_{ns} = 1$, are empty. Thus chargeless particles have short-range interactions only, where the range of the force is $g_n^2/M_\upsilon c^2$ which has an indeterminacy specified as g_n. *The chargeless massive particles predicted by this theory have the same symmetry properties as the two neutrinos* ν_e *and* ν_μ.

The symmetry of the electric charge multiplicity (i.e., invariance of the field equations under the transformation $\Gamma \to -\Gamma$) together with the symmetry $q \to -q$, imply the existence of particles with +1, -1, 0 units of electric charges and their corresponding antiparticles with -1, +1, 0 units of electric charges, respectively. In order to find other distinguishing characteristics of the particles with different electric charges we may further classify the solutions according to the *invariance properties* of the function Φ as it appears in the field equations (16)-(19). Thus from (19) we may, formally, deduce the *invariant statement*

$$\Phi^b = A + (-1)^{b+1} \int [-(2\ddot{\rho} + \dot{\rho}^2)]^{1/2}\, d\beta \ , \tag{47}$$

where A is a constant of integration and where b is the charge multiplicity number which assumes the values b=0 and b=1. From the equation (47) it is clear that we may, in principle, substitute in the field equations (16)-(18) to determine two sets of solutions ρ_b, S_b and the corresponding relations between r_o, ℓ_o, λ_o, where b=0, 1. For the solutions where $\exp(\rho)=0$ and $\exp(\rho)=\beta^2$ ($\beta \neq$constant) the constant A assumes the values 0 and $\frac{1}{2}\pi$, respectively. For these special solutions and also for the solutions where $\beta=\pm\ell_o$, $A=(s+1)\pi$, the invariant function Φ is independent of b. However, all other solutions for which the integral in (47) does not vanish will depend on b. Thus the solutions corresponding to b=0 and b=1 must, in general, represent different systems of particles. All the currents, being determined by the solutions of the field equations, will acquire a new degree of freedom defined by b=0,1. The same applies to all other derived quantities like the electric and magnetic fields and also the corresponding energy and momentum.

Based on the prediction by the theory for the existence of particles with electric charges $\pm e$, 0, the most reasonable interpretation for the *electric charge multiplicity number* b, separating leptons from baryons, is to regard it as a conservation law for electrons (b=0) and protons (b=1). For the case $e=0$, in view of the symmetries discussed in sections II and III, the natural expectation would be to assign b=0, 1 to the electron neutrino and muon neutrino, respectively.

THE FIELD EQUATIONS

The Lagrangian of the theory is given by

$$\mathcal{L}=\hat{\mathfrak{g}}^{\mu\nu}(\hat{R}_{\mu\nu}-r_o^{-2}q^{-1}F_{\mu\nu})+2r_o^{-2}[\sqrt{(-\hat{g})}-\sqrt{(-g)}], \tag{48}$$

where the nonsymmetric hermitian or non-hermitian curvature tensor $\hat{R}_{\mu\nu}$ is defined, in terms of the nonsymmetric affine connection $\Gamma^{\rho}{}_{\mu\nu}$ by,

$$\hat{R}_{\mu\nu}=-\Gamma^{\rho}_{\mu\nu,\rho}+\Gamma^{\rho}_{\mu\rho,\nu}-\Gamma^{\rho}_{\mu\nu}\Gamma^{\sigma}_{\rho\sigma}+\Gamma^{\rho}_{\mu\sigma}\Gamma^{\sigma}_{\rho\nu} \tag{49}$$

and where

$$\hat{\mathfrak{g}}^{\mu\nu}=\sqrt{(-\hat{g})}\,\hat{g}^{\mu\nu},\quad \mathrm{Det}\, g_{\mu\nu}=g, \tag{50}$$

$$\hat{g}=\mathrm{Det}\,\hat{g}_{\mu\nu}=g\,(1\pm q^{-2}\Omega-q^{-4}\Lambda^2), \tag{51}$$

where the + or - is chosen for non-hermitian or hermitian $\hat{g}_{\mu\nu}$, respectively. The auxiliary field $F_{\mu\nu}$ is defined by

$$F_{\mu\nu}=\partial_\mu A_\nu-\partial_\nu A_\mu. \tag{52}$$

Because of the presence of the extra field variables A_μ and the terms $\sqrt{(-g)}$ the Lagrangian $\mathcal{L}$ is not locally gauge invariant. Now, because of the extensive use of the various quantities in the theory we must include in this paper the following definitions.

$$\Omega=\frac{1}{2}\Phi^{\mu\nu}\Phi_{\mu\nu},\quad \Lambda=\frac{1}{4}f^{\mu\nu}\Phi_{\mu\nu},\quad f^{\mu\nu}=\frac{1}{2\sqrt{(-g)}}\epsilon^{\mu\nu\rho\sigma}\Phi_{\rho\sigma}, \tag{53}$$

$$\hat{g}^{\nu\rho}\,\hat{g}_{\mu\rho}=\delta^{\nu}_{\mu}, \tag{54}$$

$$\hat{\mathfrak{g}}^{\underline{\mu\nu}}=\hat{\mathfrak{g}}^{\underline{\nu\mu}}=\sqrt{(-g)}\,h^{\mu\nu},\quad h^{\mu\nu}=\frac{(1\pm1/2\,q^{-2}\Omega)\,g^{\mu\nu}\pm q^{-2}T^{\mu\nu}}{\sqrt{(1\pm q^{-2}\Omega-q^{-4}\Lambda^2)}}, \tag{55}$$

$$b^{\nu\rho}b_{\mu\rho}=\delta^{\nu}_{\mu},\quad b_{\mu\nu}=\frac{(1\pm 1/2\,q^{-2}\Omega)\,g^{\mu\nu}\mp q^{-2}T^{\mu\nu}}{\sqrt{(1\pm q^{-2}\Omega-q^{-4}\Lambda^{2})}}, \tag{56}$$

$$T^{\nu\rho}T_{\mu\rho}=\delta^{\nu}_{\mu}(\tfrac{1}{4}\Omega^{2}+\Lambda^{2}),\ T_{\mu\nu}=\tfrac{1}{2}\Omega\,g_{\mu\nu}-\Phi_{\mu\rho}\Phi_{\nu}{}^{\rho},\ T^{\rho}_{\rho}=g^{\mu\nu}T_{\mu\nu}=0, \tag{57}$$

$$\hat{g}^{\mu\nu}=\sqrt{(-g)}\,\frac{\Phi_{\mu\nu}\pm q^{-2}\Lambda\ f_{\mu\nu}}{\sqrt{(\ 1\pm q^{-2}\Omega-\Lambda^{2})}}, \tag{58}$$

The Lagrangian (48) in the correspondence limit $r_o = 0$ reduces to the Lagrangian of general relativity. The two supersymmetric branches can be expressed as

$$\hat{g}^{s}_{\mu\nu}=g_{\mu\nu}+i^{s}q^{-1}\Phi_{\mu\nu}, \tag{59}$$

where s = 0,1 for the non-hermitian and hermitian field variables, respectively. The nonsymmetric affine connection $\Gamma^{\rho}_{\mu\nu}$, as obtained from varying the action function

$$S=\int \mathcal{L}d^{4}x, \tag{60}$$

with respect to $\Gamma^{\rho}_{\mu\nu}$, are to be calculated from the 64 algebraic equations

$$\hat{g}_{\mu\nu,\rho}-\hat{g}_{\sigma\nu}\Gamma^{\sigma}_{\mu\rho}-\hat{g}_{\mu\sigma}\Gamma^{\sigma}_{\rho\nu}=0. \tag{61}$$

Furthermore, the curvature tensor $\hat{R}_{\mu\nu}$ is invariant under the gauge transformation

$$\Gamma^{\rho}_{\mu\nu}\rightarrow\Gamma^{\rho}_{\mu\nu}+\delta^{\rho}_{\mu}\partial_{\nu}\lambda, \tag{62}$$

which property does not hold for general relativity where one deals with symmetric connection. For real $\hat{g}_{\mu\nu}$ the curvature tensor $\hat{R}_{\mu\nu}$ is transposition invariant viz.,

$$[\tilde{\hat{R}}_{\mu\nu}]=\hat{R}_{\nu\mu}(\tilde{\Gamma})=\hat{R}_{\mu\nu}(\Gamma),$$

provided

$$\Gamma_{\rho}=\Gamma^{\sigma}_{\rho\sigma}=0, \tag{63}$$

which condition, as can be seen from the contravariant form of (61),

$$\hat{g}^{\mu\nu}{}_{,\rho}+\hat{g}^{\sigma\nu}\Gamma^{\mu}_{\sigma\rho}+\hat{g}^{\mu\sigma}\Gamma^{\nu}_{\rho\sigma}-\hat{g}^{\mu\nu}\Gamma^{\sigma}_{\underline{\rho\sigma}}=0, \tag{64}$$

entail the four field equations

$$\hat{g}^{\underset{\vee}{\mu\nu}}{}_{,\nu}=0. \tag{65}$$

The field equations (65) also follow from varying the action function S with respect to the extra field variables A_{μ}. The remaining field equations are obtained, by varying S with respect to $\hat{g}^{\mu\nu}$ as

$$\hat{R}_{\mu\nu} = r_o^{-2} (\hat{b}_{\mu\nu} - \hat{g}_{\mu\nu}), \tag{66}$$

where

$$\hat{b}^s_{\mu\nu} = b_{\mu\nu} + i^s q^{-1} F_{\mu\nu}. \tag{67}$$

By separating out symmetric and antisymmetric parts in (66) and eliminating $F_{\mu\nu}$ we obtain the final field equations as

$$\hat{R}_{\underline{\mu\nu}} = r_o^{-2} (b_{\mu\nu} - g_{\mu\nu}), \tag{68}$$

$$\hat{R}_{\underset{\vee}{\mu\nu},\rho} + \hat{R}_{\underset{\vee}{\nu\rho},\mu} + \hat{R}_{\underset{\vee}{\rho\mu},\nu} + r_o^{-2} I_{\mu\nu\rho} = 0, \tag{69}$$

$$\hat{g}^{\underset{\vee}{\mu\nu}}{}_{,\nu} = 0. \tag{70}$$

Because of the two differential identities obtainable from (69) and (70), the eighteen field equations for the sixteen field variables $\hat{g}_{\mu\nu}$ are equivalent to sixteen independent field equations. The four component fully antisymmetric conserved quantities $I_{\mu\nu\rho}$ represent magnetic current density

$$4\pi s^\mu = \frac{1}{3!} \epsilon^{\mu\nu\rho\sigma} I_{\nu\rho\sigma}, \tag{71}$$

where

$$I_{\mu\nu\rho} = \Phi_{\mu\nu,\rho} + \Phi_{\nu\rho,\mu} + \Phi_{\rho\mu,\nu}, \tag{72}$$

and where, as follows from the third expression in (53) and (71) above, the magnetic current density can also be expressed as

$$s^\mu = \frac{1}{4\pi} \frac{\partial[\sqrt{(-g)} f^{\mu\nu}]}{\partial x^\nu}. \tag{73}$$

STRATEGIC TRIAD

The triad, for most physicists, could entail the goals of: (i) finding the mass of the top quark, (ii) finding the mass of the Higgs boson, (iii) looking for CP violation in K and B decays. However, a more general requirement is the unification of interactions mediated by the exchange of the *strategic triad* of spin 0, spin 1, and spin 2 bosons between families of quarks and leptons. In order to complete the unification of all interactions we shall need, besides the classical field equations (68)-(70) and their *supersymmetric* counterparts, a generalized Dirac wave equation describing quarks and leptons interactions with spin 0, spin 1 and spin 2 bosons. We consider the role of the extremum value of the action function S_o given by

$$S_o = -\frac{q^2}{4\pi c} \int [\sqrt{(-\hat{g})} - \sqrt{(-g)}] \, d^4x, \tag{74}$$

where the square root $\sqrt{(-g)} = \sqrt{(-g)} \sqrt{(1+q^{-2}\Omega - q^{-4}\Lambda^2)}$ can be shown to be of the form

$$\sqrt{(-\hat{g})} = \sqrt{(-g)}\sqrt{[(1+\frac{1}{2}q^{-2}\Omega)^2 - q^{-4}c^2P_\mu P^\mu]}, \tag{75}$$

where

$$cP_\mu = T_{\mu\nu}\hat{v}^\nu, \quad c^2P_\mu P^\mu = \frac{1}{4}\Omega^2+\Lambda^2, \quad T_{\mu\rho}T^{\nu\rho} = \delta^\nu_\mu(\frac{1}{4}\Omega^2+\Lambda^2), \tag{76}$$

and where $T_{\mu\nu}$ is, formally, of the same form as the stress-energy-momentum tensor of the electromagnetic field viz.

$$T_{\mu\nu} = \frac{1}{2}\Omega g_{\mu\nu} - \Phi_{\mu\rho}\Phi_\nu{}^\rho. \tag{77}$$

In this case the new $T_{\mu\nu}$ contains, in addition to the stress-energy-momentum of the electromagnetic field, as will be seen, the energy and momentum densities for massive spin 0, 1, 2 fields. The unit vector $\hat{v}^\mu$ in (76) is defined by

$$\hat{v}^\mu = \frac{dx^\mu}{ds}, \quad \hat{v}^\mu\hat{v}_\mu = 1, \; ds^2 = g_{\mu\nu}dx^\mu dx^\nu. \tag{78}$$

By using the splitting of the field $\Phi_{\mu\nu}$ according to

$$\Phi_{\mu\nu} = \Phi_{o\mu\nu} + \Phi_{1\mu\nu}, \tag{79}$$

where $\Phi_{o\mu\nu}$ represents the short-range field inside the core with a neutral distribution of the magnetic charge, and $\Phi_{1\mu\nu}$ small compared to $\Phi_{o\mu\nu}$ represents pure radiation field obeying Maxwell's equations

$$\Phi_{1\mu\nu,\rho} + \Phi_{1\nu\rho,\mu} + \Phi_{1\rho\mu,\nu} = 0, \quad [\sqrt{(-g)}\,\Phi_1^{\mu\nu}]_{,\nu} = 0. \tag{80}$$

Using (79) in the definition of $T_{\mu\nu}$ by (77) we obtain the decomposition

$$T_{\mu\nu} = T_{o\mu\nu} + T_{I\mu\nu} + T_{1\mu\nu}, \tag{81}$$

$$\Omega = \Omega_o + \Omega_I + \Omega_1, \tag{82}$$

where

$$\Omega_o = \frac{1}{2}\Phi_o^{\mu\nu}\Phi_{o\mu\nu}, \quad \Omega_I = \Phi_o^{\mu\nu}\Phi_{1\mu\nu}, \quad \Omega_1 = \frac{1}{2}\Phi_1^{\mu\nu}\Phi_{1\mu\nu} \tag{83}$$

$$T_{o\mu\nu} = \frac{1}{2}\Omega_o g_{\mu\nu} - \Phi_{o\mu\rho}\Phi_{o\nu}{}^\rho, \quad T_{I\mu\nu} = \frac{1}{2}\Omega_I g_{\mu\nu} - (\Phi_{o\mu\rho}\Phi_{1\nu}{}^\rho + \Phi_{o\nu\rho}\Phi_{1\mu}{}^\rho), \tag{84}$$

$$T_{1\mu\nu} = \frac{1}{2}\Omega_1 g_{\mu\nu} - \Phi_{1\mu\rho}\Phi_{1\nu}{}^\rho, \quad [\sqrt{(-g)}T_{1\mu}{}^\nu]_{|\nu} = 0, \tag{85}$$

$$\frac{1}{4\pi}[\sqrt{(-g)}T_{o\mu}{}^\nu]_{|\nu} = \Phi_{o\mu\rho}J_e{}^\rho + f_{o\mu\rho}s^\rho, \quad J_e^\mu = \frac{1}{4\pi}\frac{\partial}{\partial x^\nu}[\sqrt{(-g)}\Phi^{\mu\nu}],$$
$$\frac{1}{4\pi}[\sqrt{(-g)}T_{I\mu}{}^\nu]_{|\nu} = \Phi_{1\mu\rho}J_e{}^\rho + f_{1\mu\rho}s^\rho, \tag{86}$$

and where the sign | indicates covariant derivative with respect to the metric $g_{\mu\nu}$.

The particle mass is generated by the scalar Ω_o according to

$$\frac{1}{4\pi c}\int_\sigma \frac{1}{2}\Omega_o\sqrt{(-g)}\,d\sigma = mc. \tag{87}$$

Expansion of the action function S_o in (74) to q^{-2} order yields

$$S_o \equiv -\frac{1}{4\pi c}\int \frac{1}{2}(\Omega_o+\Omega_1+\Omega_1)\sqrt{(-g)}\, d^4x . \tag{88}$$

Using the definitions

$$\sqrt{(-g)}\, d^4x = \sqrt{(-g)}\, d\sigma\, ds,\ d\sigma = \hat{v}^{\mu} d\sigma_{\mu},$$
$$\Phi_{1\mu\nu} = \partial_{\mu}A_{\nu} - \partial_{\nu}A_{\mu},$$

we obtain

$$S_o \equiv -mc\int ds - \frac{1}{c}\int A_{\mu}J^{\mu}_{e}\, d\sigma\, ds - \frac{1}{16\pi c}\int \Phi_1^{\mu\nu}\Phi_{1\mu\nu}\sqrt{(-g)}\, d^4x , \tag{89}$$

which is the same as the classical action function for a charged point particle moving according to Lorentz's equations of motion in an external electromagnetic field. The above is a crude approximation to a *non-linear classical action* function just to show that it contains the usual equations of motion. We shall in what follows show that the corresponding quantum action function is, in contrast to classical theory, linear.

By using the usual 4x4 Dirac matrices γ_{μ}, γ_5 defined by the anticommutation relations

$$\gamma_{\mu}\gamma_{\nu} + \gamma_{\nu}\gamma_{\mu} = -2\, g_{\mu\nu} I_o, \quad \gamma_5\gamma_{\mu}+\gamma_{\mu}\gamma_5 = 0 , \tag{90}$$

where I_o is a 4x4 unit matrix, we can obtain from (74) the two action operators

$$S^F_{o1} = -\frac{1}{4\pi c}\int_F \sqrt{(-g)}\,[c\gamma^{\mu}P_{\mu} + \tfrac{1}{2}\Omega - 2\, u_-(\tfrac{1}{2}\Omega + q^2)]\, d^4x , \tag{91}$$

$$S^F_{o2} = -\frac{1}{4\pi c}\int_F \sqrt{(-g)}\,[c\gamma_5\gamma^{\mu}P_{\mu} + \tfrac{1}{2}\Omega - 2\, u_-(\tfrac{1}{2}\Omega + q^2)]\, d^4x, \tag{92}$$

where the subscript F implies the substitution of Fermi-like solutions (i.e., the solutions of the field equations for non-hermitian $\hat{g}_{\mu\nu}$) in the integrand as compared to the substitutions of Bose-like solutions (i.e., the solutions of the field equations for the hermitian $\hat{g}_{\mu\nu}$) in the two action operators

$$S^B_{o1} = -\frac{1}{4\pi c}\int_B \sqrt{(-g)}[c\gamma^{\mu}P_{\mu} + \tfrac{1}{2}\Omega - 2\, u_-(\tfrac{1}{2}\Omega - q^2)]\, d^4x, \tag{93}$$

$$S^B_{o2} = -\frac{1}{4\pi c}\int_B \sqrt{(-g)}[c\gamma_5\gamma^{\mu}P_{\mu} + \tfrac{1}{2}\Omega - 2\, u_-(\tfrac{1}{2}\Omega - q^2)]\, d^4x, \tag{94}$$

where u_- is the projection operator

$$u_- = \frac{1}{2}(1 - i\,\gamma_5) . \tag{95}$$

Introducing a wave function Ψ the operator equations (91)-(94) can be replaced by the generalized Dirac wave equations

$$cM^F_1\Psi^F_1 = -\frac{1}{4\pi c}\int_F \sqrt{(-g)}[c\gamma^{\mu}P_{\mu} + \tfrac{1}{2}\Omega - 2u_-(\tfrac{1}{2}\Omega + q^2)]\, d\sigma\, \Psi^F_1, \tag{96}$$

$$cM_2^F\Psi_2^F = -\frac{1}{4\pi c}\int_F \sqrt{(-g)}[c\gamma_5\gamma^\mu P_\mu + \frac{1}{2}\Omega - 2u_-(\frac{1}{2}\Omega + q^2)]\, d\sigma\, \Psi_2^F, \tag{97}$$

$$cM_1^B\Psi_1^B = -\frac{1}{4\pi c}\int_B \sqrt{(-g)}[c\gamma^\mu P_\mu + \frac{1}{2}\Omega - 2u_-(\frac{1}{2}\Omega - q^2)]\, d\sigma\, \Psi_1^B, \tag{98}$$

$$cM_2^B\Psi_2^B = -\frac{1}{4\pi c}\int_B \sqrt{(-g)}[c\gamma_5\gamma^\mu P_\mu + \frac{1}{2}\Omega - 2u_-(\frac{1}{2}\Omega - q^2)]\, d\sigma\, \Psi_2^B, \tag{99}$$

where we used the substitutions

$$\frac{dS_{oj}^\ell}{ds}\Psi_j = cM_j^\ell\Psi_j\,, \tag{100}$$

and where j = 1, 2 and ℓ = F,B. The masses M^ℓ_j are free particle masses relating to quark and lepton masses and self-energy. In the mass relation (87) the mass m is generated by F- and B-type fields. Hence, the integrals in (96)-(99) each yields a momentum $p_\mu = mcv_\mu$, where m refers to the mass m_F or m_B. The *first quantization* of the equations (96)-(99) follows by the operator representation of the momentum p_μ by $p_\mu = i\hbar \frac{\partial}{dx^\mu}$, thereby obtaining *generalized Dirac wave equations* describing the coupling of photons, massive gauge bosons with spin ½ matter fields.

The appearance of $T_{\mu\nu}$ in the wave equations (96)-(99) through the definition of $P_\mu = \frac{1}{c}\, T_{\mu\nu}\hat{v}^\nu$ (momentum density) provides a *hydrodynamic picture* of the interplay between the generations of quarks and leptons unifying the mediation of forces through the spin 0, 1, 2 bosons. The energy-momentum-stress densities represented by an electromagnetic type tensor $T_{\mu\nu}$ for massless and massive bosons is not accidental but the proper way to unify all the interactions. The tensor $T_{\mu\nu}$ with nine independent components (i.e., $T^\rho{}_\rho = 0$) can be decomposed into 0, 1, 2 spin parts as

$$T_{\mu\nu} = T^{(o)}_{\mu\nu} + T^{(1)}_{\mu\nu} + T^{(2)}_{\mu\nu}\,. \tag{101}$$

By using (81)-(86) it can easily be seen that for a point charge with $J_e^\mu = e\,\hat{v}^\mu\,\delta(x-\xi)$, any one of the wave equations (96)-(99) yield terms like

$$(p_\mu - \frac{e}{c}A_\mu)\,\gamma^\mu,$$

and form the basis for the derivation of the usual gauge invariant Dirac wave equation describing the interaction of a photon with a point electric charge.

For distances large compared to Plank length (or r_o) and to order of q^{-2}, the field equations (68) and their supersymmetric counterparts, with the assumption $g_{\mu\nu} = \eta_{\mu\nu}$, reduce to

$$(\nabla^2 - \frac{\partial^2}{c^2\partial t^2} + \kappa^2)\, T^{(2)}_{\mu\nu}\,(F) = 0\,, \tag{102}$$

$$(\nabla^2 - \frac{\partial^2}{c^2\partial t^2} - \kappa^2)\, T^{(2)}_{\mu\nu}\,(B) = 0\,, \tag{103}$$

where $\kappa^2 = 2\,r_o^{-2}$ and where we used the relations (86) with $J_e^\mu = 0$, $s^\mu = 0$ to obtain four restrictions $(\sqrt{(-g)}\, T_\mu^{\ \nu})_{|\nu} = 0$ to construct the spin 2 tensors $T^{(2)}_{\mu\nu}$ (F), $T^{(2)}_{\mu\nu}$ (B) each with five independent components only. The magnetic charge current densities s^μ(F) and s^μ(B), with the same approximation, obey the equations (102) and (103), respectively. Despite *tachyon* type solutions of the linearized equation (102), its exact spherically symmetric form has no such solutions and yields magnetic charge distribution in stratified layers with alternating signs and decreasing amounts of magnetic charges g_n (n=0, 1, 2 . . .) where $\sum_0^\infty g_n = 0$. The magnetic charge distribution for s^μ(B) has the form of a dipole with no restriction of magnetic charge amounts in each opposite sign poles, i.e., $g_+ + g_- = 0$.

GENERATION OF MASS

The wave equations (96)-(99) contain important information on the masses of the *strategic triad*, leptons, and quarks. The term q^2 in the wave equations can be expressed as a function of the *mixing angle* Φ and electric and magnetic charges Ne, Ng, respectively, where N, if not set equal to 1, can vary over the interval 10^{15}- 10^{21}. By using the definition of q in

$$q^{-1} = \frac{2G}{c^4} U, \quad \text{where} \quad U = \sqrt{\left(\frac{g^2}{\cos^2\Phi} + e^2\right)} + (-1)^s g \tan\Phi, \quad s=0,1. \tag{104}$$

we obtain it as

$$q^2 = \frac{1}{N^2} \frac{M_p c^2}{V_p} q_o^2, \tag{105}$$

where $V_p = \frac{4\pi}{3} r_p^3$ is the Planck volume, and

$$q_o^2 = 3\pi \left(\frac{g^2}{e^2+g^2}\right) \alpha_p Y^2, \tag{106}$$

$$\alpha_p = M_p c^2 / \left(\frac{e^2+g^2}{r_o}\right), \tag{107}$$

$$Y = \sqrt{\left(\frac{1}{\cos^2\Phi} + \frac{e^2}{g^2}\right)} + (-1)^{s+1} \tan\Phi, \quad s=0,1\,. \tag{108}$$

In principle all the masses carried by the interactions in the wave equations (96)-(99) are calculable as functions of e, g, and Φ. Thus, instead of the Standard Theory's Higgs bosons providing, by a spontaneous symmetry breaking, the origin of mass, we have here the *Planck boson* as the generator of mass. We may also use the results in (104) and the corresponding Compton wave lengths to define the mass relations

$$M^2_\pm(r_o) = \frac{\hbar c}{g^2} M_p^2 Y_\pm^{-2}, \quad M^2_\pm(\lambda_o) = \frac{\hbar c}{eg} M_p^2 Y_\pm^{-1}, \quad M^2_\pm(\ell_o) = \frac{\hbar c}{g^2} M_p^2 Y_\pm^{-1}. \tag{109}$$

The masses $M_\pm(r_o)$ of spin 2 bosons are related to Planck mass, and spin 1 boson masses $M_\pm(\lambda_o)$, $M_\pm(\ell_o)$, as follow from (109), according to

$$M_{\pm}(r_o) = \sqrt{\left(\frac{e^2}{\hbar c}\right)}\,\frac{M_{\pm}^2(\lambda_o)}{M_p}, \quad M_{\pm}(r_o) = \sqrt{\left(\frac{g^2}{\hbar c}\right)}\,\frac{M_{\pm}^2(\ell_o)}{M_p}. \tag{110}$$

Based on the construction of Planck bosons out of spin 1 bosons satisfying the two wave equations (102), (103) we can expect these spin 2 bosons to decay into two spin 1 bosons within the span of Planck time of 10^{-43} seconds. Thus, the four wave equations (96)-(99) contain the coupling of, instead of four spin 2 bosons, eight spin 1 bosons (gluons) to the spin ½ matter fields of quarks and leptons. The bosons themselves are constituted by quark-antiquark pairs. The quarks themselves must have the constituents of an infinite number of infinitesimal monopoles with neutral distribution.

A further mass relation can be obtained by using

$$q^2 r_o^2 = \frac{c^4}{2G} \quad \text{and} \quad \ell_o^2 = g\, q^{-1}, \quad \lambda_o^2 = \ell_w r_o$$

and writing the identity

$$\lambda_o^4 + \ell_o^4 = \frac{2G}{c^4}(e^2+g^2)\, r_o^2. \tag{111}$$

Hence we obtain the mass relation

$$\frac{1}{M_{\pm}^4(\ell_o)} + \frac{1}{M_{\pm}^4(\lambda_o)} = N^2 \frac{1}{M_p^2 M_{\pm}^2(r_o)}. \tag{112}$$

By using the relation $\ell^2 = \ell_s^2 + \ell_w^2 = \frac{2G}{c^4}(e^2 + g^2)$ and the substitution $\frac{1}{2}\ell^2 = \frac{\hbar^2}{M^2(\ell)c^2}$, we obtain

$$\frac{M(\ell)}{M_p} = \sqrt{\left(\frac{\hbar c}{e^2+g^2}\right)}. \tag{113}$$

CONCLUSIONS

A glance at the wave equations (96)-(99) together with the classical field equations (68)-(70) reveals a close resemblance to quantum electrodynamics and a new way of unifying general relativity and quantum theory. We have, actually, shown that relativistic quantum theory can be derived from the generalized theory of gravitation. Some of the epistemological issues pertaining to conflict or clash between quantum theory and general relativity arising, for example, from he study of black holes as sinks of information regardless of their substance or of their fading away into oblivion, is not relevant in these discussions.

The eight spin 1 bosons, resulting from the decay of the four highly massive spin 2 bosons in the wave equations (96)-(99) despite a striking resemblance to the Standard Theory's eight gluons, they may not, because of the absence of gravity, be related to the latter. The same wave equations contain the interactions of the four massive spin 1 bosons with quarks and leptons. The masses of these four weak bosons are generated by the four scalar bosons contained in the four wave equations (96)-(99). Thus, if there exist four weak bosons then we need to theorize the existence of a *fourth generation.* The current understanding of the experiments do not include, for example, the existence of an additional neutral weak boson. The masses of the quarks and leptons, because of the q^2 term in the wave equations (96)-(99) depend on the electric and magnetic charges and the mixing angle Φ.

Just as in electrodynamics, there is a probability (though very small) proportional to strong and weak coupling for the quarks and leptons to radiate any member of the *strategic triad* of spins 0, 1, and 2. The *supersymmetry* as presented in this theory, is the only possible way to unify the descriptions of particles with different spins. It is a symmetry whose basic operation is to transform particles or fields with a known spin into other particles or fields whose spin differs by the minimal unit $\frac{1}{2}\hbar$. The process of *supersymmetrizing* transforms bosons into fermions and vice versa. The presence of discrete symmetries induced by the terms containing γ_5 and the projection operator $\frac{1}{2}(1 - i\gamma_5)$ in the wave equations (96)-(99) should be relevant in seeking to explain C*P* violating weak decays of K and B bosons. The reference 1 contains in section 7.2 a discussion of the violation of C and *P* symmetries in the absence of electric charge. This property of symmetry violation in the absence of the electric charge together with the wave equations (96)-(99) may provide a fundamental mechanism for C*P* violation.

During the instant of the big bang creation of the universe all of the symmetries were unbroken. If at an appropriate temperature a phase transition occurred from the exact symmetry to the broken unified gauge symmetries SU(3,1) × U(1) and SO(3,1) × o(2), then C*P* violation must have begun in the prenucleosynthesis era to cause the creation of surplus matter over antimatter. However, this idea of C*P* violation to explain the origin of the excess matter content of the universe by Andre Sakharov may not, like other proposals, be the final word on this important cosmological problem.

Finally, the physicists ought not to assume *ab initio* that an approach unrelated to the currently favored models cannot possibly be an acceptable effort. There are some new and useful ideas in the proposed theory. It unifies two long-range forces to create to short-range forces representing weak and strong interactions. If one pursues an idea for a long time, as is the case for the effort expended in this work, one develops a mind-set, just as others acquire their own mind-sets by investing all their time on what is fashionable for the present and ignoring other perfectly viable view points. Mind-sets are not necessarily obstacles to progress, but they can be, inherently, basic ingredients for creativity. In a conversation I had with Albert Einstein on November 19, 1953 (see the last appendix in reference 1), his parting remark was, "Your theory, because of r_o, is more general than mine, but time will show which one of us is right."

THE CURRENT STATUS OF LIGO

David H. Reitze, Qi-Ze Shu, David B. Tanner, Sanichiro Yoshida, and Guenakh Mitselmakher

Physics Department
University of Florida
P.O. Box 118440
Gainesville, FL 32611

INTRODUCTION

The Laser Interferometric Gravitational Wave Observatory (LIGO) promises to open a new window on heretofore unobservable sources of gravitational radiation from astrophysical objects such as in-spiraling binary pulsars, neutron stars, black holes, and supernovae. More significantly, it will provide a unique way of studying astrophysical events in distinct yet complementary fashion to other astrophysical techniques such as visible, x-ray, gamma ray, and radio astronomy [1]. LIGO will use ultrahigh precision laser interferometry to measure *directly* gravitational radiation by detecting the induced strain, or change in length per unit length $\Delta L/L$, as a gravitational wave passes the detector (See Figure 1). The LIGO initial design strain sensitivity goal of $10^{-21}/\sqrt{\text{Hz}}$ at 100 Hz will be reached by using a power recycled Michelson interferometer with Fabry-Perot arm cavities. Three interferometers, currently under construction, will operate in coincidence at two separate sites (Hanford, WA and Livingston, LA) to provide high confidence gravitational wave detection and to discriminate against spurious events from seismic or other types of environmental disturbances. LIGO is one of a series of gravitational wave interferometers currently under development today throughout the world. Other interferometric gravitational wave projects include VIRGO (located in Italy) [2], GEO600 (in Germany) [3], and TAMA300 (in Japan) [4]. In this paper, we give an overview of LIGO and review its present status.

GRAVITATIONAL WAVES

First predicted over 80 years ago, gravitational waves are perturbations in the space-time metric which radiate outward from massive astrophysical objects that possess time-dependent quadrapole mass moments [5]. In the weak-field limit of General Relativity, the solution to the massless Einstein field equations results in a propagating wave of the form

$$h(\mathbf{r},t) = h_0 e^{i(\mathbf{k}\cdot\mathbf{r}-\omega t)} \tag{1}$$

Confluence of Cosmology, Massive Neutrinos, Elementary Particles, and Gravitation
Edited by Kursunoglu *et al.*, Kluwer Academic / Plenum Publishers, New York, 1999.

$$h_0 \approx G\ddot{Q}/c^4 r \qquad (2)$$

where h_0 is the amplitude of the strain (metric perturbation) and Q is the second derivative of the quadrapole mass moment or, equivalently, the kinetic energy of non-axisymmetric rotations. In general, a gravitational waves is a superposition of two orthogonal polarizations, h_+ and h_x (see Figure 2).

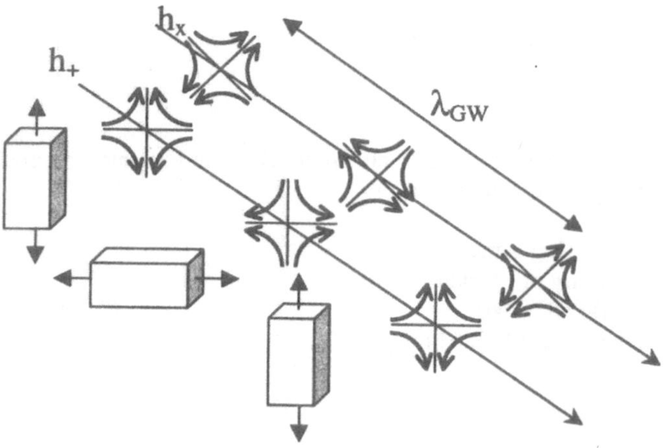

Figure 1 The induced strain as a gravitational wave of wavelength λ_{GW} passes through an object. h_+ and h_x denote the two possible orthogonal polarizations of a gravitational wave. The object is shown responding to h_+

To date, the only physical evidence for the existence of gravitational waves comes from observations of the compact binary neutron system PSR1913+16 [6,7]. One of the stars in the binary system is a pulsar whose radio emissions are Doppler shifted by the orbital velocity, providing a very accurate clock for measuring the orbital period of the system. Detailed measurements of the orbital period of PSR1913+16 over 25 years have shown an increase in orbital frequency commensurate with a decrease in the separation distance (and therefore loss of energy) of the neutron star system. According to General Relativity, the system will lose energy through the radiation of gravitational waves. Indeed, the calculated and observed decrease in orbital period agree to better than 1%.

PSR1913+16 is a member of a class of sources known as compact coalescing binary systems (consisting of neutron star/neutron star, neutron star/black hole, or black hole/black hole pairs). During their lifetime, the binary system components will gradually "spin up", spiraling inward towards each other and losing energy via radiation of gravitational waves. During the final few seconds before coalescence, the orbital frequency and gravitational wave amplitude increase significantly and are capable of detection by earth-based interferometric detectors. The resulting gravitational-wave-induced strain signal can be computed analytically and is found to be a chirped (frequency swept) signal whose frequency f and amplitude h_0 increase as $f^{11/3}$ and $f^{2/3}$ respectively [8]. The character of the emitted waveform during the merger and ringdown phases is more difficult to predict. (We note that observation of gravitational radiation from PSR1913+16 is impossible with LIGO since the coalescence will not occur for another 300,000,000 years.)

Several other astrophysical sources are possible candidates for gravitation wave emissions which could be observed using earth-based detectors. Supernovae can undergo asymmetric collapse due to density fluctuations and rotational and hydrodynamic instabilities. If these instabilities are sufficiently large, they can result in a large, time dependent quadrapole mass moment accompanied by substantial energy radiation into

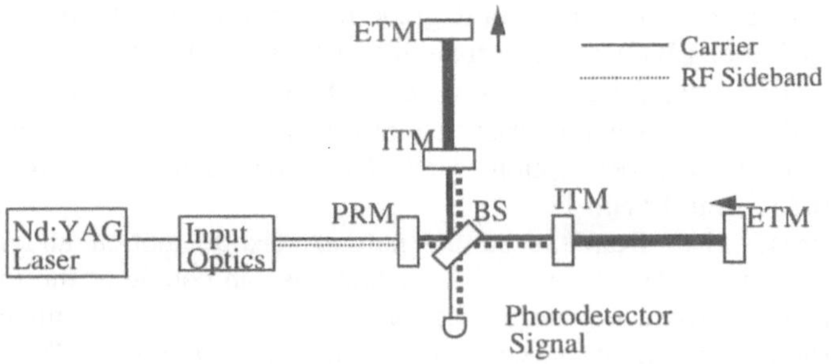

Figure 2 Schematic of the initial LIGO detector configuration. PRM=power recycling mirror, ITM=input test mass, ETM=end test mass, BS=beamsplitter. The arrows indicate the direction of the strain at a given point during the GW passage.

gravitational wave modes as the collapse progresses. Rotating neutron stars with non-axisymmetric mass distributions, neutron star *r*-modes (hydrodynamic currents), and supermassive black holes are also potential sources. Finally, it is possible that a residual stochastic background of gravitational waves exists from the dynamics of formation of the early universe.

THE LIGO INTERFEROMETER

In order to detect the infinitesimal strains caused by the passage of a gravitational wave, LIGO will employ a power-recycled, Fabry-Perot arm cavity Michelson interferometer. A conceptual layout of the interferometer is shown in Figure 2. A single frequency continuous wave Nd:YAG laser operating at 10 W output power will provide the light for the interferometer. The light is then phase-modulated (dashed line) and mode-matched into the interferometer cavities via the input optics. The phase-modulated RF sidebands on the carrier are resonant only in short Michelson cavity (consisting of the PRM and ITMs). The cavities are held in resonance to an absolute position of less than 10^{-13} m using Pound-Drever-Hall locking. Under quiescent conditions, the arm cavities have a path difference of (N+1/2) wavelengths. Thus, the light returning from each arm recombines such that the light travels back toward the laser/input optics. The power recycling mirror is positioned resonantly to reflect this light back into the interferometer. As a gravitational wave passes through the interferometer, the light in each arm cavity experiences a phase shift (equal in magnitude and opposite in sign) which results in some constructive interference of the carrier in the output photodetector port. The phase shift experienced by the carrier is compared to the RF sideband light at the photodetector to read out the gravitational wave signal.

Since the gravitational wave signal is readout as a strain ($\Delta L/L$), it is desirable to have as long an arm length as possible to increase the sensitivity of the detector. LIGO will employ interferometers with arm lengths of 2 and 4 km, less than 10^{-5} the wavelength of gravitational waves in the frequency band of interest. In order to increase the total interaction time, LIGO will employ resonant Fabry-Perot cavities in the arms of the interferometer. The Fabry-Perot arm cavities enhance the storage time of the light in the interferometer (or alternatively, the effective length of the interferometer) by a factor $\sqrt{(r_1r_2)}/[2(1-r_1r_2)]$, where $r_{1,2}$ are the amplitude reflectivities of the input test mass and end test mass, respectively. The storage time is further enhanced by the power recycling mirror. For LIGO, the overall storage time is increased approximately by approximately 1000 times compared to the transit time of the light in the 4 km or 2 km arms.

To achieve a strain sensitivity of $10^{-21}/\sqrt{Hz}$, LIGO must overcome several noise sources which reduce the performance of the detector. Weiss was the first to catalog all of the predominant noise sources which can degrade interferometer sensitivity [9]. Figure 3 shows the overall initial LIGO design sensitivity with noise floors imposed by various sources [10,11]. The predominant noise sources in the initial LIGO detectors are seismic noise (f < 50 Hz), thermal noise in the suspensions (50 Hz < f < 110 Hz), and shot noise (f > 110 Hz). A more detailed description of all of the noise sources and their relation to LIGO can be found in Refs 10 and 11.

Seismic noise arises from the earth's natural seismic motion and from ground motion induced by man-made sources. These vibrations can couple to the interferometer mirrors and produce low frequency mirror displacements many orders of magnitude above the mean displacement amplitude caused by gravitational radiation. To minimize the coupling of ground motion to the mirrors, LIGO will employ two methods. First, mirrors will be suspended from rigid mounts using a pendulum configuration. These suspensions are then mounted on multi-stage vibration isolation platforms. At frequencies greater than the resonant frequencies of the pendulum and stacks, the vibration is attenuated as $\sim f^2$ per isolation stage. We discuss this in further detail below.

Thermal excitations of the optical platform, suspension towers, suspension wires, and mirrors also couple to the motion of the mirrors and will limit the sensitivity of LIGO around 100 Hz. Objects at finite temperatures are thermally excited which induces microscopic motions. The normal modes, peak amplitudes, and lifetimes of these motions depend on the geometry of the object and viscous damping present. While the total root

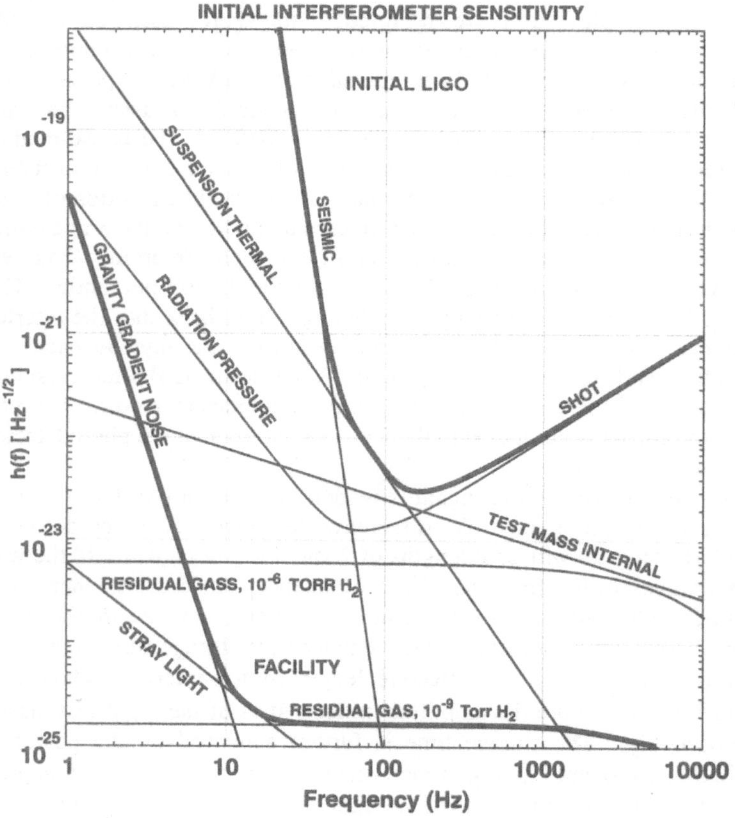

Figure 3 The initial strain sensitivity for the LIGO interferometers showing the contributions of all the noise sources.

mean square microscopic motion of an object (integrated over all frequencies) is governed by the temperature of the object through the equipartition theorem, the peak amplitude of vibration and the damping time of the resonance (or, alternatively, the Q of the resonance) are related through the fluctuation-dissipation theorem. In particular, the spectral amplitude density of the normal mode excitations can be confined to narrow regions about the resonance frequencies for sufficiently high Qs. If the damping is lowered, more of the excitation energy is concentrated in the resonance peak. Ideally, rigid objects with high internal Q's which are thermally excited have most of their motion localized in narrow region about the resonant frequency. While the resonant frequencies in the suspensions and optical platforms are lower than the LIGO detection band, the high frequency thermal 'tails' of the resonant motion can couple to mirror internal modes, exciting displacements in the LIGO band.

Shot noise in the interferometer arises from the discrete, particle-like nature of light. The probability that a single photon will impinge upon the photodetector is governed by Poisson statistics. Since the interferometer is not perfect, some carrier light is always present on the photodetector. For a gravitational wave event to be recorded, the change in detected photocurrent arising from a displacement of the end test masses (due to the GW signal) must be greater than the change in photocurrent from the detection of a random photon. For a power recycled Fabry-Perot arm cavity Michelson interferometer, this leads to a minimum detectable displacement given by [12]:

$$\tilde{x}(f) = \left(\frac{\lambda \sqrt{3E_{SB}^2 + E_S^2}}{4\pi E_C E_{SB}} \right) \left(\frac{(1 - r_1 r_2)^2}{(1 - r_1^2 - \alpha) r_2} \right) \sqrt{1 + \left(\frac{2\pi f}{f_{FP}} \right)^2} \tag{3}$$

where $E^2_{C,SB}$ are the carrier and sideband in the interferometer, E^2_S is the stray light power at the photodetector, f_{FP} is the cavity pole frequency of the arms, and α represents the interferometer losses due to scattering and absorption. For a given E_{SB} and minimal scattered light E_S, the displacement sensitivity depends inversely on the carrier amplitude E_C (alternatively, $\sqrt{P_C}$). While increasing the carrier amplitude (or power) minimizes the shot noise, the photons impinging on the test masses exert radiation pressure in discrete bursts which impart random impulsive forces. Radiation pressure noise increases with increasing interferometer power:

$$\tilde{x}_{RP}(f) = \frac{2}{\pi^2} \frac{\sqrt{r_1 r_2}}{(1 - r_1 r_2) m f^2} \sqrt{\frac{hGE_c^{\,2}}{c\lambda}} \tag{4}$$

where m is the test mass and λ is the laser wavelength. In the limit of negligible scattering, we see that equation (3) and (4) define, for a given mass, the minimum detectable strain for a given frequency.

The Current Status of LIGO Detectors

As of this writing, all of the infrastructure (in the form of laboratory buildings, beam tubes, light baffles, optics chambers, and vacuum systems) is in place at both sites. Construction on the optical beamline of the first interferometer (the 2 km arm length interferometer at Hanford) has begun with the installation of the 10 W, 1064 nm frequency Nd:YAG Pre-stabilized Laser and the LIGO Input Optics. All of the critical detector subsystems have been completed through the final design phase. Fabrication of the components of the core interferometer optics, interferometer length and alignment control systems, seismic isolation systems, and data acquisition, storage, and analysis are

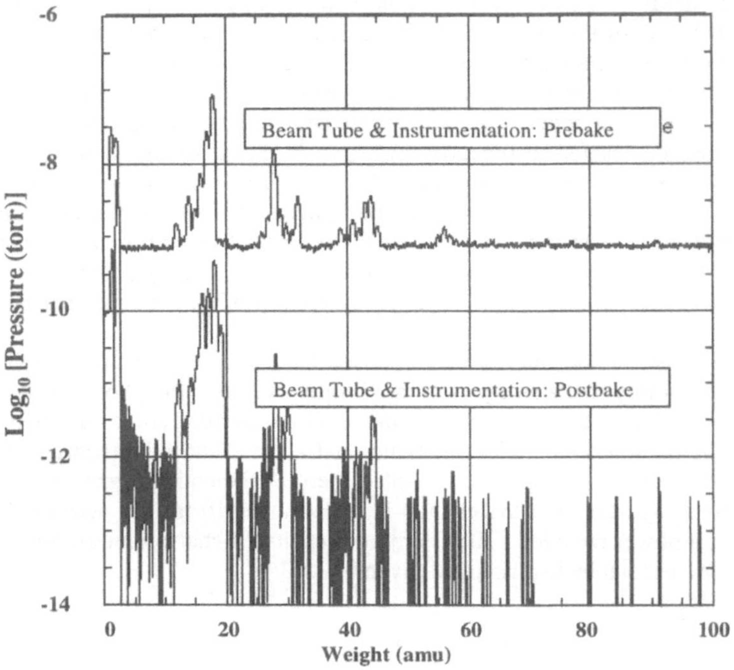

Figure 4 Pressure in the LIGO y-arm beam tube as a function of molecular weight before and after baking.

essentially complete with installation scheduled to begin in early 1999. Detector commissioning will begin in late 1999 for the first (2 km) detector. The second interferometer (the 4 km long interferometer in Livingston) will begin installation in mid 1999 and the final detector (4 km interferometer in Hanford) will follow in late 1999. A two year long LIGO science run at full sensitivity will commence in 2002. In this section, we highlight the status of some representative LIGO subsystems.

LIGO Vacuum System

For a variety of reasons, the LIGO detectors must be operated in a vacuum. The presence of residual gas limits the sensitivity of the interferometer in two ways. First, the microscopic molecular collisions with mirrors and suspensions can excite motions in the LIGO detection band. Second, gas density fluctuations along the propagation length of the laser beam modulate the index of refraction, inducing a time dependent phase shift in the beam which can be interpreted as a gravitational wave signal. They can also scatter light out of the laser beam which can reflect off vacuum tubes and be 'upconverted' into the LIGO detection band by the acoustic motion of the tube. In addition, hydrocarbons present in the beam tubes can contaminate the mirrors by chemically bonding on the surface of the mirror through photochemical reactions on the mirror surface.

In order to minimize these effects, the LIGO interferometers will operate in 1.2 m diameter vacuum tubes which extend for the entire 4 km length of both arms of the interferometer (a total volume of ~ 9000 m^3). The initial LIGO sensitivity goal requires operational pressures of 10^{-9} torr. Hydrocarbon partial pressures must be kept below 10^{-12} torr to minimize contamination.

The initial pump down of the LIGO beam tubes has been completed. Figure 4 shows the pressure of the LIGO Hanford Y-arm beam tube before as a function of

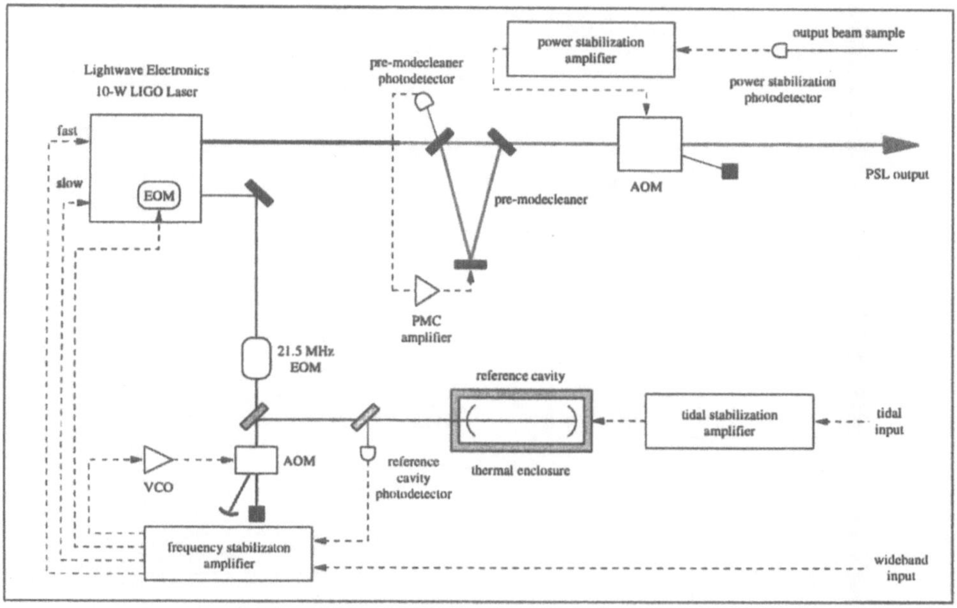

Figure 5 Schematic layout of the LIGO Pre-Stabilized Laser.

molecular weight. The predominant contribution at 18 amu, H_2O, has reduced from ~ 10^{-7} torr to less than 10^{-9} torr after baking. Hydrocarbon contributions in the 40 amu range have been reduced to close to 10^{-12} torr.

LIGO Pre-Stabilized Laser

The LIGO Pre-stabilized Laser (PSL) will provide a frequency-stabilized laser source for the LIGO interferometers. A schematic of the PSL layout is shown in Figure 5. At the heart of the PSL is a Lightwave Electronics single longitudinal mode Nd:YAG laser operating at 1064 nm. The Lightwave laser is configured as a master oscillator power amplifier (MOPA) producing 10 W of output power and nominal TEM_{00} (Gaussian) beam mode. The laser beam then travels though a pre-mode cleaner (PMC) optical cavity that provides spatial filtering and reduces high-frequency intensity fluctuations [13]. Additional intensity stabilization is provided by an acousto-optic modulator positioned after the PMC. In Figure 6, a plot of the relative intensity noise in the frequency range 100 Hz – 100 kHz is shown. The LIGO requirement of $10^{-6}/\sqrt{Hz}$ is also shown for comparison.

In order to stabilize the laser frequency, a small portion of the laser is split off and locked to an external reference cavity. The error signal from the cavity is fed back into the laser oscillator via fast and slow path frequency inputs. This feedback allows the laser to reach a frequency noise level of < 10^{-2} Hz/$\sqrt{Hz}$ in the LIGO detection band (See Figure 7). Further frequency noise suppression to < 10^{-7} Hz/$\sqrt{Hz}$ is provided by control inputs derived from the locking of the mode cleaner and the LIGO arm cavities.

LIGO Optical Components

The development of optical components (input test masses, end test masses, recycling mirrors, and beamsplitters) for the LIGO interferometers has pushed the state-of-

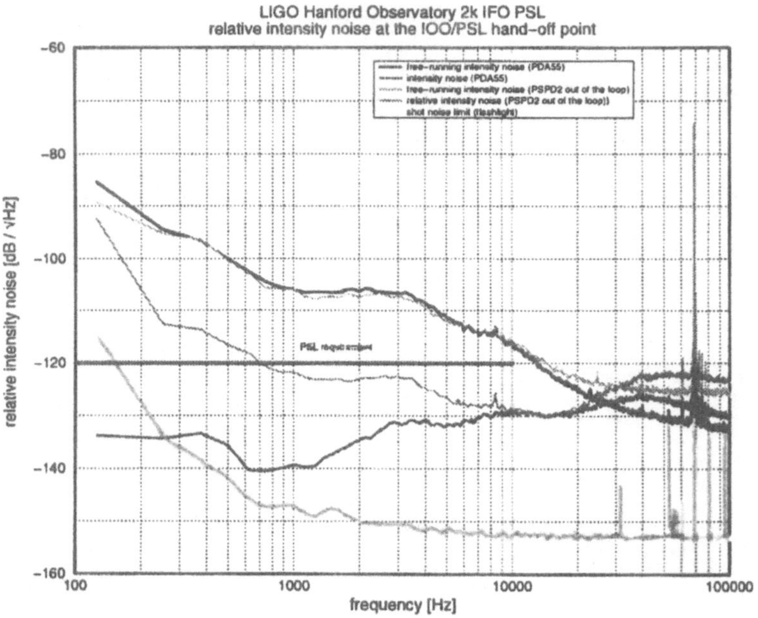

Figure 6 The relative intensity noise of the LIGO Pre-Stabilized Laser for under free running and locked operation. The requirement of 120 dB/√Hz is also shown.

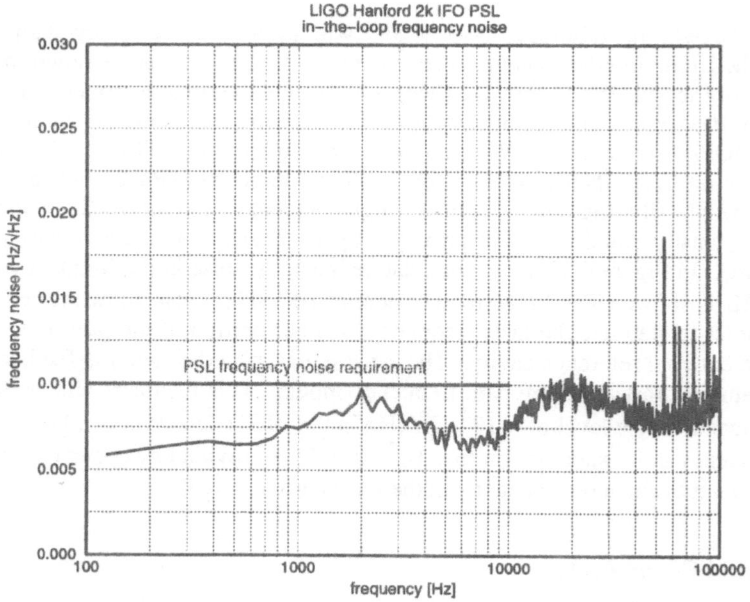

Figure 7 Frequency noise of the LIGO Pre-Stabilized Laser when locked to an external reference cavity. The requirement is shown for comparison.

the-art in optical materials, optics processing (polishing and coating), and optical metrology. The LIGO core optical components are made from high-quality, low-absorption fused silica. To collect and reflect the cavity laser light over 4 km propagation lengths, the mirror dimensions are 250 mm in diameter. Thermal noise considerations dictate a thickness of 100 mm.

The initial LIGO strain sensitivity places stringent restrictions on all aspects of the optical fabrication:

- Absorption – The absorption of light in optical components and coatings can alter the spatial profile of the laser beam and reduce the amount of laser light stored in the interferometer. For initial LIGO, absorption levels must be in the range of 50 – 100 ppm or less for all of the LIGO core optics. Requirements on the surface absorption are particularly severe for the input and end test masses (0.6 and 2 ppm respectively).
- Scattering – As noted above, laser light scattered from the surface or the bulk matter of the optic can be interpreted as a gravitational wave signal if it is upconverted into the LIGO detection band. For initial LIGO, scattering levels must be in the range of 100 – 200 ppm or less for all of the LIGO core optics.
- High Mechanical Q's – The location and Q's of the resonances in the LIGO core optics must be such that they minimize the thermal noise in the 100-1000 Hz frequency range. This requires resonance frequencies greater than 3 kHz with internal mechanical Q's of 5 x 10^6.
- Polishing – Wave front distortions in the LIGO laser beams couple to increased shot noise through the enhancement of contrast defects at the dark port of the interferometer. For the input and end test masses, the distortions in the surface figure of the optics must be kept at root mean square levels of $\lambda/1200$ (equivalently, < 1 nm) or better. In addition, small spatial scale distortions (microroughness) lead to increased surface scattering. LIGO core optics must be maintained at root mean square levels of 0.6 nm or better.
- Coating – Mismatches in arm cavity reflectivities can lead to a loss of contrast at the dark port. In addition, the coatings cannot distort the surface figure or enhance the scattering losses of the optic.
- Metrology – Methods of sufficient sensitivity must be developed to determine that the above requirements are met.

Two examples of high precision metrology on the LIGO core optics are displayed in Figures 8 and 9. Figure 8 displays the absorption (in ppm/cm) as a function of

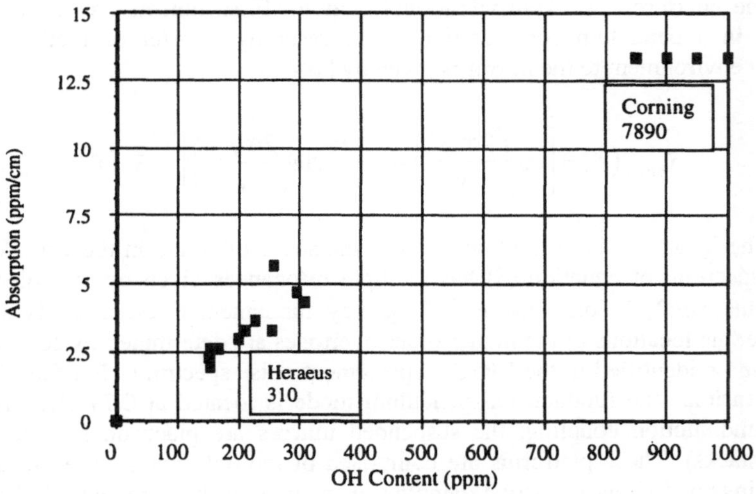

Figure 8 Absorption as a function of OH content for two different types of fused silica glasses.

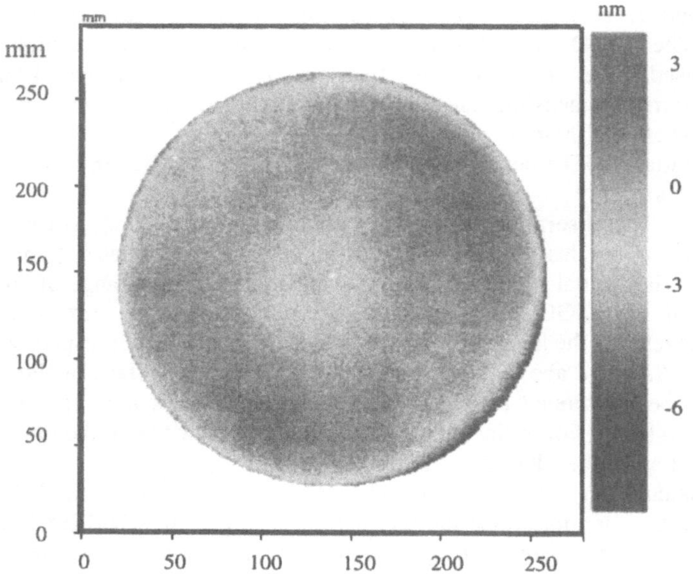

Figure 9 **Phase map of a LIGO folding mirror after polishing. Over the central 10 cm of the optic, the mean surface deviation is < 2 nm.**

molecular OH content in the fused silica for two different types of glass. This data provides strong evidence that the absorption of 1064 nm light is predominantly due to the presence of OH. In Figure 9 we present a phase map of the surface of a LIGO folding mirror for the 2 km interferometer. This measurement was performed at CSIRO in Australia using a high resolution Fizeau interferometer. Over the central 10 cm aperture of the optic, the surface deviates from an ideal flat surface by less than 2 nm.

LIGO Vibration Isolation

As discussed above, minimizing seismic coupling to the LIGO optical components is critical for achieving optimal low frequency strain sensitivity. LIGO uses two methods to isolate the interferometer from vibrational motion. First, mirrors are suspended from steel wires in a pendulum configuration. The resulting transfer of motion from the surrounding environment to the mirror is given by [14]:

$$\tilde{x}_{mirror}(f) = \left[\cos\left(\frac{2\pi f l}{v}\right) - \frac{2\pi f v}{g}\sin\left(\frac{2\pi f l}{v}\right)\right]^{-1} \tilde{x}_{in}(f) \qquad (5)$$

where l is the length of the pendulum and v is the speed of sound in the wire. While the amplitude spectrum of equation (5) has multiple resonances given by the solutions of $f = \{2\pi v/[g \tan(2\pi f l/v)]\}^{-1}$, only the low frequency fundamental resonance is amplified. Furthermore, the locations of the higher order overtones are determined by equation 5 and therefore can be identified in the LIGO displacement noise spectrum. For the LIGO core optics suspensions, the fundamental pendulum mode is located at 0.74 Hz. To further reduce ground motion coupling, the suspended mirrors are place on seismic isolation platforms (stacks). These platforms are comprised of multiple mass stages separated by damped springs and allow for active damping of the motion though feedback. Each stage acts like a damped harmonic oscillator and severely attenuate motion above its resonant

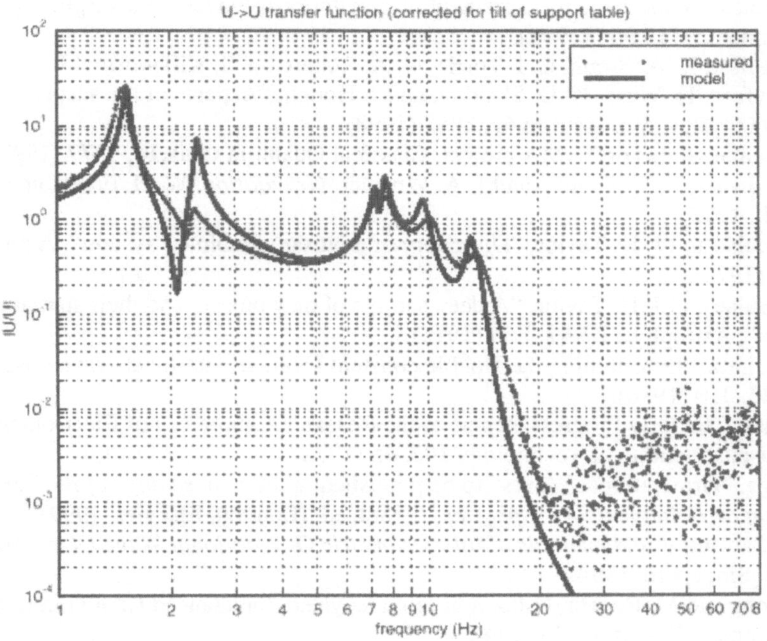

Figure 10 **Transfer function of the seismic isolation platform for transverse excitations.**

frequency [11]. Resonant frequencies are typically below 10 Hz. Figure 10 shows the calculated and measured horizontal-horizontal transfer function for a prototype LIGO suspension platform. In accordance with the stack design, multiple resonances appear below 10 Hz. Above 10 Hz, the transfer function decreases rapidly, falling off as $\sim f^8$.

CONCLUSIONS

LIGO promises to open a new window on heretofore unobservable astrophysical processes in the universe through its ability to detect *directly* gravitational waves. The detection of gravitational radiation, in itself a formidable scientific challenge, will also allow us to observe the dynamics of astrophysical events such as binary star coalescences and supernovae explosions in new ways. The LIGO detectors will become operational in 2000.

ACKNOWLEDGMENTS

The authors thank Mark Barton, Garilynn Billingsley, Jordan Camp, Janeen Hazel-Romie, Peter King, Daniel Sigg, Rai Weiss, and Stan Whitcomb for their contributions to this article. This work is supported by the National Science Foundation through grant PHY-9722114.

REFERENCES

1. A. Abramovici, W. E. Althouse, R. Drever, Y. Gursel, S. Kawamura, F. Raab, D. Shoemaker, L. Sievers, R. Spero, K. Thorne, R. Vogt, R. Weiss, S. Whitcomb, and M. E. Zucker, "LIGO - the laser-interferometer-gravitational-wave-observatory" Science **256**, 325 (1992).

2.A. Giazotto, "Wide-band measurement of gravitational-waves - the Virgo Project", Nuovo Cimento C**15** 955 (1992).
3. M. V. Plissi, K. A. Strain, C. I. Torrie, N. A. Robertson, S. Killbourn, S.Rowan, S. M. Twyford, H. Ward, K. D. Skeldon, and J. Hough, "Aspects of the suspension system for GEO 600" Rev. Sci. Instrum.**69**, 3055 (1998).
4. K. Kawabe, Status of TAMA project, Classical Quantum Grav.**14**, 1477 (1997).
5. C. W. Misner, K. S. Thorne, and J. A. Wheeler, *Gravitation*, W. H. Freeman and Co., New York, 1973.
6. R. A. Hulse and J. H. Taylor, "Discovery of a pulsar in a binary system", Astrophys. J. **195**, L51 (1975).
7. R. A. Hulse and J. H. Taylor, "A deep sample of new pulsars and their spatial extent in the galaxy", Astrophys. J. **201**, L55 (1975).
8. B. F. Schutz, "Determining the Hubble constant From gravitational-wave observations", Nature **323**, 310 (1986).
9. R. Weiss, Quarterly Progress Report of the Research Laboratory of Electronics of the Massachusetts Institute of Technology **105**, 45 (1972).
10. D. Sigg, "Gravitational Waves" to be published in the *Proceedings of the Texas Symposium on Relativistic Astrophysics*, J. Paul, ed. (London, Elsevier, 1999).
11. P. R. Saulson, *Fundamentals of Interferometric Gravitational Wave Detectors*, World Scientific, Singapore, 1994.
12. T. Lyons and M. Regehr, "Shot noise in a recycled, unbalanced LIGO interferometer", LIGO Technical Report #113 (1994).
13. B. Wilke, N. Uehara, E. K. Gustafson, R. L. Byer, P. J. King, S. U. Seel, and R. L. Savage, Jr. "Spatial and temporal filtering of a 10-W Nd:YAG laser with a Fabry-Perot ring-cavity premode cleaner" Opt. Lett. **23**, 1704 (1998).
14. N. A. Robertson, in *The Detection of Gravitational Waves,* ed. D. G. Blair, Cambridge University Press, Cambridge, 1991.

SOLAR NEUTRINOS: AN OVERVIEW

J. N. Bahcall[1]

Institute for Advanced Study
Olden Lane
Princeton, NJ 08540

INTRODUCTION

The most important result from solar neutrino research is, in my view, that solar neutrinos have been detected experimentally with fluxes and energies that are qualitatively consistent with solar models that are constructed assuming that the sun shines by nuclear fusion reactions. The first experimental result, obtained by Ray Davis and his collaborators in 1968[1, 2], has now been confirmed by four other beautiful experiments, Kamiokande[3], SAGE[4], GALLEX[5], and SuperKamiokande[6].

The observation of solar neutrinos with approximately the predicted energies and fluxes establishes empirically the theory[7] that main sequence stars derive their energy from nuclear fusion reactions in their interiors and has inaugurated what we all hope will be a flourishing field of observational neutrino astronomy. The detections of solar neutrinos settle experimentally the debate over the age and energy source of the sun that raged for many decades, beginning in the middle of the 19th century. The leading theoretical physicists of the 19th century argued convincingly that the sun could not be more than 10^7 years old because that was the maximum lifetime that could be fueled by gravitational energy("No other natural explanation, except chemical action, can be conceived.", Lord Kelvin[8]). On the other hand, geologists and evolutionary biologists argued that the sun must be $> 10^9$ years old in order to account for observed geological features and for evolutionary processes[9].[2] Today we know that the biologists and geologists were right and the theoretical physicists were wrong, which may be a historical lesson to which we physicists should pay attention.

I will discuss predictions of the combined standard model in the main part of this review. By 'combined' standard model, I mean the predictions of the standard solar model and the predictions of the minimal electroweak theory. We need a solar model to tell us how many neutrinos of what energy are produced in the sun and we need electroweak theory to tell us how the number and flavor content of the neutrinos are changed as they make their way from the center of the sun to detectors on earth. For all practical purposes, standard electroweak theory states that nothing happens

[1]e-mail: jnb@sns.ias.edu

[2]The arguments of Lord Kelvin and his theoretical physics associates were so persuasive that in later editions Darwin removed all mention of time scales from "The Origin of the Species."

to solar neutrinos after they are created in the deep interior of the sun.

Using standard electroweak theory and fluxes from the standard solar model, one can calculate the rates of neutrino interactions in different terrestrial detectors with a variety of energy sensitivities. The combined standard model also predicts that the energy spectrum from a given neutrino source should be the same for neutrinos produced in terrestrial laboratories and in the sun and that there should not be measurable time-dependences (other than the seasonal dependence caused by the earth's orbit around the sun). The spectral and temporal departures from standard model expectations are expected to be small in all currently operating experiments[10] and have not yet yielded definitive results. Therefore, I will concentrate here on inferences that can be drawn by comparing the total rates observed in solar neutrino experiments with the combined standard model predictions.

I will begin by reviewing the quantitative predictions of the combined standard solar model and describing the three solar neutrino problems that are established by the chlorine, Kamiokande, SAGE, GALLEX, and SuperKamiokande experiments. I then detail the uncertainties in the standard model predictions and then show that helioseismological measurements indicate that the standard solar model predictions are accurate for our purposes. I next discuss the implications for solar neutrino research of the precise agreement between helioseismological measurements and the predictions of standard solar models. Ignoring all knowledge of the sun, I then cite analyses in that show that one cannot fit the existing experimental data with neutrino fluxes that are arbitrary parameters, unless one invokes new physics to change the shape or flavor content of the neutrino energy spectrum. Finally, I summarize the characteristics of the best-fitting neutrino oscillation descriptions of the experimental data, and discuss and summarize the results.

If you want to obtain numerical data or subroutines that are discussed in this talk, or to see relevant background information, you can copy them from my Web site: http://www.sns.ias.edu/~jnb .

STANDARD MODEL PREDICTIONS

Table 1 gives the neutrino fluxes and their uncertainties for our best standard solar model, hereafter BP98[11]. Figure 1 shows the predicted neutrino fluxes from the dominant *p-p* fusion chain.

The BP98 solar model includes diffusion of heavy elements and helium, makes use of the nuclear reaction rates recommended by the expert workshop held at the Institute of Nuclear Theory[12], recent (1996) Livermore OPAL radiative opacities[13], the OPAL equation of state[14], and electron and ion screening as determined by the recent density matrix calculation[15, 16]. The neutrino absorption cross sections that are used in constructing Table 1 are the most accurate values available[17, 18] and include, where appropriate, the thermal energy of fusing solar ions and improved nuclear and atomic data. The validity of the absorption cross sections has recently been confirmed experimentally using intense radioactive sources of ^{51}Cr. The ratio, R, of the capture rate measured (in GALLEX and SAGE) to the calculated ^{51}Cr capture rate is $R = 0.95 \pm 0.07$ (exp) + $^{+0.04}_{-0.03}$ (theory) and was discussed extensively at Neutrino 98 by Gavrin and by Kirsten. The neutrino-electron scattering cross sections, used in interpreting the Kamiokande and SuperKamiokande experiments, now include electroweak radiative corrections[19].

Figure 2 shows for the chlorine experiment all the predicted rates and the estimated uncertainties(1σ) published by my colleagues and myself since the first mea-

Table 1. Standard Model Predictions (BP98): solar neutrino fluxes and neutrino capture rates, with 1σ uncertainties from all sources (combined quadratically).

Source	Flux $(10^{10}\ \mathrm{cm^{-2}s^{-1}})$	Cl (SNU)	Ga (SNU)
pp	$5.94\left(1.00^{+0.01}_{-0.01}\right)$	0.0	69.6
pep	$1.39\times10^{-2}\left(1.00^{+0.01}_{-0.01}\right)$	0.2	2.8
hep	2.10×10^{-7}	0.0	0.0
^{7}Be	$4.80\times10^{-1}\left(1.00^{+0.09}_{-0.09}\right)$	1.15	34.4
^{8}B	$5.15\times10^{-4}\left(1.00^{+0.19}_{-0.14}\right)$	5.9	12.4
^{13}N	$6.05\times10^{-2}\left(1.00^{+0.19}_{-0.13}\right)$	0.1	3.7
^{15}O	$5.32\times10^{-2}\left(1.00^{+0.22}_{-0.15}\right)$	0.4	6.0
^{17}F	$6.33\times10^{-4}\left(1.00^{+0.12}_{-0.11}\right)$	0.0	0.1
Total		$7.7^{+1.2}_{-1.0}$	129^{+8}_{-6}

surement by Ray Davis and his colleagues in 1968. This figure should give you some feeling for the robustness of the solar model calculations. Many hundreds and prob-

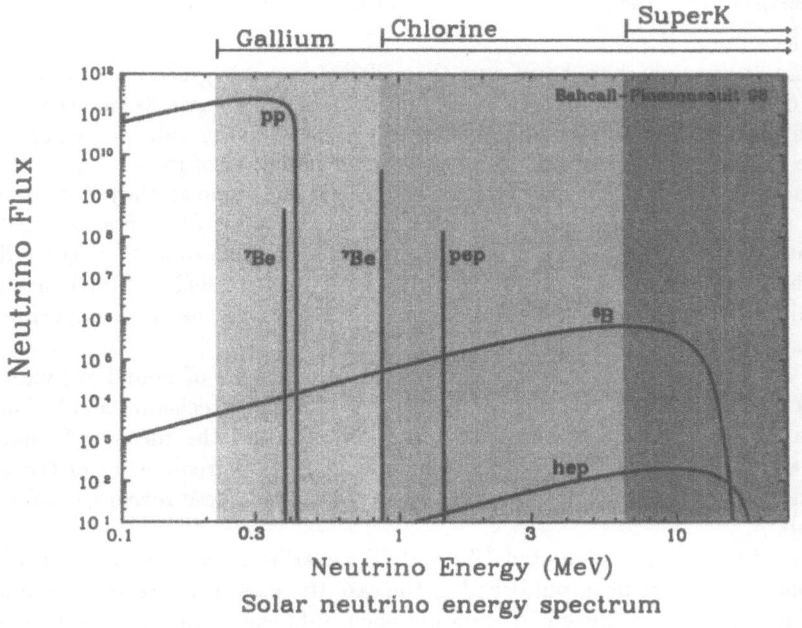

Figure 1. The energy Spectrum of neutrinos from the pp chain of interactions in the Sun, as predicted by the standard solar model. Neutrino fluxes from continuum sources (such as pp and 8B) are given in the units of counts per cm2 per second. The pp chain is responsible for more than 98% of the energy generation in the standard solar model. Neutrinos produced in the carbon-nitrogen-oxygen CNO chain are not important energetically and are difficult to detect experimentally. The arrows at the top of the figure indicate the energy thresholds for the ongoing neutrino experiments.

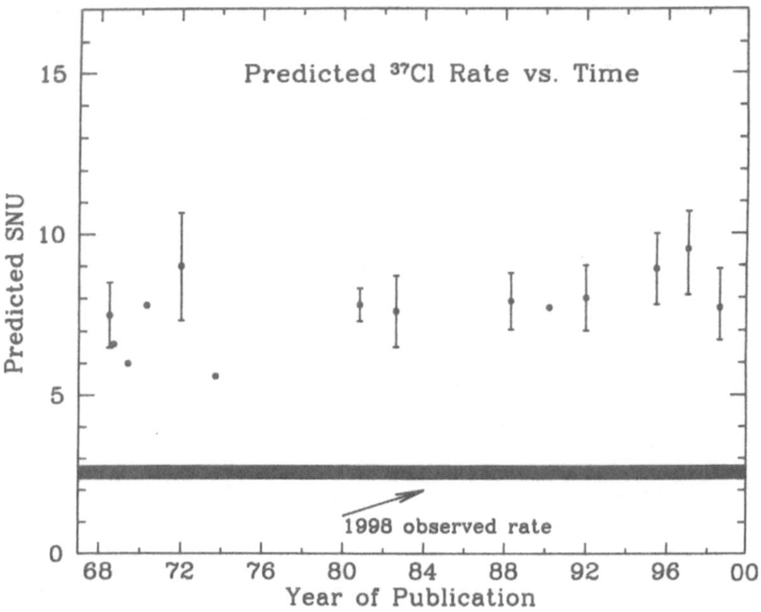

Figure 2. The predictions of John Bahcall and his collaborators of neutrino capture rates in the ^{37}Cl experiment are shown as a function of the date of publication(since the first experimental report in 1968[1]). The event rate SNU is a convenient product of neutrino flux times interaction cross section, 10^{-36} interactions per target atom per sec. The format is from Figure 1.2 of the book Neutrino Astrophysics[20]. The predictions have been updated through 1998.

ably thousands of researchers have, over three decades, made great improvements in the input data for the solar models, including nuclear cross sections, neutrino cross sections, measured element abundances on the surface of the sun, the solar luminosity, the stellar radiative opacity, and the stellar equation of state. Nevertheless, the most accurate predictions of today are essentially the same as they were in 1968 (although now they can be made with much greater confidence). For the gallium experiments, the neutrino fluxes predicted by standard solar models, corrected for diffusion, have been in the range 120 SNU to 141 SNU since 1968[17]. A SNU is a convenient unit with which to describe the measured rates of solar neutrino experiments: 10^{-36} interactions per target atom per second.

There are three reasons that the theoretical calculations of neutrino fluxes are robust: 1) the availability of precision measurements and precision calculations of input data; 2) the connection between neutrino fluxes and the measured solar luminosity; and 3) the measurement of the helioseismological frequencies of the solar pressure-mode (p-mode) eigenfrequencies. I have discussed these reasons in detail in another talk[21].

Figure 3 displays the calculated ^{7}Be and ^{8}B neutrino fluxes for all 19 standard solar models which have been published in the last 10 years in refereed science journals. The fluxes are normalized by dividing each published value by the flux from the BP98 solar model[11]; the abscissa is the normalized ^{8}B flux and the ordinate is the normalized ^{7}Be neutrino flux. The rectangular box shows the estimated 3σ uncertainties in the predictions of the BP98 solar model.

All of the solar model results from different groups fall within the estimated 3σ uncertainties in the BP98 analysis (with the exception of the Dar-Shaviv model whose

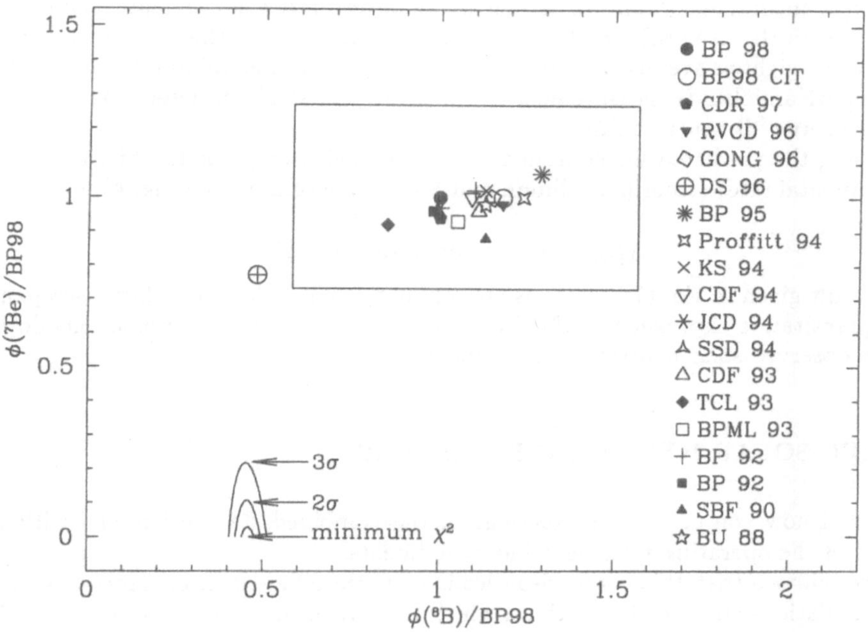

Figure 3. **Predictions of standard solar models since 1988. This figure, which is Fig. 1 of Ref. 10, shows the predictions of 19 standard solar models in the plane defined by the ^{7}Be and ^{8}B neutrino fluxes. The abbreviations that are used in the figure to identify different solar models are defined in the bibliographical item, Ref. 22. The figure includes all standard solar models with which I am familiar that were published in refereed journals in the decade 1988-1998. All of the fluxes are normalized to the predictions of the Bahcall-Pinsonneault 1998 solar model, BP98[11]. The rectangular error box defines the 3σ error range of the BP98 fluxes. The best-fit ^{7}Be neutrino flux is negative. At the 99% C.L., there is no solution[10] with all positive neutrino fluxes (see discussion in the section on "Fits Without Solar Models"). All of the standard model solutions lie far from the best-fit solution, even far from the 3σ contour.**

results have not been reproduced by other groups). This agreement demonstrates the robustness of the predictions since the calculations use different computer codes (which achieve varying degrees of precision) and involve a variety of choices for the nuclear parameters, the equation of state, the stellar radiative opacity, the initial heavy element abundances, and the physical processes that are included.

The largest contributions to the dispersion in values in Figure 3 are due to the choice of the normalization for S_{17} (the production cross-section factor for ^{8}B neutrinos) and the inclusion, or non-inclusion, of element diffusion in the stellar evolution codes. The effect in the plane of Fig. 3 of the normalization of S_{17} is shown by the difference between the point for BP98 (1.0,1.0), which was computed using the most recent recommended normalization[12], and the point at (1.18,1.0) which corresponds to the BP98 result with the earlier (CalTech) normalization[23].

Helioseismological observations have shown[11, 24] that element diffusion is occurring and must be included in solar models, so that the most recent models shown in Fig. 3 now all include helium and heavy element diffusion. By comparing a large number of earlier models, it was shown that all published standard solar models give the same results for solar neutrino fluxes to an accuracy of better than 10% if the same input parameters and physical processes are included[25, 26].

Bahcall, Krastev, and Smirnov[10] have compared the observed rates with the calculated, standard model values, combining quadratically the theoretical solar model and experimental uncertainties, as well as the uncertainties in the neutrino cross sections. Since the GALLEX and SAGE experiments measure the same quantity, we treat the weighted average rate in gallium as one experimental number. We adopt the SuperKamiokande measurement as the most precise direct determination of the higher-energy ^8B neutrino flux.

Using the predicted fluxes from the BP98 model, the χ^2 for the fit to the three experimental rates (chlorine, gallium, and SuperKamiokande, see Fig. 4) is

$$\chi^2_{\mathrm{SSM}}(\text{3 experimental rates}) = 61 . \tag{1}$$

The result given in Eq. (1), which is approximately equivalent to a 20σ discrepancy, is a quantitative expression of the fact that the standard model predictions do not fit the observed solar neutrino measurements.

THREE SOLAR NEUTRINO PROBLEMS

I will now compare the predictions of the combined standard model with the results of the operating solar neutrino experiments.

We will see that this comparison leads to three different discrepancies between the calculations and the observations, which I will refer to as the three solar neutrino problems.

Figure 4 shows the measured and the calculated event rates in the five ongoing solar neutrino experiments. This figure reveals three discrepancies between the experimental results and the expectations based upon the combined standard model. As we shall see, only the first of these discrepancies depends in an important way upon the predictions of the standard solar model.

Calculated Versus Observed Absolute Rate

The first solar neutrino experiment to be performed was the chlorine radiochemical experiment[2], which detects electron-type neutrinos that are more energetic than 0.81 MeV. After more than a quarter of a century of operation of this experiment, the measured event rate is 2.56 ± 0.23 SNU, which is a factor of three less than is predicted by the most detailed theoretical calculations, $7.7^{+1.2}_{-1.0}$ SNU[11]. Most of the predicted rate in the chlorine experiment is from the rare, high-energy ^8B neutrinos, although the ^7Be neutrinos are also expected to contribute significantly. According to standard model calculations, the *pep* neutrinos and the CNO neutrinos (for simplicity not discussed here) are expected to contribute less than 1 SNU to the total event rate.

This discrepancy between the calculations and the observations for the chlorine experiment was, for more than two decades, the only solar neutrino problem. I shall refer to the chlorine disagreement as the "first" solar neutrino problem.

Incompatibility of Chlorine and Water Experiments

The second solar neutrino problem results from a comparison of the measured event rates in the chlorine experiment and in the Japanese pure-water experiments, Kamiokande[3] and SuperKamiokande[6]. The water experiments detect higher-energy neutrinos, most easily above 7 MeV, by by observing the Cerenkov radiation from neutrino-electron scattering: $\nu + e \longrightarrow \nu' + e'$. According to the standard solar

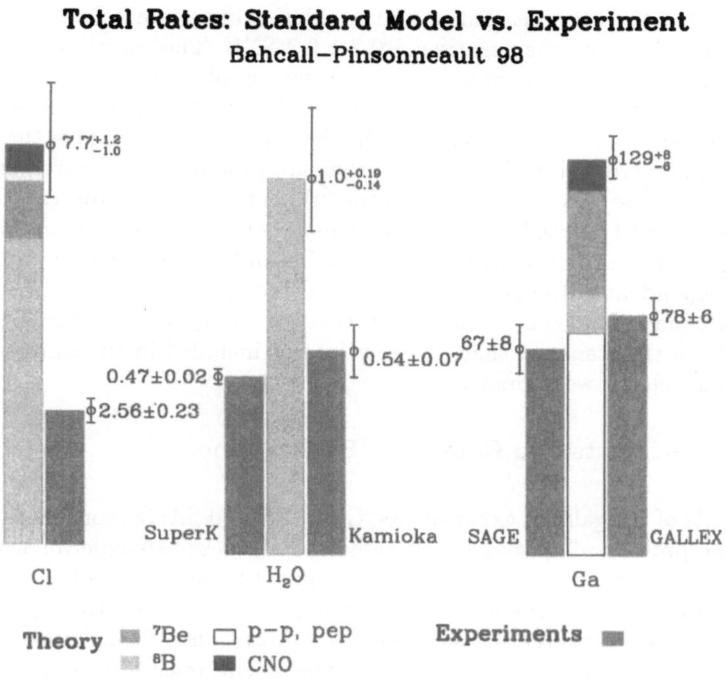

Figure 4. Comparison of measured rates and standard-model predictions for five solar neutrino experiments, [2, 3, 4, 5, 6] The unit for the radiochemical experiments (chlorine and gallium) is SNU (see Fig. 2 for a definition); the unit for the water-Cerenkov experiments(Kamiokande and SuperKamiokande) is the rate predicted by the standard solar model plus standard electroweak theory[11].

model, ^{8}B beta decay, and possibly the *hep* reaction[27], are the only important source of these higher-energy neutrinos.

The Kamiokande and SuperKamiokande experiments show that the observed neutrinos come from the sun. The electrons that are scattered by the incoming neutrinos recoil predominantly in the direction of the sun-earth vector; the relativistic electrons are observed by the Cerenkov radiation they produce in the water detector. In addition, the water Cerenkov experiments measure the energies of individual scattered electrons and therefore provide information about the energy spectrum of the incident solar neutrinos.

The total event rate in the water experiments, about 0.5 the standard model value (see Fig. 4), is determined by the same high-energy ^{8}B neutrinos that are expected, on the basis of the combined standard model, to dominate the event rate in the chlorine experiment. I have shown elsewhere[28] that solar physics changes the shape of the ^{8}B neutrino spectrum by less than 1 part in 10^5 . Therefore, we can calculate the rate in the chlorine experiment (threshold 0.8 MeV) that is produced by the ^{8}B neutrinos observed in the Kamiokande and SuperKamiokande experiments at an order of magnitude higher energy threshold.

If no new physics changes the shape of the ^{8}B neutrino energy spectrum, the chlorine rate from ^{8}B alone is 2.8 ± 0.1 SNU for the SuperKamiokande normalization (3.2 ± 0.4 SNU for the Kamiokande normalization), which exceeds the total observed chlorine rate of 2.56 ± 0.23 SNU.

Comparing the rates of the SuperKamiokande and the chlorine experiments, one finds–assuming that the shape of the energy spectrum of ^{8}B ν_e's is not changed by new physics–that the net contribution to the chlorine experiment from the *pep*, ^{7}Be, and CNO neutrino sources is negative: -0.2 ± 0.3 SNU. The contributions from the *pep*, ^{7}Be, and CNO neutrinos would appear to be completely missing; the standard model prediction for the combined contribution of *pep*, ^{7}Be, and CNO neutrinos is a relatively large 1.8 SNU(see Table 1). On the other hand, we know that the ^{7}Be neutrinos must be created in the sun since they are produced by electron capture on the same isotope(^{7}Be) which gives rise to the ^{8}B neutrinos by proton capture.

Hans Bethe and I pointed out[29] that this apparent incompatibility of the chlorine and water-Cerenkov experiments constitutes a "second" solar neutrino problem that is almost independent of the absolute rates predicted by solar models. The inference that is usually made from this comparison is that the energy spectrum of ^{8}B neutrinos is changed from the standard shape by physics not included in the simplest version of the standard electroweak model.

Gallium Experiments: No Room for ^{7}Be Neutrinos

The results of the gallium experiments, GALLEX and SAGE, constitute the third solar neutrino problem. The average observed rate in these two experiments is 73 ± 5 SNU, which is accounted for in the standard model by the theoretical rate of 72.4 SNU that is calculated to come from the basic *p-p* and *pep* neutrinos (with only a 1% uncertainty in the standard solar model *p-p* flux). The ^{8}B neutrinos, which are observed above 6.5 MeV in the Kamiokande experiment, must also contribute to the gallium event rate. Using the standard shape for the spectrum of ^{8}B neutrinos and normalizing to the rate observed in Kamiokande, ^{8}B contributes another 6 SNU. (The contribution predicted by the standard model is 12 SNU, see Table 1.) Given the measured rates in the gallium experiments, there is no room for the additional 34 ± 3 SNU that is expected from ^{7}Be neutrinos on the basis of standard solar models(see Table 1).

The seeming exclusion of everything but *p-p* neutrinos in the gallium experiments is the "third" solar neutrino problem. This problem is essentially independent of the previously-discussed solar neutrino problems, since it depends strongly upon the *p-p* neutrinos that are not observed in the other experiments and whose theoretical flux can be calculated accurately.

The missing ^{7}Be neutrinos cannot be explained away by a change in solar physics. The ^{8}B neutrinos that are observed in the Kamiokande experiment are produced in competition with the missing ^{7}Be neutrinos; the competition is between electron capture on ^{7}Be versus proton capture on ^{7}Be. Solar model explanations that reduce the predicted ^{7}Be flux generically reduce much more (too much) the predictions for the observed ^{8}B flux.

The flux of ^{7}Be neutrinos, $\phi(^7\mathrm{Be})$, is independent of measurement uncertainties in the cross section for the nuclear reaction $^7\mathrm{Be}(p,\gamma)^8\mathrm{B}$; the cross section for this proton-capture reaction is the most uncertain quantity that enters in an important way in the solar model calculations. The flux of ^{7}Be neutrinos depends upon the proton-capture reaction only through the ratio

$$\phi(^7\mathrm{Be}) \;\propto\; \frac{R(e)}{R(e)+R(p)}, \tag{2}$$

where $R(e)$ is the rate of electron capture by ^{7}Be nuclei and $R(p)$ is the rate of proton capture by ^{7}Be. With standard parameters, solar models yield $R(p) \approx 10^{-3} R(e)$. Therefore, one would have to increase the value of the $^7\mathrm{Be}(p,\gamma)^8\mathrm{B}$ cross section by more than two orders of magnitude over the current best-estimate (which has an estimated uncertainty of $\sim 10\%$) in order to affect significantly the calculated ^{7}Be solar neutrino flux. The required change in the nuclear physics cross section would also increase the predicted neutrino event rate by more than 100 in the Kamiokande experiment, making that prediction completely inconsistent with what is observed.

I conclude that either: 1) at least three of the five operating solar neutrino experiments (the two gallium experiments plus either chlorine or the two water Cerenkov experiments, Kamiokande and SuperKamiokande) have yielded misleading results, or 2) physics beyond the standard electroweak model is required to change the energy spectrum of ν_e after the neutrinos are produced in the center of the sun.

UNCERTAINTIES IN THE FLUX CALCULATIONS

I will now discuss uncertainties in the solar model flux calculations.

Table 2 summarizes the uncertainties in the most important solar neutrino fluxes and in the Cl and Ga event rates due to different nuclear fusion reactions (the first four entries), the heavy element to hydrogen mass ratio (Z/X), the radiative opacity, the solar luminosity, the assumed solar age, and the helium and heavy element diffusion coefficients. The ^{14}N $+ p$ reaction causes a 0.2% uncertainty in the predicted pp flux and a 0.1 SNU uncertainty in the Cl (Ga) event rates.

The predicted event rates for the chlorine and gallium experiments use recent improved calculations of neutrino absorption cross sections[17, 18]. The uncertainty in the prediction for the gallium rate is dominated by uncertainties in the neutrino absorption cross sections, +6.7 SNU (7% of the predicted rate) and −3.8 SNU (3% of the predicted rate). The uncertainties in the chlorine absorption cross sections cause an error, ±0.2 SNU (3% of the predicted rate), that is relatively small compared to other uncertainties in predicting the rate for this experiment. For non-standard neutrino energy spectra that result from new neutrino physics, the uncertainties in the predictions for currently favored solutions (which reduce the contributions from the least well-determined ^{8}B neutrinos) will in general be less than the values quoted here for standard spectra and must be calculated using the appropriate cross section uncertainty for each neutrino energy[17, 18].

The nuclear fusion uncertainties in Table 2 were taken from Adelberger *et al.*[12], the neutrino cross section uncertainties from[17, 18], the heavy element uncertainty was

Table 2. Average uncertainties in neutrino fluxes and event rates due to different input data. The flux uncertainties are expressed in fractions of the total flux and the event rate uncertainties are expressed in SNU. The ^{7}Be electron capture rate causes an uncertainty of ±2%[30] that affects only the ^{7}Be neutrino flux. The average fractional uncertainties for individual parameters are shown.

<Fractional uncertainty>	pp	^{3}He^{3}He	^{3}He^{4}He	^{7}Be $+ p$	Z/X	opac	lum	age	diffuse
	0.017	0.060	0.094	0.106	0.033		0.004	0.004	
Flux									
pp	0.002	0.002	0.005	0.000	0.002	0.003	0.003	0.0	0.003
^{7}Be	0.0155	0.023	0.080	0.000	0.019	0.028	0.014	0.003	0.018
^{8}B	0.040	0.021	0.075	0.105	0.042	0.052	0.028	0.006	0.040
SNUs									
Cl	0.3	0.2	0.5	0.6	0.3	0.4	0.2	0.04	0.3
Ga	1.3	0.9	3.3	1.3	1.6	1.8	1.3	0.20	1.5

taken from helioseismological measurements[31], the luminosity and age uncertainties were adopted from BP95[26], the 1σ fractional uncertainty in the diffusion rate was taken to be 15%[32], which is supported by helioseismological evidence[24], and the opacity uncertainty was determined by comparing the results of fluxes computed using the older Los Alamos opacities with fluxes computed using the modern Livermore opacities[25]. To include the effects of asymmetric errors, the now publicly-available code for calculating rates and uncertainties (see discussion in previous section) was run with different input uncertainties and the results averaged. The software contains a description of how each of the uncertainties listed in Table 2 were determined and used.

The low energy cross section of the $^7\mathrm{Be} + p$ reaction is the most important quantity that must be determined more accurately in order to decrease the error in the predicted event rates in solar neutrino experiments. The ^{8}B neutrino flux that is measured by the Kamiokande[3], Super-Kamiokande[6], and SNO[33] experiments is, in all standard solar model calculations, directly proportional to the $^7\mathrm{Be} + p$ cross section. If the 1σ uncertainty in this cross section can be reduced by a factor of two to 5%, then it will no longer be the limiting uncertainty in predicting the crucial ^{8}B neutrino flux (cf. Table 2).

HOW LARGE AN UNCERTAINTY DOES HELIOSEISMOLOGY SUGGEST?

Could the solar model calculations be wrong by enough to explain the discrepancies between predictions and measurements for solar neutrino experiments? Helioseismology, which confirms predictions of the standard solar model to high precision, suggests that the answer is probably "No."

Figure 5 shows the fractional differences between the most accurate available sound speeds measured by helioseismology[34] and sound speeds calculated with our best solar model (with no free parameters). The horizontal line corresponds to the hypothetical case in which the model predictions exactly match the observed values. The rms fractional difference between the calculated and the measured sound speeds is 1.1×10^{-3} for the entire region over which the sound speeds are measured, $0.05 R_\odot < R < 0.95 R_\odot$. In the solar core, $0.05 R_\odot < R < 0.25 R_\odot$ (in which about 95% of the solar energy and neutrino flux is produced in a standard model), the rms fractional difference between measured and calculated sound speeds is 0.7×10^{-3}.

Helioseismological measurements also determine two other parameters that help characterize the outer part of the sun (far from the inner region in which neutrinos are produced): the depth of the solar convective zone (CZ), the region in the outer part of the sun that is fully convective, and the present-day surface abundance by mass of helium (Y_{surf}). The measured values, $R_{\mathrm{CZ}} = (0.713 \pm 0.001) R_\odot$[35], and $Y_{\mathrm{surf}} = 0.249 \pm 0.003$[31], are in satisfactory agreement with the values predicted by the solar model BP98, namely, $R_{\mathrm{CZ}} = 0.714 R_\odot$, and $Y_{\mathrm{surf}} = 0.243$. However, we shall see below that precision measurements of the sound speed near the transition between the radiative interior (in which energy is transported by radiation) and the outer convective zone (in which energy is transported by convection) reveal small discrepancies between the model predictions and the observations in this region.

If solar physics were responsible for the solar neutrino problems, how large would one expect the discrepancies to be between solar model predictions and helioseismological observations? The characteristic size of the discrepancies can be estimated using the results of the neutrino experiments and scaling laws for neutrino fluxes and

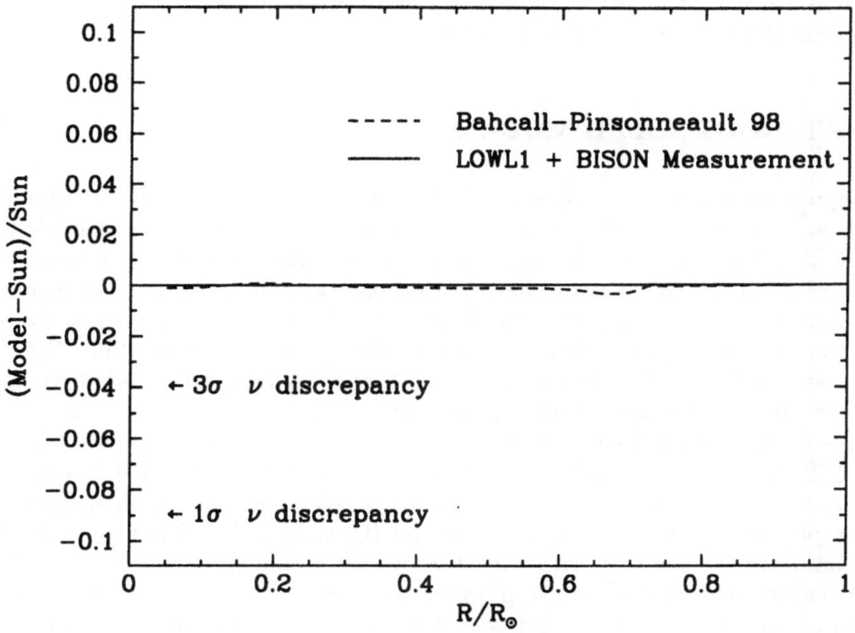

Figure 5. Predicted versus Measured Sound Speeds. This figure shows the excellent agreement between the calculated (solar model BP98, Model) and the measured (Sun) sound speeds, a fractional difference of 0.001 rms for all speeds measured between $0.05R_\odot$ and $0.95R_\odot$. The vertical scale is chosen so as to emphasize that the fractional error is much smaller than generic changes in the model, 0.04 to 0.09, that might significantly affect the solar neutrino predictions.

sound speeds.

All recently published solar models predict essentially the same fluxes from the fundamental pp and pep reactions (amounting to 72.4 SNU in gallium experiments, cf. Table 1), which are closely related to the solar luminosity. Comparing the measured gallium rates and the standard predicted rate for the gallium experiments, the ^{7}Be flux must be reduced by a factor N if the disagreement is not to exceed n standard deviations, where N and n satisfy $72.4 + (34.4)/N = 72.2 + n\sigma$. For a 1σ (3σ) disagreement, $N = 6.1(2.05)$. Sound speeds scale like the square root of the local temperature divided by the mean molecular weight and the ^{7}Be neutrino flux scales approximately as the 10th power of the temperature[36]. Assuming that the temperature changes are dominant, agreement to within 1σ would require fractional changes of order 0.09 in sound speeds (3σ could be reached with 0.04 changes), if all model changes were in the temperature[3]. This argument is conservative because it ignores the ^{8}B and CNO neutrinos which contribute to the observed counting rate (cf. Table 1) and which, if included, would require an even larger reduction of the ^{7}Be flux.

I have chosen the vertical scale in Fig. 5 to be appropriate for fractional differences between measured and predicted sound speeds that are of order 0.04 to 0.09 and that might therefore affect solar neutrino calculations. Fig. 5 shows that the characteristic agreement between solar model predictions and helioseismological measurements is

[3] I have used in this calculation the GALLEX and SAGE measured rates reported by Kirsten and Gavrin at Neutrino 98. The experimental rates used in BP98 were not as precise and therefore resulted in slightly less stringent constraints than those imposed here. In BP98, we found that agreement to within 1σ with the then available experimental numbers would require fractional changes of order 0.08 in sound speeds (3σ could be reached with 0.03 changes.)

more than a factor of 40 better than would be expected if there were a solar model explanation of the solar neutrino problems.

FITS WITHOUT SOLAR MODELS

Suppose (following the precepts of Hata *et al.*[37], Parke[38], and Heeger and Robertson[39]) we now ignore everything we have learned about solar models over the last 35 years and allow the important *pp*, ^{7}Be, and ^{8}B fluxes to take on any non-negative values. What is the best fit that one can obtain to the solar neutrino measurements assuming only that the luminosity of the sun is supplied by nuclear fusion reactions among light elements (the so-called 'luminosity constraint'[40])?

The answer is that the fits are bad, even if we completely ignore what we know about the sun. I quote the results from Ref. 10.

If the CNO neutrino fluxes are set equal to zero, there are no acceptable solutions at the 99% C. L. ($\sim 3\sigma$ result). The best-fit is worse if the CNO fluxes are not set equal to zero. All so-called 'solutions' of the solar neutrino problems in which the astrophysical model is changed arbitrarily (ignoring helioseismology and other constraints) are inconsistent with the observations at much more than a 3σ level of significance. No fiddling of the physical conditions in the model can yield the minimum value, quoted above, that was found by varying the fluxes independently and arbitrarily.

Figure 3 shows, in the lower left-hand corner, the best-fit solution and the 1σ -3σ contours. The 1σ and 3σ limits were obtained by requiring that $\chi^2 = \chi^2_{\rm min} + \delta\chi^2$, where for 1σ $\delta\chi^2 = 1$ and for 3σ $\delta\chi^2 = 9$. All of the standard model solutions lie far from the best-fit solution and even lie far from the 3σ contour.

Since standard model descriptions do not fit the solar neutrino data, we will now consider models in which neutrino oscillations change the shape of the neutrino energy spectra.

NEUTRINO OSCILLATIONS

The experimental results from all five of the operating solar neutrino experiments (chlorine, Kamiokande, SAGE, GALLEX, and SuperKamiokande) can be fit well by descriptions involving neutrino oscillations, either vacuum oscillations(as originally suggested by Gribov and Pontecorvo[41])or resonant matter oscillations(as originally discussed by Mikeyhev, Smirnov, and Wolfenstein(MSW)[42]).

Table 3. Neutrino Oscillation Solutions.

Solution	Δm^2	$\sin^2 2\theta$
SMA	5×10^{-6} eV^2	5×10^{-3}
LMA	2×10^{-5} eV^2	0.8
LOW	8×10^{-8} eV^2	0.96
VAC	8×10^{-11} eV^2	0.7

Table 3 summarizes the four best-fit solutions that are found in the two-neutrino approximation[10, 27]. Only the SMA MSW solution fits well all the data–including the recoil electron energy spectrum measured in the SuperKamiokande experiment–if the standard value for the *hep* production reaction cross section(^{3}He+$p \rightarrow$ ^{4}He+$e^+ + \nu_e$) is used[10]. However, for over a decade I have not given an estimated uncertainty for this cross section[20]. The transition matrix element is essentially forbidden and the actual quoted value for the production cross section depends upon a delicate cancellation between two comparably sized terms that arise from very different and hard to evaluate nuclear physics. I do not see anyway at present to determine from experiment or from first principles theoretical calculations a relevant, robust upper limit to the *hep* production cross section (and therefore the *hep* solar neutrino flux).

The possible role of *hep* neutrinos in solar neutrino experiments is discussed extensively in Ref. 27. The most important unsolved problem in theoretical nuclear physics related to solar neutrinos is the range of values allowed by fundamental physics for the *hep* production cross section.

DISCUSSION AND CONCLUSION

When the chlorine solar neutrino experiment was first proposed[43], the only stated motivation was "...to see into the interior of a star and thus verify directly the hypothesis of nuclear energy generation in stars." This goal has now been achieved,

The focus has shifted to using solar neutrino experiments as a tool for learning more about the fundamental characteristics of neutrinos as particles. Experimental effort is now concentrated on answering the question: What are the probabilities for transforming a solar ν_e of a definite energy into the other possible neutrino states? Once this question is answered, we can calculate what happens to ν_e's that are created in the interior of the sun. Armed with this information from weak interaction physics, we can return again to the original motivation of using neutrinos to make detailed, quantitative tests of nuclear fusion rates in the solar interior. Measurements of the flavor content of the dominant low energy neutrino sources, *p-p* and ^{7}Be neutrinos, will be crucial in this endeavor and will require another generation of superb solar neutrino experiments.

Three decades of refining the input data and the solar model calculations has led to a predicted standard model event rate for the chlorine experiment, 7.7 SNU, which is very close to 7.5 SNU, the best-estimate value obtained in 1968[44]. The situation regarding solar neutrinos is, however, completely different now, thirty years later. Four experiments have confirmed the original chlorine detection of solar neutrinos. Helioseismological measurements are in excellent agreement with the standard solar model predictions and very strongly disfavor (by a factor of 40 or more) hypothetical deviations from the standard model that are require to fits the neutrino data(cf. Fig. 5). Just in the last two years, improvements in the helioseismological measurements have resulted in a five-fold improvement in the agreement between the calculated standard solar model sound speeds and the measured solar velocities(cf. Figure 2 of the Neutrino 96 talk[45] with Figure 5 of this talk).

I acknowledge support from NSF grant #PHY95-13835.

REFERENCES

1. Davis, R., Jr., Harmer, D. S. and Hoffman, K. C., Phys. Rev. Lett. **20**, 1205 (1968).

2. Davis, R., Jr., Prog. Part. Nucl. Phys. **32**, 13 (1994); Cleveland, B. T., Daily, T., Davis, R., Jr., Distel, J. R., Lande, K., Lee, C. K., Wildenhain, P. S. and Ullman, J., Astrophys. J. **496**, 505 (1998).

3. KAMIOKANDE Collaboration, Fukuda, Y. *et al.*, Phys. Rev. Lett. **77**, 1683 (1996).

4. SAGE Collaboration, Gavrin, V. *et al.*, in "Neutrino '96," Proceedings of the 17th International Conference on Neutrino Physics and Astrophysics (Helsinki). Edited by K. Enqvist, K. Huitu and J. Maalampi (World Scientific, Singapore 1997), p. 14; SAGE collaboration, Abdurashitov, J. N. *et al.*, Phys. Rev. Lett. **77**, 4708 (1996).

5. GALLEX Collaboration, Anselmann, P. *et al.*, Phys. Lett. B **342**, 440 (1995); GALLEX Collaboration, Hampel, W. *et al.*, Phys. Lett. B **388**, 364 (1996).

6. SuperKamiokande Collaboration, Y. Suzuki, in "Neutrino 98." Proceedings of the XVIII International Conference on neutrino Physics and Astrophysics, Takayama, Japan, 4-9 June 1998. Edited by Y. Suzuki and Y. Totsuka. To be published in Nucl. Phys. B (Proc. Suppl.); the Super-Kamiokande Collaboration, Fukuda, Y. *et al.*, Phys. Rev. Lett. **81**, 1562 (1998); Fukuda, Y. *et al.*, hep-ex/9812009; Totsuka, Y., in the Proceedings of the 18th Texas Symposium on Relativistic Astrophysics and Cosmology, December 15–20, 1996, Chicago, Illinois. Edited by A. Olinto, J. Frieman and D. Schramm (World Scientific, Singapore 1998) p. 114.

7. Bethe, H. A., Phys. Rev. **55**, 434 (1939).

8. Thompson, W., "On the Age of the Sun's Heat," MacMilan's Magazine **5**, 288 (1862); reprinted in "Treatise on Natural Philosphy," W. Thompson and P. Gutherie, **Vol. 1**, appendix E, (Cambridge Univ. Press, Cambridge 1883), p. 485.

9. Darwin, C., "On the Origin of the Species," 1859 (Reprinted: Harvard University Press, Cambridge, MA, 1964).

10. Bahcall, J. N., Krastev, P. I. and Smirnov, A. Yu., Phys. Rev. D **58**, 096016-1 (1998).

11. Bahcall, J. N., Basu, S. and Pinsonneault, M. H., Phys. Lett. B **433**, 1 (1998).

12. Adelberger, E. *et al.*, Rev. Mod. Phys. **70**, 1265 (1998).

13. Iglesias, C. A. and Rogers, F. J., Astrophys. J. **464**, 943 (1996); Alexander, D. R. and Ferguson, J. W., Astrophys. J. **437**, 879 (1994). These references describe the different versions of the OPAL opacities.

14. Rogers, F. J., Swenson, F. J. and Iglesias, C. A.. Astrophys. J. **456**, 902 (1996).

15. Gruzinov, A. V. and Bahcall, J. N., Astrophys. J. **504**, 996 (1998).

16. Salpeter, E. E., Australian J. Phys. **7**, 373 (1954).

17. Bahcall, J. N., Phys. Rev. C **56**, 3391 (1997).

18. Bahcall, J. N., Lisi, E., Alburger, D. E., De Braeckeleer, L., Freedman, S. J. and Napolitano, J., Phys. Rev. C **54**, 411 (1996).

19. Bahcall, J. N., Kamionkowski, M. and Sirlin, A., Phys. Rev. D **51**, 6146 (1995).

20. Bahcall, J. N., "Neutrino Astrophysics" (Cambridge University Press, Cambridge 1989).

21. Bahcall, J. N., Astrophys. J. **467**, 475 (1996).

22. (GONG) Christensen-Dalsgaard, J. *et al.*, GONG Collaboration, Science **272**, 1286 (1996); (BP95) Bahcall, J. N. and Pinsonneault, M. H., Rev. Mod. Phys. **67**, 781 (1995); (KS94) Kovetz, A. and Shaviv, G., Astrophys. J. **426**, 787 (1994); (CDF94) Castellani, V., Degl'Innocenti, S., Fiorentini, G., Lissia, L. M. and Ricci, B., Phys. Lett. B **324**, 425 (1994); (JCD94) Christensen-Dalsgaard, J., Europhys. News **25**, 71 (1994); (SSD94) Shi, X., Schramm, D. N. and Dearborn, D. S. P., Phys. Rev. D **50**, 2414 (1994); (DS96) Dar, A. and Shaviv, G., Astrophys. J. **468**, 933 (1996); (CDF93) Castellani, V., Degl'Innocenti, S. and Fiorentini, G.. Astron. Astrophys. **271**, 601 (1993); (TCL93) Turck-Chièze, S. and Lopes, I., Astrophys. J. **408**, 347 (1993); (BPML93) Berthomieu, G., Provost, J., Morel, P. and Lebreton, Y., Astron. Astrophys. **268**, 775 (1993); (BP92) Bahcall, J. N. and Pinsonneault, M. H., Rev. Mod. Phys. **64**, 885 (1992); (SBF90) Sackman, I.-J., Boothroyd, A. I. and Fowler, W. A., Astrophys. J. **360**, 727 (1990); (BU88) Bahcall, J. N. and Ulrich, R. K., Rev. Mod. Phys. **60**, 297 (1988); (RVCD96) Richard, O., Vauclair, S., Charbonnel, C. and Dziembowski, W. A., Astron. Astrophys. **312**, 1000 (1996); (CDR97) Ciacio, F., Degl'Innocenti, S. and Ricci, B., Astron. Astrophys. Suppl. Ser. **123**, 449 (1997).

23. Johnson, C. W., Kolbe, E., Koonin, S. E. and Langanke, K., Astrophys. J. **392**, 320 (1992).

24. Bahcall, J. N., Pinsonneault, M. H., Basu, S. and Christensen-Dalsgaard, J., Phys. Rev. Lett. **78**, 171 (1997).

25. Bahcall, J. N. and Pinsonneault, M. H., Rev. Mod. Phys. **64**, 885 (1992).

26. Bahcall, J. N. and Pinsonneault, M. H., Rev. Mod. Phys. **67**, 781 (1995).

27. Bahcall, J. N. and Krastev, P. I., Phys. Lett. B **436**, 243 (1998).

28. Bahcall, J. N., Phys. Rev. D **44**, 1644 (1991).

29. Bahcall, J. N. and Bethe, H. A., Phys. Rev. Lett. **65**, 2233 (1990).

30. Gruzinov, A. V. and Bahcall, J. N., Astrophys. J. **490**, 437 (1997).

31. Basu, S. and Antia, H. M., Mon. Not. R. Astron. Soc. **287**, 189 (1997).

32. Thoul, A. A., Bahcall, J. N. and Loeb, A., Astrophys. J. **421**, 828 (1994).

33. McDonald, A. B., in the Proceedings of the 9th Lake Louise Winter Institute. Edited by A. Astbury *et al.* (World Scientific, Singapore 1994), p. 1.

34. Basu, S. *et al.*, Mon. Not. R. Astron. Soc. **292**, 234 (1997).

35. Basu, S. and Antia, H. M., Mon. Not. R. Astron. Soc. **276**, 1402 (1995).

36. Bahcall, J. N. and Ulmer, A., Phys. Rev. D **53**, 4202 (1996).

37. Hata, N., Bludman, S. and Langacker, P., Phys. Rev. D **49**, 3622 (1994).

38. Parke, S., Phys. Rev. Lett. **74**, 839 (1995).

39. Heeger, K. M. and Robertson, R. G. H., Phys. Rev. Lett. **77**, 3720 (1996).

40. Bahcall, J. N. and Krastev, P. I., Phys. Rev. D **53**, 4211 (1996).

41. Gribov, V. N. and Pontecorvo, B. M., Phys. Lett. B **28**, 493 (1969); Pontecorvo, B., Sov. Phys. JETP **26**, 984 (1968).

42. Wolfenstein, L., Phys. Rev. D **17**, 2369 (1978); Mikheyev, S. P. and Smirnov, A. Yu., Yad. Fiz. **42**, 1441 (1985) [Sov. J. Nucl. Phys. **42**, 913 (1985)]; Nuovo Cimento C **9**, 17 (1986).

43. Bahcall, J. N., Phys. Rev. Lett. **12**, 300 (1964); Davis, R., Jr., Phys. Rev. Lett. **12**, 303 (1964); Bahcall, J. N. and Davis, R., Jr., in "Stellar Evolution." Edited by R. F. Stein and A. G. Cameron (Plenum Press, New York 1966), p. 241 [proposal first made in 1963 at this conference].

44. Bahcall, J. N., Bahcall, N. A. and Shaviv, G., Phys. Rev. Lett. **20**, 1209 (1968).

45. Bahcall, J. N. and Pinsonneault, M. H. in "Neutrino '96," Proceedings of the 17th International Conference on Neutrino Physics and Astrophysics (Helsinki). Edited by K. Enqvist, K. Huitu and J. Maalampi (World Scientific, Singapore 1997), p. 56.

THREE–NEUTRINO VACUUM OSCILLATION SOLUTIONS TO THE SOLAR AND ATMOSPHERIC ANOMALIES

Kerry Whisnant

Department of Physics and Astronomy
Iowa State University
Ames, IA 50011

INTRODUCTION

There has long been speculation that the solar[1,2] and atmospheric[3,4] neutrino anomalies may be the result of neutrinos oscillating from one species to another.[5,6] The recent results from the Super–Kamiokande (Super–K) experiment[7,8,9] provide further evidence for oscillations, and indicate that different mass–squared difference scales are required to explain both effects. There is also data from the LSND experiment,[10] which can be explained by oscillations at a third mass–squared difference scale. Since three separate mass–squared difference scales are apparently necessary to describe all three types of experiment, a fourth neutrino is needed. However, there has not yet been a confirmation of the LSND results, and recent data from the KARMEN experiment[11] excludes much of the LSND allowed region.

In this talk I will take the conservative approach that vacuum neutrino oscillations are only responsible for the solar and atmospheric anomalies, and discuss the possible three–neutrino models that are consistent with the data. Four–neutrino models that describe all the data, including the LSND results, will be reviewed elsewhere.[12] I will then discuss the implications for neutrino mass, and possible future oscillation tests of three–neutrino models.

REVIEW OF SOLAR AND ATMOSPHERIC NEUTRINO DATA

Table 1 shows the results of the solar neutrino experiments.[1,7] Each experiment has a different neutrino energy threshhold (E_ν^{min}), and therefore is sensitive to a different part of the solar neutrino spectrum. The last column in Table 1 shows the experimental rate compared to the latest prediction of the Standard Solar Model (SSM).[13] Furthermore, Super–Kamiokande (also Kamiokande, although with much less data) measures the recoil electron energy spectrum; electrons with energy above

Confluence of Cosmology, Massive Neutrinos, Elementary Particles, and Gravitation
Edited by Kursunoglu *et al.*, Kluwer Academic / Plenum Publishers, New York, 1999.

Table 1. Solar neutrino data

Experiment	Reaction	E_ν^{min} (GeV)	Data/SSM
Homestake	$\nu_e\ ^{37}\mathrm{Cl} \to\ ^{37}\mathrm{Ar}\ e$	0.81	0.33 ± 0.03
GALLEX	$\nu_e\ ^{71}\mathrm{Ga} \to\ ^{71}\mathrm{Ge}\ e$	0.25	0.60 ± 0.06
SAGE	$\nu_e\ ^{71}\mathrm{Ga} \to\ ^{71}\mathrm{Ge}\ e$	0.25	0.52 ± 0.06
Kamiokande	$\nu\ e \to \nu\ e$	6.5	0.52 ± 0.06
Super–Kamiokande	$\nu\ e \to \nu\ e$	6.5	0.47 ± 0.02

13 MeV have a somewhat lesser suppression compared to the SSM. Super–K also measures the direction of the detected electron, which shows a very strong correlation with the vector from the sun, thereby indicating that they are indeed seeing solar neutrinos.

The results of the atmospheric neutrino experiments[3,8,14] are shown in Table 2. The theoretical fluxes are thought to be known to 20% uncertainty, while different estimates for the ν_μ/ν_e ratio agree to about 5% over the range of energies appropriate for these experiments.[15] Therefore the results are usually quoted in terms of the measured μ/e ratio compared to the theoretical expectation, so that the overall normalization drops out.

Table 2. Atmospheric neutrino data

Data sample	$R = (N_\mu/N_e)_{data}/(N_\mu/N_e)_{theory}$
Kamiokande sub–GeV	$0.60^{+0.06}_{-0.05} \pm 0.05$
Kamiokande multi–GeV	$0.57^{+0.08}_{-0.07} \pm 0.07$
Irvine–Michigan–Brookhaven (IMB)	$0.54 \pm 0.05 \pm 0.11$
SOUDAN II	$0.61 \pm 0.11^{+0.06}_{-0.05}$
MACRO	$0.74 \pm 0.036 \pm 0.046 \pm 0.13$
Super–K sub–GeV	$0.63 \pm 0.025 \pm 0.05$
Super–K multi–GeV	$0.65 \pm 0.05 \pm 0.08$

Super–K also measures the zenith angle distribution and energy of detected muons and electrons. These results[8] are shown in Table 3 as a function of L/E, where L is the inferred distance traveled and E is the inferred neutrino energy, which are determined from the direction and energy, respectively, of the detected charged lepton. Although the normalization is not correct, the electron data is consistent with a constant ratio for data/theory (within experimental errors), while the muon data has a large suppression of the high L/E data compared to the low L/E data. Therefore, if one adjusts the overall normalization upward to be in agreement with the electron data, it appears that the dominant effect is that ν_μ at larger L/E are disappearing (but *not* into ν_e).

Table 3. Super–Kamiokande atmospheric neutrino data versus L/E

L/E (km/GeV)	$R_e = (N_e)_{data}/(N_e)_{theory}$	$R_\mu = (N_\mu)_{data}/(N_\mu)_{theory}$
5.6	1.13 ± 0.18	1.00 ± 0.18
18	1.24 ± 0.11	0.99 ± 0.08
56	1.22 ± 0.11	0.91 ± 0.07
180	1.13 ± 0.14	0.84 ± 0.10
560	1.13 ± 0.15	0.63 ± 0.09
1800	1.19 ± 0.12	0.68 ± 0.07
5600	1.18 ± 0.10	0.61 ± 0.05
18000	1.34 ± 0.15	0.59 ± 0.07

THREE–NEUTRINO OSCILLATION MODELS

In a three–neutrino model the charged–current eigenstates $(\nu_e, \nu_\mu, \nu_\tau)$ can be written in terms of the mass eigenstates (ν_1, ν_2, ν_3) as

$$\begin{pmatrix} \nu_e \\ \nu_\mu \\ \nu_\tau \end{pmatrix} = U \begin{pmatrix} \nu_1 \\ \nu_2 \\ \nu_3 \end{pmatrix} = \begin{pmatrix} c_1c_3 & c_1s_3 & s_1e^{-i\delta} \\ -c_2s_3 - s_1s_2c_3e^{i\delta} & c_2c_3 - s_1s_2s_3e^{i\delta} & c_1s_2 \\ s_2s_3 - s_1c_2c_3e^{i\delta} & -s_2c_3 - s_1c_2s_3e^{i\delta} & c_1c_2 \end{pmatrix} \begin{pmatrix} \nu_1 \\ \nu_2 \\ \nu_3 \end{pmatrix}, \qquad (1)$$

where $c_j \equiv \cos\theta_j$ and $s_j \equiv \sin\theta_j$. Without loss of generality we assume $m_1 < m_2 < m_3$. The general off–diagonal oscillation probabilities in vacuum are

$$P(\nu_\alpha \to \nu_\beta) = -\sum_{j<k} \left[4\ \mathrm{Re}(U_{\alpha j}U^*_{\beta j}U^*_{\alpha k}U_{\beta k})\sin^2\Delta_{jk} - 2\ \mathrm{Im}(U_{\alpha j}U^*_{\beta j}U^*_{\alpha k}U_{\beta k})\sin 2\Delta_{jk}\right], \qquad (2)$$

where $\Delta_{jk} \equiv 1.27(\delta m^2_{jk}/\mathrm{eV}^2)(L/\mathrm{km})(E/\mathrm{GeV})^{-1}$ and $\delta m^2_{jk} \equiv m_j^2 - m_k^2$. To fit both the solar and atmospheric data, we assume $\delta m^2_{sun} = |\delta m^2_{21}| \ll |\delta m^2_{31}| \simeq |\delta m^2_{32}| = \delta m^2_{atm}$ (the alternate possibility with δm^2_{32} much smaller gives similar results). Then for atmospheric and long–baseline experiments ($1\ \mathrm{km/GeV} \ll L/E \leq 10^4\ \mathrm{km/GeV}$), only δm^2_{atm} contributes appreciably to oscillations, and

$$P(\nu_e \to \nu_\mu) = \sin^2\theta_2 \sin^2 2\theta_1 \sin^2\Delta_{atm}, \qquad (3)$$

$$P(\nu_e \to \nu_\tau) = \cos^2\theta_2 \sin^2 2\theta_1 \sin^2\Delta_{atm}, \qquad (4)$$

$$P(\nu_\mu \to \nu_\tau) = \cos^4\theta_1 \sin^2 2\theta_2 \sin^2\Delta_{atm}. \qquad (5)$$

The corresponding expressions for anti–neutrinos are the same; there are no CP–violating effects in the leading oscillation.[16] For solar neutrino experiments ($L/E \sim 10^{10}$ km/GeV), the leading oscillation term averages ($\langle\sin^2\Delta_{atm}\rangle \to \frac{1}{2}$) and we have

$$P(\nu_e \to \nu_e) = 1 - \frac{1}{2}\sin^2 2\theta_1 - \cos^4\theta_1 \sin^2 2\theta_3 \sin^2\Delta_{sun}. \qquad (6)$$

It is important to note that if both the solar and atmospheric data are to be completely described by oscillations (including the spectrum distortion of solar neutrinos and zenith angle dependence of atmospheric neutrinos), then there can be no measurable oscillations at short baselines ($L/E \sim 1$ km/GeV), since the δm^2 scales are too small.

The only parameter common to both solar and atmospheric oscillations is θ_1. This means that if $\theta_1 = 0$ (i.e., $U_{e3} = 0$), then the solar and atmospheric oscillations decouple[17,18] (the solar oscillations are determined by θ_3 and δm^2_{sun}, and the atmospheric oscillations are determined by θ_2 and δm^2_{atm}), and each case reduces to a simple two–neutrino–like fit. However, the oscillations still involve three neutrinos: the atmospheric oscillations are $\nu_\mu \to \nu_\tau$ and the solar oscillations are $\nu_e \to \cos^2\theta_2\ \nu_\mu + \sin^2\theta_2\ \nu_\tau$.

TWO–NEUTRINO FITS

Assuming $\theta_1 = 0$, we can find the two–neutrino–like oscillation solutions to the solar and atmospheric data. In our analysis of the atmospheric data we used the Super–K fully contained data versus L/E in Table 3, and also included as a free parameter an overall flux normalization correction α.[19,20] Table 4 shows the overall best two–neutrino fit, the best fit with fixed normalization ($\alpha = 1$), and for comparison the best fit for $\nu_\mu \leftrightarrow \nu_e$ oscillations. We see that $\nu_\mu \to \nu_\tau$ oscillations are

Table 4. Best two–neutrino fits to Super–K atmospheric data.

Oscillation	δm^2_{atm} (eV²)	$\sin^2 2\theta_2$	α	χ^2/DOF	Goodness–of-fit
$\nu_\mu \to \nu_\tau$	2.8×10^{-3}	1.00	1.16	7/13	90%
$\nu_\mu \to \nu_\tau$	1.1×10^{-3}	0.85	–	23/14	6%
$\nu_\mu \leftrightarrow \nu_e$	1.1×10^{-3}	0.88	0.71	82/13	4×10^{-12}

very strongly favored over $\nu_\mu \leftrightarrow \nu_e$; also, $\nu_\mu \leftrightarrow \nu_e$ oscillations with $\delta m^2_{atm} > 10^{-3}$ eV^2 are excluded by the CHOOZ reactor experiment.[21] At 95% C.L., the allowed range of parameters is approximately $5.5 \times 10^{-4} < \delta m^2_{atm}/\text{eV}^2 < 9.5 \times 10^{-3}$ and $0.8 < \sin^2 2\theta_2 < 1.0$. In our analysis of the solar neutrino data,[18,22] we used the Homestake, GALLEX and SAGE capture rates,[1] and the Super–K 504-day detected electron spectrum.[7] Because the number of ^{8}B neutrinos in the SSM is relatively uncertain, we also included as a free parameter an overall flux normalization β for the ^{8}B neutrinos. Table 5 shows the best fits to the solar neutrino data, which occur in four regions of parameter space that correspond roughly to having the Earth–Sun distance $\frac{1}{2}$, $\frac{3}{2}$, $\frac{5}{2}$, and $\frac{7}{2}$ wavelengths for a typical ^{8}B neutrino. The 95% C.L. allowed regions are very narrow in δm^2_{sun} (lying near the values listed in Table 5), and extend from 0.6 to nearly 1.0 in $\sin^2 2\theta_3$ for Solution A and from 0.85 to 1.0 in $\sin^2 2\theta_3$ for Solution D. Also shown for comparison is the best fit when the normalization of the ^{8}B neutrinos is fixed ($\beta = 1$). Solution C gives the best fit to the Super–K spectrum alone, reproducing fairly well the smaller suppression at higher energies.

We see that both the atmospheric and solar data are well–described by large–amplitude vacuum neutrino oscillations. A more extensive list of references on other fits to the solar and atmospheric data may be found in Ref. 18.

GLOBAL THREE–NEUTRINO FIT

A full three–neutrino fit may be done by allowing θ_1 to vary. The best three–neutrino fit to the atmospheric data is

$$\delta m^2_{atm} = 2.8 \times 10^{-3}\ \text{eV}^2, \qquad \sin^2 2\theta_2 = 1.00, \qquad \sin\theta_1 = 0.00, \qquad \alpha = 1.16, \tag{7}$$

and the best three–neutrino fit to the solar data is

$$\delta m^2_{sun} = 7.5 \times 10^{-11}\ \text{eV}^2, \qquad \sin^2 2\theta_3 = 0.91, \qquad \sin\theta_1 = 0.00, \qquad \beta = 1.62. \tag{8}$$

We see that in each case the best fit occurs when the solar and atmospheric parameters are decoupled. The atmospheric data provides the strongest constraint on θ_1, and places the limit $\sin\theta_1 < 0.3$ at 95% C.L. Allowed regions for the other parameters

Table 5. Best two–neutrino fits to solar neutrino data.

Solution	δm^2_{sun} (eV²)	$\sin^2 2\theta_3$	β	χ^2/DOF	Goodness–of-fit
A	7.5×10^{-11}	0.91	1.62	21.6/16	16%
B	2.5×10^{-10}	0.86	0.84	26.3/16	5%
C	4.4×10^{-10}	0.97	0.80	23.5/16	10%
D	6.4×10^{-10}	1.00	0.80	27.6/16	4%
	6.5×10^{-11}	0.74	–	26.5/17	7%

are similar to those obtained for the two–neutrino fits, although they do tend to shrink for larger values of $\sin\theta_1$.

BI–MAXIMAL MIXING

Since both the solar and atmospheric data favor maximal, or nearly maximal, mixing, it is interesting to ask what happens if both have maximal mixing ("bi–maximal mixing"). In fact, there is unique solution for the mixing matrix (up to trivial sign changes) which gives maximal mixing in both the solar and atmospheric sectors:[23,24]

$$U = \begin{pmatrix} \frac{1}{\sqrt{2}} & -\frac{1}{\sqrt{2}} & 0 \\ \frac{1}{2} & \frac{1}{2} & -\frac{1}{\sqrt{2}} \\ \frac{1}{2} & \frac{1}{2} & \frac{1}{\sqrt{2}} \end{pmatrix}, \tag{9}$$

which corresponds to $\theta_1 = 0$, $\sin^2 2\theta_2 = 1$, and $\sin^2 2\theta_3 = 1$. Then atmospheric oscillations are $\nu_\mu \to \nu_\tau$, and solar ν_e oscillate into an equal mixture of ν_μ and ν_τ (although this could not be verified since solar neutrino experiments do not distinguish between ν_μ and ν_τ).

Although the mixing matrix is fixed for bi–maximal mixing, since neutrino oscillations measure only mass–squared differences, the absolute values of the masses are not determined. Many possible bi–maximal mixing scenarios have been discussed in the literature: $m_1 \ll m_2 \ll m_3$ (hierarchical masses),[23,25] $m_1 \simeq m_2 \ll m_3$ (one large mass),[26] $m_1 \ll m_2 \simeq m_3$ (two large, nearly degenerate masses),[27] $m_1 \simeq m_2 \simeq m_3$ (three nearly degenerate masses).[28] The last case is especially interesting since the overall mass scale is not fixed, and can be adjusted, e.g., to give a neutrino contribution to hot dark matter. There are also other cases (sometimes mistakenly called bi–maximal in the literature) which might be called "quasi bi–maximal." where one or both mixings are not maximal but are nevertheless nearly maximal. The most interesting such example can be obtained from the so–called "democratic" lepton mass matrix, and gives $\theta_1 = 0$, $\sin^2 2\theta_2 = \frac{8}{9}$, and $\sin^2 2\theta_3 = 1$.[29] More general models with decoupling ($\sin\theta_1 = 0$) that are nearly bi–maximal have also been studied.[30]

SEPARATE SCALES FOR SOLAR AND ATMOSPHERIC OSCILLATIONS?

One may ask if two separate δm^2 scales are required for the solar and atmospheric data. If they were not, then one could use the second δm^2 to fit the LSND results and a fourth neutrino is not needed to describe all the data. For example, if the two independent δm^2 scales were chosen to fit the LSND and atmospheric data, how bad then is the fit to the solar data? Since in this case all δm^2_{jk} are much larger than 10^{-10} eV^2. the solar oscillations all reach their average values, which for three neutrinos means that the solar oscillation probability is energy–independent and must lie in the range $\frac{1}{3} \le P(\nu_e \to \nu_e) \le 1$. After allowing the ^{8}B normalization to vary, we find that the best such energy–independent fit to the solar data gives a χ^2/DOF of 48/17, which is excluded at 99.98% C.L.[18,31] The poor fit occurs mainly because the Homestake result is significantly lower than the others and because the Super–K spectrum shows an energy–dependent suppression when compared to the SSM. Even if the Homestake result is ignored. the resulting best energy–independent fit has χ^2/DOF = 25/16, which is excluded at 93% C.L. Therefore a separate δm^2 scale for the solar oscillations is strongly favored.

Similarly, if the two independent δm^2 scales were chosen to fit the LSND and solar data, how bad is the fit to the atmospheric data? In this case the L/E is not large enough for the δm^2_{sun} oscillation to have any effect, but is more than large enough to cause the oscillations at the LSND δm^2 scale to reach their average values, so that any atmospheric ν_μ suppression should be independent of L/E. The best L/E-independent fit to the Super–K data has $\chi^2/\mathrm{DOF} = 33/13$, which is excluded at 99.8% C.L.[18] The poor fit is due to the strong L/E dependence exhibited by the Super–K data (see Table 3). Therefore a separate δm^2 scale for the atmospheric oscillations is also strongly favored. The conclusion is that separate scales for the solar and atmospheric oscillations are clearly favored by the data, and that a complete description of all data (including the LSND results) requires three independent δm^2 and hence four neutrinos.[12]

INFERRED LIMITS ON NEUTRINO MASSES

The current direct limits on neutrino masses from endpoint spectra are[32]

$$m_{\nu_e} < 4.4\ \mathrm{eV}, \qquad m_{\nu_\mu} < 170\ \mathrm{keV}, \qquad m_{\nu_\tau} < 18\ \mathrm{MeV}. \tag{10}$$

If we assume that $\delta m^2_{sun} \sim 10^{-10}\ \mathrm{eV}^2$ and $\delta m^2_{atm} \sim 10^{-3}\ \mathrm{eV}^2$, then on the scale of the measurements in Eq. (10), all three neutrinos must be nearly degenerate ($m_1 \simeq m_2 \simeq m_3 \equiv m$). Then it is not hard to show[33] that $\langle m^2_{\nu_e}\rangle = \sum_j |U_{ej}|^2 m_j^2 < (4.4\ \mathrm{eV})^2$ implies

$$m < 4.4\ \mathrm{eV}, \qquad \text{and} \qquad \langle m_{\nu_\alpha}\rangle < 4.4\ \mathrm{eV}, \quad \alpha = e, \mu, \tau. \tag{11}$$

These inferred bounds on the masses of ν_μ and ν_τ are up to six orders of magnitude better than the corresponding direct limits. The same argument can be made if the solar oscillations are of the MSW type ($\delta m^2_{sun} \sim 10^{-5}\ \mathrm{eV}^2$).

FUTURE TESTS OF THREE–NEUTRINO PARAMETERS

The parameters θ_1, θ_2 and δm^2_{atm} that determine the atmospheric oscillations can also be tested in long–baseline experiments. Currently plans are being made for the following experiments: FNAL to SOUDAN (MINOS), KEK to Kamioka (K2K), and CERN to GRAN SASSO (AQUA–RICH, ICARUS, NICE, NOE, OPERA, and Super–ICARUS).[34] Generally these experiments will test ν_μ survival, ν_μ appearance as ν_τ, and ν_μ disappearance (via neutral current events); typical sensitivities to oscillation parameters are $\delta m^2_{atm} > 2\times10^{-3}\ \mathrm{eV}^2$ for maximal mixing and $\sin^2 2\theta_2 > 10^{-2}$ for large δm^2_{atm}. More extensive tests may be done with muon storage rings,[35] which will have the options of simultaneous ν_μ and $\bar\nu_e$ beams (from μ^- decay), or simultaneous $\bar\nu_\mu$ and ν_e beams (from μ^+ decay). Oscillations between ν_μ and ν_e (or $\bar\nu_\mu$ and $\bar\nu_e$) therefore give wrong–sign charged leptons in the detector. Oscillations to ν_τ are detected via the $\tau \to \ell\nu\bar\nu$ decay, where the sign of the detected lepton determines whether it came from an initial ν_μ or $\bar\nu_e$ ($\bar\nu_\mu$ or ν_e) for an original μ^- (μ^+). In particular, strong limits on $\nu_e \to \nu_\mu$ and $\nu_e \to \nu_\tau$ oscillations are possible, which should strongly constrain θ_1.[36]

The parameters θ_1, θ_3 and δm^2_{sun} can be further constrained by future solar neutrino experiments. The Sudbury Neutrino Observatory (SNO),[37] and to a lesser extent future runs at Super–K, can make neutral–current (NC) as well as charged–current (CC) measurements; a measurement of $(CC/NC)_{data}/(CC/NC)_{theory} < 1$ will be evidence for ν_e oscillations into ν_μ and/or ν_τ (as opposed to a sterile neutrino).

They can also make improved measurements of the day/night ratio (a deviation from unity signals MSW rather than vacuum oscillations) and ^{8}B neutrino energy spectrum (the shape of which could help distinguish between the vacuum and MSW scenarios, and perhaps even distinguish[18] between the different vacuum scenarios in Table 5). Super–K, GALLEX, SAGE, and BOREXINO[38] can also in principle detect a seasonal variation in the ^{7}Be neutrinos due to the variation in Earth–Sun distance, which can occur only for vacuum oscillations.[39] The BOREXINO detector does the best job here since its primary signal comes from the ^{7}Be neutrinos. Although there are many ways to analyze the seasonal data, a simple but effective method is to look at the two asymmetries

$$A_1 = \frac{2(N_W - N_S)}{(N_W + N_S + N_F + N_{SP})}, \qquad A_2 = \frac{(N_W + N_S - N_F - N_{SP})}{(N_W + N_S + N_F + N_{SP})}, \tag{12}$$

where N_W, N_{SP}, N_S, N_F are the number of events in the winter, spring, summer, and fall, respectively; A_1 tends to be nonzero if the oscillation probability is monotonic with L, while A_2 tends to be nonzero if the oscillation probability has a local extremum in the middle of the L range. The eccentricity of the Earth's orbit alone gives $A_1 = 0.03$ (and $A_2 = 0$). Vacuum oscillations tend to give a larger value for A_1 in Super–K for recoil electron energies above 10 MeV.[18] A simultaneous measurement of A_1 and A_2 in GALLEX, SAGE, or especially BOREXINO, could help distinguish between vacuum solar solutions; for more details, see Ref. 20.[40]

SUMMARY

Full three–neutrino fits to the solar and atmospheric neutrino data seem to favor a decoupling of the parameters associated with the solar and atmospheric oscillations, and also indicate maximal or nearly maximal mixing for each type of experiment. There are many scenarios that could give bi–maximal or quasi bi–maximal mixing. Furthermore, the data strongly favors separate δm^2 scales for the solar and atmospheric oscillations, which means that apparently four neutrinos are required to also explain the LSND result. Once the solar and atmospheric oscillation scales are assumed, current beta decay limits on m_{ν_e} may be extended to include m_{ν_μ} and m_{ν_τ} in a three–neutrino model. Finally, future long–baseline and solar neutrino experiments should be able to greatly constrain the allowed ranges of the three–neutrino parameters.

ACKNOWLEDGMENTS

I gratefully acknowledge collaboration with V. Barger, S. Pakvasa, and T.J. Weiler.

REFERENCES

1. B.T. Cleveland *et al.*, Nucl. Phys. Proc. Suppl. B 38: 47 (1995); Kamiokande collaboration, Y. Fukuda *et al.*, Phys. Rev. Lett. 77: 1683 (1996); GALLEX Collaboration, W. Hampel *et al.*, Phys. Lett. B 388: 384 (1996); SAGE collaboration, J.N. Abdurashitov *et al.*, Phys. Rev. Lett. 77: 4708 (1996).
2. J.N. Bahcall and M.H. Pinsonneault, Rev. Mod. Phys. 67: 781 (1995).
3. Kamiokande collaboration, K.S. Hirata *et al.*, Phys. Lett. B 280: 146 (1992);

Y. Fukuda *et al.*, Phys. Lett. B335: 237 (1994); IMB collaboration, R. Becker-Szendy *et al.*, Nucl. Phys. Proc. Suppl. B 38: 331 (1995); Soudan-2 collaboration, W.W.M. Allison *et al.*, Phys. Lett. B 391: 491 (1997); MACRO collaboration, M. Ambrosio *et al.*, Phys. Lett. B 434: 451 (1998).

4. G. Barr, T.K. Gaisser, and T. Stanev, Phys. Rev. D 39: 3532 (1989); M. Honda, T. Kajita, K. Kasahara, and S. Midorikawa, Phys. Rev. D 52: 4985 (1995); V. Agrawal, T.K. Gaisser, P. Lipari, and T. Stanev, Phys. Rev. D 53: 1314 (1996); T.K. Gaisser and T. Stanev, Phys. Rev. D 57: 1977 (1998).
5. V. Barger, R.J.N. Phillips, and K. Whisnant, Phys. Rev. D 24: 538 (1981); L. Wolfenstein, Phys. Rev. D 17: 2369 (1978); S.P. Mikheyev and A. Smirnov, Yad. Fiz. 42: 1441 (1985); Nuovo Cim. 9 C: 17 (1986); H. Bethe, Phys. Rev. Lett. 56: 1305 (1986); S.P. Rosen and J.M. Gelb, Phys. Rev. D 34: 969 (1986); V. Barger, R.J.N. Phillips, and K. Whisnant, Phys. Rev. D 34: 980 (1986); S.J. Parke, Phys. Rev. Lett. 57: 1275 (1986); S.J. Parke and T.P. Walker, Phys. Rev. Lett. 57: 2322 (1986); W.C. Haxton, Phys. Rev. Lett. 57: 1271 (1986); S.L. Glashow and L.M. Krauss, Phys. Lett. B 190: 199 (1987). T.K. Kuo and J. Pantaleone, Rev. Mod. Phys. 61: 937 (1989); A. Acker, S. Pakvasa, and J. Pantaleone, Phys. Rev. D 43: 1754 (1991); V. Barger, R.J.N. Phillips, and K. Whisnant, Phys. Rev. Lett. 69: 3135 (1992); P.I. Krastev and S.T. Petcov, Phys. Lett. B 285: 85 (1992); Phys. Rev. Lett. 72: 1960 (1994); Phys. Rev. D 53: 1665 (1996); N. Hata and P. Langacker, Phys. Rev. D 56: 6116 (1997);
6. J.G. Learned, S. Pakvasa, and T.J. Weiler, Phys. Lett. B 207: 79 (1988); V. Barger and K. Whisnant, Phys. Lett. B 209: 365 (1988); K. Hidaka, M. Honda, and S. Midorikawa, Phys. Rev. Lett. 61: 1537 (1988).
7. Super–Kamiokande Collaboration, talk by Y. Suzuki at *Neutrino-98*, Takayama, Japan, June 1998.
8. Super–Kamiokande Collaboration, Y. Fukuda et al., Phys. Lett. B 433: 9 (1998); Phys. Lett. B 436: 33 (1998); Phys. Rev. Lett. 81: 1562 (1998).
9. See talk by J. Goodman, these proceedings.
10. Liquid Scintillator Neutrino Detector (LSND) collaboration, C. Athanassopoulos *et al.*, Phys. Rev. Lett. 75: 2650 (1995); *ibid.* 77: 3082 (1996); *ibid.* 81: 1774 (1998); talk by H. White at *Neutrino-98*, Takayama, Japan, June 1998.
11. KARMEN collaboration, K. Eitel and B. Zeitnitz, talk at *Neutrino-98*, Takayama, Japan, June 1998, hep-ex/9809007; future tests of the LSND results will be made by the KARMEN experiment and also by the mini–BooNE experiment, E. Church *et al.*, nucl-ex/9706011; see also J.M. Conrad, talk at ICHEP–98, hep-ex/9811009.
12. See talk by K. Whisnant on four–neutrino models, these proceedings.
13. J.N. Bahcall, S. Basu, and M.H. Pinsonneault, Phys. Lett. B 433: 1 (1998).
14. Early, low–statistics experiments which did not indicate a depletion of ν_μ (Frejus and NUSEX) are not included in Table 2.
15. T.K. Gaisser *et al.*, Phys. Rev. D 54: 5578 (1996).
16. V. Barger, K. Whisnant, and R.J.N. Phillips, Phys. Rev. Lett. 45: 2084 (1980).
17. S.M. Bilenky and C. Giunti, hep-ph/9802201.
18. V. Barger and K. Whisnant, hep-ph/9812273.
19. V. Barger, T.J. Weiler, and K. Whisnant, Phys. Lett. B 440: 1 (1998).
20. For other discussions of the recent atmospheric neutrino data, see J.W. Flanagan, J.G. Learned, and S. Pakvasa, Phys. Rev. D 57: 2649 (1998); M.C. Gonzalez–Garcia, H. Nunokawa, O. Peres, T. Stanev, and J.W.F. Valle, Phys.

Rev. D 58: 033004 (1998); M.C. Gonzalez–Garcia, H. Nunokawa, O. Peres, and J.W.F. Valle, hep-ph/9807305; C.H. Albright, K.S. Babu, and S.M. Barr, Phys. Rev. Lett. 81: 1167 (1998); J.G. Learned, S. Pakvasa, and J.L. Stone, Phys. Lett. B 435: 131 (1998); L.J. Hall and H. Murayama, Phys. Lett. B 436: 323 (1998).
21. CHOOZ collaboration, M. Apollonio *et al.*, Phys. Lett. B 420: 397 (1998).
22. Similar fits have been done in J. Bahcall, P. Krastev, and A. Smirnov, Phys. Rev. D 58: 096016 (1998); see also talk by J. Bahcall, these proceedings.
23. V. Barger, S. Pakvasa, T.J. Weiler, and K. Whisnant, Phys. Lett. B 437: 107 (1998).
24. A.J. Baltz, A.S. Goldhaber, and M. Goldhaber, Phys. Rev. Lett. 81: 5730 (1998).
25. Y. Nomura and T. Yanagida, Phys. Rev. D 59: 017303 (1999); S. Davidson and S.F. King, Phys. Lett. B 445: 191 (1998).
26. G. Altarelli and F. Feruglio, Phys. Lett. B 439: 112 (1998); M. Jezabek and Y. Sumino, Phys. Lett. B 440: 327 (1998).
27. C. Jarlskog, M. Matsuda, S. Skadhauge, and M. Tanimoto, hep-ph/9812282.
28. H. Georgi and S. Glashow, hep-ph/9808293; R.N. Mohapatra and S. Nussinov, Phys. Lett. B 441: 299 (1998); hep-ph/9809415; S.K. Kang and C.S. Kim, hep-ph/9811379; E. Ma, hep-ph/9812344.
29. H. Fritzsch and Z. Xing, Phys. Lett. B 372: 265 (1996); Phys. Lett. B 440: 313 (1998); M. Fukugita, M. Tanimoto, and T. Yanagida, Phys. Rev. D 57: 4429 (1998); R.N. Mohapatra and S. Nussinov, Phys. Lett. B 441: 299 (1998); hep-ph/9809415; C. Jarlskog, M. Matsuda, S. Skadhauge, and M. Tanimoto, hep-ph/9812282.
30. See the first paper of Ref. 26 and L.J. Hall and D. Smith, hep-ph/9812308.
31. See also P.I. Krastev and S.T. Petcov, Phys. Lett. B 395: 69 (1997).
32. See, e.g., the Particle Data Group's 1998 Review of Particle Properties, C. Caso *et al.*, Euro. Phys. J. 3: 1 (1998).
33. V. Barger, T.J. Weiler, and K. Whisnant, Phys. Lett. B 442: 255 (1998).
34. Links to long–baseline experiments may be found at: http://www.hep.anl.gov/NDK/Hypertext/nuindustry.html
35. S. Geer, Phys. Rev. D 57: 6989 (1998).
36. See, e.g., the discussions in Refs. 18 and 19.
37. SNO Collaboration, E. Norman *et al.*, in proc. of *The Fermilab Conference: DPF 92*, November 1992, Batavia, IL, ed. by C. H. Albright, P.H. Kasper, R. Raja, and J. Yoh (World Scientific, Singapore, 1993), p. 1450.
38. BOREXINO Collaboration, C. Arpesella *et al.*, "INFN Borexino proposal," Vols. I and II, edited by G. Bellini, R. Raghavan *et al.* (Univ. of Milan, 1992); J. Benziger, F.P. Calaprice *et al.*, "Proposal for Participation in the Borexino Solar Neutrino Experiment," (Princeton University, 1996).
39. V. Barger, R.J.N. Phillips, and K. Whisnant, Phys. Rev. Lett. 65: 3084 (1990); Phys. Rev. D 43: 1110 (1991); S. Pakvasa and J. Pantaleone, Phys. Rev. Lett. 65: 2479 (1990).
40. For other recent discussions of seasonal effects in solar neutrinos, see S.P. Mikheyev and A.Yu. Smirnov, Phys. Lett. B 429: 343 (1998). B. Faid, G.L. Fogli, E. Lisi, and D. Montanino, Astropart. Phys. 10: 93 (1999); S.L. Glashow, P.J. Kernan, and L.M. Krauss, Phys. Lett. B 445: 412 (1999); J.M. Gelb and S.P. Rosen, hep-ph/9809508; V. Berezinsky, G. Fiorentini, and M. Lissia, hep-ph/9811352.

ARE THERE FOUR OR MORE NEUTRINOS?

Kerry Whisnant

Department of Physics and Astronomy
Iowa State University
Ames, IA 50011

INTRODUCTION

It now appears that the solar[1,2] and atmospheric[3,4] neutrino anomalies could very well be the result of neutrinos oscillating from one species to another.[5] The recent results from the Super–Kamiokande (Super–K) experiment[6,7,8] provide very strong evidence for oscillations, and indicate that different mass–squared difference (δm^2) scales are required to explain both effects. Data from the LSND experiment[9] can be explained by oscillations, but this requires a third δm^2 scale. Since only two independent δm^2 scales are possible with three neutrinos, a fourth neutrino is needed. Recent data from the KARMEN experiment[10] excludes much of the LSND allowed region, but there still remains an area in the oscillation parameter space that is consistent with all the data.

In this talk I will review the evidence that there are four (or more) neutrinos, and that at least one of the extra neutrinos must be sterile. I will discuss the mass spectrum required in a four–neutrino model, and analyze the phenomenology of a large class of models that contains most of the models currently being discussed in the literature. I will then discuss the new phenomenology possible in these models, especially in long–baseline experiments and CP violation. I will conclude with the implications of four–neutrino models for neutrino mass.

REVIEW OF NEUTRINO OSCILLATION DATA

The evidence for vacuum neutrino oscillations in the solar and atmospheric neutrino data is reviewed in Ref. 5; a discussion of both vacuum and MSW solar solutions is given in Ref. 11. The conclusion is that if all of the data are taken at face value, atmospheric ν_μ are oscillating into a neutrino species other than ν_μ or ν_e, and solar ν_e are oscillating into a neutrino species other than ν_e. The δm^2 scale associated with the solar oscillation is $\delta m^2_{sun} \sim 10^{-10}\,\mathrm{eV}^2$ $(10^{-5}\,\mathrm{eV}^2)$ for

Confluence of Cosmology, Massive Neutrinos, Elementary Particles, and Gravitation
Edited by Kursunoglu *et al.*, Kluwer Academic / Plenum Publishers, New York, 1999.

vacuum (MSW) oscillations; for atmospheric oscillations it is $\delta m^2_{atm} \sim 10^{-3}$ eV^2. All oscillations have large (perhaps maximal) oscillation amplitude except for the MSW solar solution, which favors a small amplitude of order 10^{-2}.

The LSND experiment also has evidence for neutrino oscillations.[9] Their measurements of neutrinos from μ^+ decay at rest (π^+ decay in flight) indicate that $\bar{\nu}_\mu$ (ν_μ) are oscillating into $\bar{\nu}_e$ (ν_e) with an oscillation probability of $0.31^{+0.11}_{-0.10} \pm 0.05\%$ ($0.26 \pm 0.10 \pm 0.05\%$). The KARMEN experiment[10] places an upper limit on the $\bar{\nu}_\mu \to \bar{\nu}_e$ oscillation probability of 0.07% at 90% C.L. at a different value of L/E. The degree to which the two experiments agree (or disagree) depends on the exact statistical method used to analyze the data.[12] Regardless of the method used, there is a range of two–neutrino parameters (assuming $\nu_\mu \to \nu_e$ oscillations) that is consistent with both experiments, given approximately by the line segment defined by

$$0.3 \text{ eV}^2 \leq \delta m^2_{LSND} = \frac{0.030 \text{ eV}^2}{(\sin^2 2\theta)^{0.7}} \leq 2.0 \text{ eV}^2. \tag{1}$$

The lower bound in Eq. (1) comes from the BUGEY reactor experiment[13] while the upper bound comes from KARMEN and BNL E776.[14]

As discussed in Ref. 5, to simultaneously explain the solar, atmospheric and LSND data requires three independent δm^2 scales, which implies there must be (at least) four neutrinos. Each type of neutrino data is consistent with simple two–neutrino oscillations:

$$\begin{aligned} \nu_e &\to \nu_x && \text{for solar,} \\ \nu_\mu \to \nu_y \quad &(y \neq e) && \text{for atmospheric,} \\ \nu_\mu &\to \nu_e && \text{for LSND.} \end{aligned} \tag{2}$$

NEED FOR ONE (OR MORE) STERILE NEUTRINO(S)

The invisible width of the Z boson measures the number of light neutrinos with standard weak interactions. The LEP measurement[15] is $N_\nu = 2.993 \pm 0.011$, which is consistent with the standard model with only ν_e, ν_μ, and ν_τ. Any extra neutrino that participates in oscillations must be light, but since it does not contribute to the Z width it must be "sterile," i.e., have little or no weak interactions. In this talk a sterile neutrino will be labeled ν_s.

Sterile neutrinos have often been considered in the literature.[16] The simplest possibility is to have either the solar ν_e or atmospheric ν_μ oscillate into ν_s, while the other oscillates into ν_τ. However, there are constraints from Big Bang nucleosynthesis (BBN) on the mixing of ν_s with an active neutrino species. If the mixing is sizable, then the effective number of neutrinos that contribute during BBN would be larger than three; current limits[17] indicate $N_\nu^{eff} \leq 3$. This would appear to rule out all sterile–active mixing except for the small–angle MSW solar solution. However, some recent estimates of N_ν^{eff} using a higher inferred abundance of ^{4}He yield a considerably weaker bound.[18] Therefore sizable sterile–active mixing may still be possible; in this talk I will consider both large and small mixings with sterile neutrinos.

There are many recent models in the literature in which there is one additional light sterile neutrino[19–24] In most of the models the solar (atmospheric) oscillation is $\nu_e \to \nu_s$ ($\nu_\mu \to \nu_\tau$), although $\nu_e \to \nu_\tau$ ($\nu_\mu \to \nu_s$) has also been considered.[19,24] Another interesting possibility is that *both* solar ν_e and atmospheric ν_μ oscillate into sterile neutrinos;[25] the two sterile states must be distinct since three independent δm^2 are still required. In the latter models ν_τ (and in some cases a third sterile neutrino) do not participate in the visible oscillations, although they may help contribute to hot dark matter.

NEUTRINO MASS SPECTRUM

Sterile–active neutrino mass terms are of the Dirac form, but active–active and sterile–sterile mass terms are Majorana; therefore the most general neutrino mass matrix involving a sterile neutrino is Majorana.[26] The most general Majorana mass matrix is complex and symmetric and can be diagonalized by a unitary matrix U as follows: $M_D = U^T M U$.[27] If we assume the neutrino flavor basis $(\nu_x, \nu_e, \nu_\mu, \nu_y)$, where solar oscillations are $\nu_e \to \nu_x$ and atmospheric oscillations are $\nu_\mu \to \nu_y$, then U is a 4×4 matrix that rotates the flavor states to the mass eigenstates $(\nu_0, \nu_1, \nu_2, \nu_3)$.

Given that three different mass-squared differences scales ($\delta m^2_{sun} \ll \delta m^2_{atm} \ll \delta m^2_{LSND}$) are required to explain all of the neutrino oscillation data, there are in principle two possibilities for the mass spectrum in four–neutrino models: one mass eigenstate separated from three nearly degenerate states (which might be called the 1+3 or 3+1 scenario, depending on the relative masses), or two pairs of nearly degenerate mass eigenstates separated from each other (the 2+2 scenario). An example of the 1+3 scenario is $m_0 \simeq m_1 \simeq m_2 < m_3$ with $\delta m^2_{sun} = \delta m^2_{10}$, $\delta m^2_{atm} = \delta m^2_{21}$, and $\delta m^2_{LSND} = \delta m^2_{32}$. An example of the 2+2 scenario is $m_0 \simeq m_1 < m_2 \simeq m_3$ with $\delta m^2_{sun} = \delta m^2_{10}$, $\delta m^2_{LSND} = \delta m^2_{21}$, and $\delta m^2_{atm} = \delta m^2_{32}$. Then because of the different ordering of δm^2 values, the various δm^2 contribute to different experiments in these two cases (see Table 1), which leads to different constraints on the neutrino mixing.

Table 1. Contribution of δm^2 to oscillation phenomena in different scenarios.

Experiment	1+3	2+2
LSND	δm^2_{32}, δm^2_{31}, δm^2_{30}	δm^2_{31}, δm^2_{30}, δm^2_{21}, δm^2_{20}
Atmospheric	δm^2_{21}, δm^2_{20}	δm^2_{32}
Solar	δm^2_{10}	δm^2_{10}

For example, the constraints from the atmospheric, LSND and BUGEY experiments at $\delta m^2_{LSND} = 0.3\ \text{eV}^2$ in the 1+3 and 2+2 cases are listed in Table 2. It is obvious from the table that the 1+3 scenario is ruled out, but that the 2+2 scenario is (barely, in this example) allowed. In fact in can be shown for all δm^2_{LSND} that the null results of the BUGEY reactor and CDHS[28] accelerator experiments, when combined with the allowed regions from the solar, atmospheric, and LSND data, completely excludes all 1+3 (or 3+1) scenarios, so that only the 2+2 scenario is allowed.[29,19] The overall mass scale is not determined, although the splitting of the mass eigenstates implies $m_3 \geq \sqrt{\delta m^2_{LSND}} \simeq 0.55$ eV.

Table 2. Some constraints on U at $\delta m^2_{LSND} = 0.3\ \text{eV}^2$.

Experiment	1+3	2+2
Atmospheric	$\lvert U_{\mu3}\rvert^2 \leq 0.11$	$\lvert U_{\mu2}\rvert^2 + \lvert U_{\mu3}\rvert^2 \geq 0.89$
Bugey	$\lvert U_{e3}\rvert^2 \leq 0.01$	$\lvert U_{e2}\rvert^2 + \lvert U_{e3}\rvert^2 \leq 0.01$
LSND	$\lvert U_{e3}\rvert^2 \lvert U_{\mu3}\rvert^2 \geq 0.01$	$(\lvert U_{e2}\rvert^2 + \lvert U_{e3}\rvert^2)(\lvert U_{\mu2}\rvert^2 + \lvert U_{\mu3}\rvert^2) \geq 0.01$

There is an additional restriction if we require MSW oscillations to describe the solar data. The so-called resonance condition for maximal mixing in the sun (which is needed for the MSW effect to occur) is $2\sqrt{2} G_F E N_e = \delta m^2_{01} \cos 2\theta$, where N_e is the number density of electrons and θ is the mixing angle between ν_e and ν_x.[30]

Therefore, a resonance can only occur if $\delta m_{01}^2 \cos 2\theta$ is positive, which implies that the mass eigenstate in vacuum mostly closely associated with ν_x must be heavier than the one mostly closely associated with ν_e. This is usually (but not always) satisfied when M_{xx}, the diagonal mass matrix element for ν_x, is larger than M_{ee}. There is no such restriction for vacuum oscillations since they do not depend on the sign of δm_{01}^2.

MASS MATRIX MODELS AND NEUTRINO MIXING

Most explicit four–neutrino mass matrices proposed in the literature may be put into the form (in the $\nu_x, \nu_e, \nu_\mu, \nu_y$ basis)

$$M = \begin{pmatrix} \epsilon_1 & \epsilon_2 & 0 & 0 \\ \epsilon_2 & 0 & \epsilon_5 & \epsilon_3 \\ 0 & \epsilon_5 & \delta & m \\ 0 & \epsilon_3 & m & \delta' \end{pmatrix}. \tag{3}$$

Examples include the cases $m \ll \delta \simeq \delta'$ and $\epsilon_3 = 0$,[20] $\delta = \delta' \ll m$ and $\epsilon_5 = 0$,[19] $\delta = -\delta' \ll m$ and $\epsilon_5 = 0$,[21] $\delta' \ll m$ and $\epsilon_1 = \epsilon_3 = \delta = 0$,[23] and $\delta, \delta' \ll m$.[24] In all of these models, the LSND effect is of order ϵ^2, where ϵ_3 and/or $\epsilon_5 \sim \epsilon \sim 0.05 - 0.20$. Also, U has the form

$$U = \begin{pmatrix} U_1 & U_2 \\ U_3 & U_4 \end{pmatrix}, \tag{4}$$

where the U_j are all 2×2 matrices, and U_2 and U_3 are of order ϵ in magnitude. The only large mixings are $\nu_e \to \nu_x$ and $\nu_\mu \to \nu_y$; the former may also be small if the solar oscillation is of the MSW type.

The most general 4×4 U has 6 mixing angles and 6 phases; only 3 of the phases may be determined in neutrino oscillations. Assuming the form of Eq. (4), U may be written approximately as[24]

$$U = \begin{pmatrix} c_{01} & s_{01}^* & s_{02}^* & s_{03}^* \\ -s_{01} & c_{01} & s_{12}^* & s_{13}^* \\ \begin{array}{c} -c_{01}(s_{23}^* s_{03} + c_{23} s_{02}) \\ +s_{01}(s_{23}^* s_{13} + c_{23} s_{12}) \end{array} & \begin{array}{c} -s_{01}^*(s_{23}^* s_{03} + c_{23} s_{02}) \\ -c_{01}(s_{23}^* s_{13} + c_{23} s_{12}) \end{array} & c_{23} & s_{23}^* \\ \begin{array}{c} c_{01}(s_{23} s_{02} - c_{23} s_{03}) \\ -s_{01}(s_{23} s_{12} - c_{23} s_{13}) \end{array} & \begin{array}{c} s_{01}^*(s_{23} s_{02} - c_{23} s_{03}) \\ +c_{01}(s_{23} s_{12} - c_{23} s_{13}) \end{array} & -s_{23} & c_{23} \end{pmatrix}, \tag{5}$$

where $c_{jk} = \cos\theta_{jk}$, $s_{jk} = \sin\theta_{jk} e^{i\delta_{jk}}$, and s_{02}, s_{03}, s_{12}, and s_{13} are small (of order ϵ or less). The mixing in Eq. (5) is the most general possible if no neutrino has large mixing with more than one other neutrino. We see that in the approximation given by Eq. (5), U_1 (U_4) is a simple 2×2 rotation matrix involving θ_{01} (θ_{23}), and is therefore approximately unitary. The elements of U_2 and U_3 are all small (of order ϵ or less), as required.

NEW PHENOMENOLOGY

The mixing matrix of Eq. (5) can imply possible new physics. The signatures and future experiments where they might be detected are briefly listed below (for a more detailed discussion, see Ref. 24).

Short–Baseline Experiments

In short–baseline experiments ($L/E \sim 1$ km/GeV) only δm^2_{LSND} is probed. Oscillations of $\nu_e \to \nu_y$ and $\nu_\mu \to \nu_x$ with amplitude of order ϵ^2 are possible in many of the models, but CP violation is suppressed in the leading oscillation.[31] Future short–baseline experiments will test $\nu_\mu \to \nu_e$ (mini–BooNE at FNAL and ORLaND at Oak Ridge) and $\nu_\mu \to \nu_\tau$ (COSMOS at FNAL and TOSCA at CERN).[32]

Long–Baseline Experiments

In long–baseline experiments ($L/E \geq 100$ km/GeV) the δm^2_{LSND} oscillations reach their average values and δm^2_{atm} is probed. Oscillations of $\nu_e \to \nu_y$ and $\nu_\mu \to \nu_x$ are possible here, too, as are new contributions to $\nu_\mu \leftrightarrow \nu_e$; all of these channels can have an oscillation amplitude of order ϵ^2. Also, visible CP violation is possible. Defining the CP–violation parameter

$$A^{CP}_{\alpha\beta} \equiv \frac{P(\nu_\alpha \to \nu_\beta) - P(\bar{\nu}_\alpha \to \bar{\nu}_\beta)}{P(\nu_\alpha \to \nu_\beta) + P(\bar{\nu}_\alpha \to \bar{\nu}_\beta)}, \tag{6}$$

then $A^{CP}_{e\mu} = -A^{CP}_{\mu e} = -A^{CP}_{ey} = A^{CP}_{ye} \sim \epsilon^2$ to leading order in ϵ for the mixing in Eq. (5); CP violation of order ϵ^2 is also possible in the $\nu_\mu \to \nu_x$ and $\nu_\mu \to \nu_y$ channels (although the latter will be difficult to measure since $\nu_\mu - \nu_y$ mixing is nearly maximal).

Currently plans are being made for the following long–baseline experiments: FNAL to SOUDAN (MINOS), KEK to Kamioka (K2K), and CERN to GRAN SASSO (AQUA-RICH, ICARUS, NICE, NOE, OPERA, and Super–ICARUS).[32] Generally these experiments will test ν_μ survival, ν_μ appearance as ν_τ, and ν_μ disappearance (via neutral current events); typical sensitivities to oscillation parameters are $\delta m^2_{atm} > 2 \times 10^{-3}$ eV^2 for maximal mixing and $\sin^2 2\theta > 10^{-2}$ for large δm^2_{atm}. More extensive tests, including tests for CP violation, may be done with muon storage rings,[33] which will have the options of simultaneous ν_μ and $\bar{\nu}_e$ beams (from μ^- decay), or simultaneous $\bar{\nu}_\mu$ and ν_e beams (from μ^+ decay). Oscillations between ν_μ and ν_e (or $\bar{\nu}_\mu$ and $\bar{\nu}_e$) therefore give wrong–sign charged leptons in the detector. Oscillations to ν_τ are detected via the $\tau \to \ell\nu\bar{\nu}$ decay, where the sign of the detected lepton determines whether it came from an initial ν_μ or $\bar{\nu}_e$ ($\bar{\nu}_\mu$ or ν_e) for an original μ^- (μ^+).

All new signals involving ν_τ are potentially observable in appearance experiments. Measurements of small–amplitude oscillations with a sterile neutrino will be very difficult since they require measuring a disappearance amplitude of order ϵ^2, which means that the initial flux must be known to within a very small uncertainty. If $\nu_x = \nu_\tau$, then all 6 angles and 2 of the 3 independent phases that can be determined by oscillation experiments will be accessible; on the other hand, if $\nu_x = \nu_s$, then only 4 angles and 1 phase can be easily measured.[24] To fully determine the accessible parameters, it is important to measure both short– and long–baseline oscillations since some oscillation channels have distinct contributions from more than one δm^2 scale.

Although s_{12} and s_{13} are always constrained to be small, s_{02} and s_{03} are not. In this case U no longer has the form of Eq. (4), and the parametrization in Eq. (5) becomes more complicated[24] (no specific models of this type have been proposed, however). Furthermore, in this case atmospheric ν_μ oscillate to a linear combination of ν_x and ν_y, and solar ν_e oscillate into the orthogonal combination. Vacuum CP–violation effects involving ν_e are still of order ϵ^2, but there are potentially large CP–violation effects in long–baseline $\nu_\mu \to \nu_y$ oscillations.[34]

DISTINGUISHING ν_τ FROM ν_s

In order to determine which one (if any) of ν_x and ν_y is ν_τ, it will be necessary to distinguish ν_τ from ν_s in solar and atmospheric neutrino experiments. The Sudbury Neutrino Observatory (SNO),[35] and to a lesser extent future runs at Super–K, can make neutral–current (NC) as well as charged–current (CC) measurements; a measurement of $(CC/NC)_{data}/(CC/NC)_{theory} = 1$ will be evidence for $\nu_e \to \nu_s$ oscillations. For atmospheric neutrinos, the NC/CC ratio,[36] the up/down ratio for NC,[37] evidence for ν_μ disappearance in conjunction with extra events without muons,[38] and matter effects in $\nu_\mu \to \nu_s$[39] can all be used to determine if $\nu_y = \nu_s$. Also possible is direct detection of taus in atmospheric[40] and, of course, long–baseline experiments.

OTHER IMPLICATIONS

Massive neutrinos will contribute to the mass density of the universe; the Sloan Digital Sky Survey should be sensitive to m_ν as low as $0.2 - 0.9$ eV for two nearly degenerate neutrinos,[41] which covers the LSND range in 2+2 scenarios. Neutrinoless double beta decay puts a upper limit of about 0.5 eV the magnitude of the M_{ee} element of the Majorana mass matrix;[42] although $M_{ee} = 0$ in most explicit models, it can put strong constraints on more general models, especially those where the mass eigenstate most closely associated with ν_e is in the upper pair in 2+2 scenarios. Finally, beta decay limits involving ν_e put an upper limit on the largest mass eigenstate: $m_3^2 \le (4.4 \text{ eV})^2 + \delta m^2_{LSND} \simeq (5 \text{ eV})^2$.[43]

SUMMARY

At least four neutrinos are required in order to completely explain the solar, atmospheric, and LSND neutrino data. A 2+2 mass spectrum (with two pairs of nearly degenerate mass eigenstates) is favored over a 1+3 or 3+1 spectrum. Models exist with three active and one sterile neutrino, or two active and two sterile neutrinos. If no neutrino has large mixing with more than one other neutrino, then the four–neutrino mixing matrix reduces to a set of 2×2 blocks, where the off–diagonal blocks are small in magnitude; possible new physics in this case includes new oscillation channels and CP violation that can be tested in future long–baseline experiments. Finally, the largest neutrino mass has a lower limit of 0.55 eV, which may have implications for the mass density of the universe and beta decay physics.

ACKNOWLEDGMENTS

I gratefully acknowledge collaboration with V. Barger, Y.–B. Dai, S. Pakvasa, T.J. Weiler, and B.–L. Young.

REFERENCES

1. B.T. Cleveland *et al.*, Nucl. Phys. Proc. Suppl. B 38: 47 (1995); Kamiokande collaboration, Y. Fukuda *et al.*, Phys. Rev. Lett. 77: 1683 (1996); GALLEX Collaboration, W. Hampel *et al.*, Phys. Lett. B 388: 384 (1996); SAGE

collaboration, J.N. Abdurashitov *et al.*, Phys. Rev. Lett. 77: 4708 (1996).
2. J.N. Bahcall and M.H. Pinsonneault, Rev. Mod. Phys. 67: 781 (1995); J.N. Bahcall, S. Basu, and M.H. Pinsonneault, Phys. Lett. B 433: 1 (1998).
3. Kamiokande collaboration, K.S. Hirata *et al.*, Phys. Lett. B 280: 146 (1992); Y. Fukuda *et al.*, Phys. Lett. B335: 237 (1994); IMB collaboration, R. Becker-Szendy *et al.*, Nucl. Phys. Proc. Suppl. B 38: 331 (1995); Soudan-2 collaboration, W.W.M. Allison *et al.*, Phys. Lett. B 391: 491 (1997); MACRO collaboration, M. Ambrosio *et al.*, Phys. Lett. B 434: 451 (1998).
4. G. Barr, T.K. Gaisser, and T. Stanev, Phys. Rev. D 39: 3532 (1989); M. Honda, T. Kajita, K. Kasahara, and S. Midorikawa, Phys. Rev. D 52: 4985 (1995); V. Agrawal, T.K. Gaisser, P. Lipari, and T. Stanev, Phys. Rev. D 53: 1314 (1996); T.K. Gaisser and T. Stanev, Phys. Rev. D 57: 1977 (1998).
5. See K. Whisnant talk on three–neutrino models, these proceedings, and references therein.
6. Super-Kamiokande Collaboration, talk by Y. Suzuki at *Neutrino–98*, Takayama, Japan, June 1998.
7. Super–Kamiokande Collaboration, Y. Fukuda et al., Phys. Lett. B 433: 9 (1998); Phys. Lett. B 436: 33 (1998); Phys. Rev. Lett. 81: 1562 (1998).
8. See J. Goodman, these proceedings.
9. Liquid Scintillator Neutrino Detector (LSND) collaboration, C. Athanassopoulos *et al.*, Phys. Rev. Lett. 75: 2650 (1995); *ibid.* 77: 3082 (1996); *ibid.* 81: 1774 (1998); talk by H. White at *Neutrino–98*, Takayama, Japan, June 1998.
10. KARMEN collaboration, K. Eitel and B. Zeitnitz, talk at *Neutrino–98*, Takayama, Japan, June 1998, hep-ex/9809007; future tests of the LSND results will be made by the KARMEN experiment and also by the mini–BooNE experiment, E. Church *et al.*, nucl-ex/9706011.
11. J. Bahcall, P. Krastev, and A. Smirnov, Phys. Rev. D 58: 096016 (1998); see also J. Bahcall, these proceedings.
12. See, e.g., J.M. Conrad, talk at ICHEP-98, hep-ex/9811009.
13. Y. Declais *et al.*, Nucl. Phys. B 434: 503 (1995).
14. L. Borodovsky *et al.*, Phys. Rev. Lett. 68: 274 (1992).
15. LEP Electroweak Working Group and SLD Heavy Flavor Group, D. Abbaneo *et al.*, CERN-PPE-96-183, December 1996.
16. See, e.g., V. Barger, P. Langacker, J. Leveille, and S. Pakvasa, Phys. Rev. Lett. 45: 692 (1980); V. Barger, N. Deshpande, P.B. Pal, R.J.N. Phillips, and K. Whisnant, Phys. Rev. D 43: 1759 (1991); S. Bludman, D.C. Kennedy, and P. Langacker, Nucl. Phys. B 374: 373 (1992); Z.G. Berezhiani and R.N. Mohapatra, Phys. Rev. D 52: 6607 (1995); E.J. Chun, A.S. Joshipura and A.Y. Smirnov, Phys. Lett. B 357: 608 (1995); Phys. Rev. D 54: 4654 (1996); K. Benakli and A.Y. Smirnov, Phys. Rev. Lett 79: 4314 (1997); G. Cleaver, M. Cvetic, J.R. Espinosa, L. Everett, and P. Langacker, Phys. Rev. D 57: 2701 (1998).
17. R. Barbieri and A. Dolgov, Phys. Lett. B 237: 440 (1990); K. Enqvist, K. Kainulainen, and M. Thomson, Nucl. Phys. B 373: 498 (1992); X. Shi, D.N. Schramm, and B.D. Fields, Phys. Rev. D 48: 2563 (1993); C.Y. Cardall and G.M. Fuller, Phys. Rev. D 54: 1260 (1996); D.P. Kirilova and M.V. Chizhov, hep–ph/9707282; S.M. Bilenky, C. Giunti, W. Grimus and T. Schwetz, hep–ph/9804421.
18. P.J. Kernan and S. Sarkar, Phys. Rev. 54: R3681 (1996); S. Sarkar, Rept. Prog. Phys. 59: 1 (1996); K.A. Olive, talk at 5th International Workshop on Topics in Astroparticle and Underground Physics (TAUP 97), Gran Sasso,

Italy, 1997; K.A. Olive, proc. of 5th International Conference on Physics Beyond the Standard Model, Balholm, Norway, 1997; C.J. Copi, D.N. Schramm and M.S. Turner, Phys. Rev. Lett. 75: 3981 (1995); K.A. Olive and G. Steigman, Phys. Lett. B 354: 357 (1995).

19. V. Barger, T.J. Weiler, and K. Whisnant, Phys. Lett. B 427: 97 (1998); V. Barger, S. Pakvasa, T.J. Weiler, and K. Whisnant, Phys. Rev. D 58: 093016 (1998).
20. D.O. Caldwell and R.N. Mohapatra, Phys. Rev. D 48: 3259 (1993).
21. S.C. Gibbons, R.N. Mohapatra, S. Nandi, and A. Raychaudhuri, Phys. Lett. B 430: 296 (1998).
22. B. Brahmachari and R.N. Mohapatra, Phys. Lett. B 437: 100 (1998); A. Joshipura and A.Yu. Smirnov, Phys. Lett. B 439: 103 (1998).
23. S. Mohanty, D.P. Roy, and U. Sarkar, Phys. Lett. B 445:185 (1998).
24. V. Barger, Y.-B. Dai, K. Whisnant, and B.-L. Young, hep-ph/9901388.
25. R. Foot and R.R. Volkas, Phys. Rev. D 52: 6595 (1995); A. Geiser, Phys. Lett. B 444: 358 (1999); D. Suematsu, hep-ph/9805305; W. Krolikowski, hep-ph/9808207; hep-ph/9808307.
26. See, e.g., G. Gelmini and E. Roulet, Rept. Prog. Phys. 58: 1207 (1995); P. Langacker, Phys. Rev. D 58: 093017 (1998).
27. R.N. Mohapatra and P.B. Pal, *Massive Neutrinos in Physics and Astrophysics*, (World Scientific, Singapore, 1991), pp. 218.
28. F. Dydak et al., Phys. Lett. **B134**, 281 (1984).
29. S. M. Bilenky, C. Giunti and W. Grimus, Eur. Phys. J. C 1: 247 (1998).
30. V. Barger, R.J.N. Phillips, and K. Whisnant, Phys. Rev. D 24: 538 (1981); P. Langacker, J.P. Leveille, and J. Sheiman, Phys. Rev. D 27: 1228 (1983).
31. This was discussed, in the context of three–neutrino models, in V. Barger, K. Whisnant, and R.J.N. Phillips, Phys. Rev. Lett. 45: 2084 (1980).
32. Links to long–baseline experiments may be found at: http://www.hep.anl.gov/NDK/Hypertext/nuindustry.html
33. S. Geer, Phys. Rev. D 57: 6989 (1998).
34. S. M. Bilenky, C. Giunti and W. Grimus, Phys. Rev. D 58: 033001 (1998).
35. SNO Collaboration, E. Norman *et al.*, in proc. of *The Fermilab Conference: DPF 92*, November 1992, Batavia, IL, ed. by C. H. Albright, P.H. Kasper, R. Raja, and J. Yoh (World Scientific, Singapore, 1993), p. 1450.
36. F. Vissani and A. Smirnov, Phys. Lett. B 432: 376 (1998).
37. J. G. Learned, S. Pakvasa and J. Stone, Phys. Lett. B 435: 131 (1998); L.J. Hall and H. Murayama, Phys. Lett. B 436: 323 (1998).
38. A. Curioni *et al.*, hep-ph/9805249.
39. Q.Y. Liu and A.Yu. Smirnov, Nucl. Phys. B 524: 505 (1998).
40. L.J. Hall and H. Murayama, hep-ph/9810468.
41. W. Hu, D.J. Eisenstein, and M. Tegmark, Phys. Rev. Lett. 80: 5255 (1998).
42. H.V. Klapdor–Kleingrothaus, Int. J. Mod. Phys. A 13: 3953 (1998).
43. V. Barger, T.J. Weiler, and K. Whisnant, Phys. Lett. B 442: 255 (1998).

NEUTRINO REACTIONS IN NUCLEI AND NEUTRINO BACKGROUNDS

S.L. Mintz
Physics Department
Florida International University
Miami, Florida 33199

M. Pourkaviani
MP Consulting Associates
Altamonte Springs,FL 32701

INTRODUCTION

In the last several years two neutrino experiments have produced for the first time a relatively large body of data for neutrino reactions on ^{12}C. The first group to undertake experiments of this kind was the Los Alamos or LAMPF[1] group.They looked at two reactions, both involving electron neutrinos produced primarily by the decay of muons at rest. These were obtained from π^+'s produced by the interactions of an accelator proton beam on a target. The decay of the π^+'s produced μ^+'s which were stopped and decayed producing electron neutrinos. Thus the result was a Michel spectrum which peaks at about 37 MeV and which has a maximal energy of near 52 MeV. The reactions all made use of ^{12}C as a target. The Los Alamos group was able to run two reactions,$\nu_e + {}^{12}C \longrightarrow e^- + {}^{12}N_{(g.s.)}$ and $\nu_e + {}^{12}C \longrightarrow e^- + X$ and to find Michel spectrum averaged cross sections for both of them.

A few years later a collaboration between Karlesruhe and the Rutherford Appleton Laboratory called KARMEN[2] undertook a similar set of measurements. This collaboration also obtained its spectrum in the same way as the Los Alamos group and hence made use of a Michel spectrum of neutrinos. Thus the results of the KARMEN and Los Alamos groups are directly comparable. The KARMEN group was able to run the reactions $\nu_e + {}^{12}C \longrightarrow e^- + {}^{12}N_{(g.s.)}$ and $\nu_e + {}^{12}C \longrightarrow e^- + X$ but was also able to run the neutral current reaction $\nu_e + {}^{12}C \longrightarrow \nu_e + {}^{12}C^*$.

Both groups achieved reasonable agreement for the exclusive reaction to the $^{12}N_{(g.s.)}$ state,$\nu_e + {}^{12}C \longrightarrow e^- + {}^{12}N_{(g.s.)}$. LAMPF[1] obtained:

$$< \sigma >= 1.04 \pm \ .1 \ stat \ \pm .1 \ sys \ \times 10^{-41} \ cm^2 \tag{1}$$

and KARMEN[2] obtained:

$$< \sigma >= .93 \pm \ .08 \ stat \ \pm .075 \ sys \ \times 10^{-41} \ cm^2. \tag{2}$$

Confluence of Cosmology, Massive Neutrinos, Elementary Particles, and Gravitation
Edited by Kursunoglu *et al.*, Kluwer Academic / Plenum Publishers, New York, 1999.

These values were in good agreement with each other and with various theoretical calculations[3,4,5,6] which yielded results in the range of:

$$< \sigma > \simeq .8 \; to \; .95 \; \times 10^{-41} \; cm^2. \tag{3}$$

The KARMEN[2] collaboration was also able to measure the neutral current reaction, $\nu + {}^{12}C \longrightarrow \nu + {}^{12}C^*$ to which both muon and electron neutrinos contribute. They obtained:

$$< \sigma > = 9.1 \pm \; 2.3 \; stat \; \pm \; 1.3 \; sys \; \times 10^{-42} \; cm^2 \tag{4}$$

which was in very good agreement with theoretical prediction[7,8,9]. Thus neutrino reactions in ${}^{12}C$ to exclusive final states appear to be well understood.

The situation for inclusive neutrino reactions in ${}^{12}C$ is less clear. LAMPF[1] measured the cross section for the reaction, $\nu_e + {}^{12}C \longrightarrow e^- + X$. They subtracted off transitions to ${}^{12}N_{(g.s.)}$ and obtained for the reaction, $\nu_e + {}^{12}C \longrightarrow e^- + {}^{12}N^*$:

$$< \sigma > = (3.6 \pm \; 20 \; \%) \; \times 10^{-42} \; cm^2. \tag{5}$$

The situation at KARMEN has changed over time. Their early measurements were very large, of the order of 12 $\times$ 10^{-42} cm^2 but over the last 5 or 6 years this number has steadily declined and at present is:

$$< \sigma > = (6.4 \pm \; 1.3 \; stat \; \pm \; 1.2 \; sys) \; \times \; 10^{-42} \; cm^2. \tag{6}$$

At the time that the LAMPF measurement was made only one theoretical calculation, shell model based, was available[8]:

$$< \sigma > \simeq 3.7 \; \times \; 10^{-42} \; cm^2 \tag{7}$$

which was in good agreement with the LAMPF result. However another calculation[10] based on a random phase shell model yielded a substantially larger result which at the time was still much smaller than the first announced KARMEN results but is now in reasonable agreement with them. A third calculation[11,12] based on a semi-phenomenological model has yielded a result of:

$$< \sigma > \simeq 8.27 \; \times \; 10^{-42} \; cm^2 \tag{8}$$

in closer agreement to the KARMEN result than the LAMPF result. Thus the situation for the inclusive case at low energy is still not well understood.

The situation for inclusive muon neutrino reactions in nuclei is even less well understood. This process, $\nu_\mu + {}^{12}C \longrightarrow \mu^- + X$, has been studied by the Los Alamos group[13], making use of muon neutrinos from decay in flight pions. The spectrum here is very different than the Michel spectrum and is of generally higher energy. It peaks at around 50 MeV but is still relatively large at 150 MeV and rapidly falls to near zero at 250 MeV. LAMPF initially obtained a value of $< \sigma > = (8.3 \pm \; 0.7 \; stat \; \pm \; 1.6 \; sys) \times \; 10^{-40} \; cm^2$ but their present value is:

$$< \sigma > \simeq 12.4 \; \times \; 10^{-40} \; cm^2 \tag{9}$$

with similar errors. There are several theoretical calculations available for this process.The first, a Fermi gas model[13] calculation, yielded a result of:

$$< \sigma > \simeq 24 \; \times \; 10^{-40} \; cm^2. \tag{10}$$

The second, a random phase approximation10 yielded:

$$< \sigma > \simeq 20 \times 10^{-40} cm^2. \tag{11}$$

A third calculation[14] based on a semi-phenomenological tensor approach yields:

$$< \sigma > \simeq 11.7\ 10^{-40}\ cm^2 \tag{12}$$

in very good agreement with experimental results. We shall describe this model in detail in section III of this paper.

One of the problems in measuring the neutrino cross sections in ^{12}C stems from background processes. There are many possible kinds of background but one which has attracted some interest is that of ^{13}C. This isotope of carbon is stable and is a 1.1% impurity in carbon targets. Because the transitions of interest are mirror transitions it was feared that the backgrounds might be large. As we shall see neither the exclusive processes, $\nu_e +^{13} C \longrightarrow e^- +^{13} N_{gs}$ and $\nu_\mu +^{13} C \longrightarrow \mu^- +^{13} N_{gs}$ nor the inclusive processes $\nu_e +^{13} C \longrightarrow e^- + X$ and $\nu_\mu +^{13} C \longrightarrow \mu^- + X$ are enormous versus the present uncertainties in ^{12}C measurements. In section II of this paper we will discus the exclusive reactions $\nu_e +^{13} C \longrightarrow e^- +^{13} N_{gs}$ and $\nu_\mu +^{13} C \longrightarrow \mu^- +^{13} N_{gs}$ and in section III we discus the inclusive cases,$\nu_e +^{13} C \longrightarrow e^- + X$ and $\nu_\mu +^{13} C \longrightarrow \mu^- + X$. Finally in section IV we present some results and conclusions.

EXCLUSIVE NEUTRINO REACTIONS IN ^{13}C

We consider the exclusive neutrino reactions,$\nu_e +^{13} C \longrightarrow e^- +^{13} N_{gs}$ and $\nu_\mu +^{13} C \longrightarrow \mu^- +^{13} N_{gs}$ .These are interesting because they are mirror transitions which might in principle be large. Some early estimates placed their cross sections at 8 to 10 times that of the reaction $\nu_e + {}^{12}C \longrightarrow e^- + {}^{12}N_{(g.s.)}$. Furthermore because ^{13}C is a 1.1 % impurity in ^{12}C if such estimates were correct, these neutrino reactions would be a significant background for the corresponding neutrino reactions in ^{12}C,$\nu_e + {}^{12}C \longrightarrow e^- + {}^{12}N_{(g.s.)}$ and $\nu_\mu + {}^{12}C \longrightarrow \mu^- + {}^{12}N_{(g.s.)}$ which are currently being studied at LAMPF and KARMEN. Finally there have been proposals at LAMPF to actually study these reactions and although this is presently not likely to happen these reactions could be measured elsewhere.

We may write the matrix element for these reactions in the general form:

$$M_{fi} = \frac{G\cos(\theta_C)}{\sqrt{2}} \bar{u}\gamma^\mu(1-\gamma_5)u_\nu <^{13} N_{gs}|J_\mu(0)|^{13}C > \tag{13}$$

where $J_\mu(0) = V_\mu(0) - A_\mu(0)$ and where u stands for the charged lepton and u_ν stands for the neutrino. We calculate the cross section by making use of the elementary particle model to describe the nucleus. In this case the two nuclei are spin 1/2 particles so that the matrix elements look exactly like those for a proton or neutron:

$$<^{13} N_{gs}|V_\mu(0)|^{13}C > = \bar{u}_f[\gamma_\mu F_V(q^2) + \frac{i\sigma_{\nu\mu}q^\nu}{2m_p} F_M(q^2)]u_i \tag{14a}$$

$$<^{13} N_{gs}|A_\mu(0)|^{13}C > = \bar{u}_f[\gamma_\mu\gamma_5 F_A(q^2) + \frac{q_\mu\gamma_5}{m_\pi} F_P(q^2)]u_i. \tag{14b}$$

Thus all of the nuclear structure is contained in the form factors $F_V(q^2)$, $F_M(q^2)$,$F_A(q^2)$, and $F_P(q^2)$ which must be determined. One can make use of the CVC relation:

$$[I^+, J_\mu^{em}(0)] = -V_\mu(0) \tag{15}$$

to obtain the relations:

$$F_V(q^2) = F_{1f}(q^2) - F_{1i}(q^2) \tag{16a}$$

$$F_M(q^2) = F_{2f}(q^2) - F_{2i}(q^2) \tag{16b}$$

where the $F_{1i}, F_{2i}, F_{1f}, and F_{2f}$ are the standard electromagnetic form factors of the initial and final state nuclei, namely ^{13}C and $^{13}N_{gs}$.

Thus one needs electron scattering data to obtain the weak vector form factors. Such data is readily available for the reaction,$e^- +^{13} C \longrightarrow e^- +^{13} C$, but $^{13}N_{gs}$ is very unstable and no electron scattering data exists. Thus other arguements must be used. We may obtain some help from charge symmetry which implies that:

$$<^{13} C|J^3_\mu(0)|^{13}C >= - <^{13} N_{gs}|J^3_\mu(0)|^{13}N_{gs} > \tag{17}$$

which leads to:

$$F_f^{1,2\ V} = -F_i^{1,2\ V} \tag{18a}$$

$$F_f^{1,2\ S} = F_i^{1,2\ S} \tag{18b}$$

where V and S denote isovector and isoscalar respectively. Thus only the isovector part of the electromagnetic current contributes to F_V and F_M. With the present data there is however no way to isolate the isoscalar and isovector parts of the current form factors. However from experience,all parts of the form factors have similar q^2 dependence. With this assumption we can find F_V and F_M. From the available data we obtain:

$$F_V(q^2) = \frac{\cos(-q^2/4.85\ m_\pi^2)}{(1 - q^2/4.24\ m_\pi^2)^2} \tag{19a}$$

for $|q^2| \leq 7.0\ m_\pi^2$ and

$$F_V(q^2) = \frac{\sin[(-q^2/m_\pi^2 - 7.0)(\pi/28.5)]}{\exp(-q^2/3.6m_\pi^2)} \tag{19b}$$

for $7.0\ m_\pi^2 \leq |q^2| \leq 24.0\ m_\pi^2$ and

$$F_M(q^2) = \frac{F_M(0)(1 + q^2/m_\pi^2)}{(1 - q^2/m_\pi^2)^2} \tag{20a}$$

for $|q^2| \leq 2.0\ m_\pi^2$ and

$$F_M(q^2) = \frac{8.1\ F_M(0)\sin[(-q^2/m_\pi^2 - 3.51)(\pi/40.0)]}{(1 - q^2/42.0\ m_\pi^2 + q^4/41.0\ m_\pi^4)^2} \tag{20b}$$

for $2.0m_\pi^2 \leq |q^2| \leq 24m_\pi^2$. The two branches of F_V and F_M are necessary due to diffractive minima in the data.

We still need the axial current form factors F_A and F_P to completely determine the matrix elements, Eq.(13a) and Eq.(13b). The axial current form factor,$F_A(q^2)$ can be determined from beta decay data at $q^2 \simeq 0$ via the reaction $^{13}N_{gs} \longrightarrow^{13} C+e^+ +\nu_e$. From an analysis of this data one obtains:

$$F_A(0) = 0.31. \tag{21}$$

This still leaves the q^2 dependence of F_A to be determined. To obtain this we use a result by Kim and Primakoff[15], namely that:

$$\frac{F_A(q^2)}{F_A(0)} \simeq \frac{F_M(q^2)}{F_M(0)}. \tag{22}$$

Thus from Eq.(20a) and from Eq.(20b) as well as Eq.(21), $F_A(q^2)$ is determined. We still need F_P and to obtain it we use a well known PCAC result[15]:

$$F_P(q^2) = \frac{-m_\pi(M_i + M_f)F_A(q^2)}{(q^2 + m_\pi^2)}(1 + \xi(q^2)). \tag{23}$$

We assume that $\xi(q^2) \sim 0$ based on experience with other nuclei. In any case the cross sections are not sensitive to F_P.

We are now in a position to determine the transition matrix element squared. This is given by:

$$\begin{aligned}
|M|^2 = {} & 8F_V^2 M_i\nu[\mu_o(M_i - \delta) + M_f|\vec{\mu}|\cos(\theta) + (\nu - \mu_o)A_\mu - m_\mu^2] + \\
& 8F_V F_M M_i \nu[2A_\mu\delta(\nu - \mu_o) + 4\nu A_\mu^2 - 3\mu^2 A_\mu - \delta m_\mu^2] \\
& 4M_i\nu F_M^2\{A_\mu[4\nu M_f A_\mu - 3\mu^2 M_f - 4\mu_o(M_i\mu_o + \nu A_\mu - \mu^2) \\
& - 4(M_i\nu - \nu A_\mu) + m_\mu^2(M_i + \nu - \mu_o)] + 4m_\mu^2(M_i\nu - \nu A_\mu\} \\
& 16M_i(M_i + M_f)\nu F_A F_M[A_\mu(\mu_o + \nu) - m_\mu^2] \\
& 8F_A^2 M_i^2\nu[(2 + M_f/M - i) - (M_f/M_i)|\vec{\mu}|\cos(\theta) + \\
& ((\nu - \mu_o)/M_i)A_\mu - m_\mu^2/M_i] \\
& - \frac{8F_A F_P}{m_\pi^2} M_i m_\mu^2 \nu(A_\mu + \delta) + \frac{4F_P^2}{m_\pi^2} m_\mu^2 M_i \nu A_\mu(\nu - \mu_o - \delta) \\
& 16F_V F_A M_i \nu[\nu\mu_o + |\vec{\mu}|^2 - A_\mu(\mu_o + \nu)]
\end{aligned} \tag{24}$$

where $A_\mu = \mu_o - |\vec{\mu}|\cos(\theta)$ and $\delta = M_f - M_i$. From Eq.(23) we may calculate the cross sections in the usual manner. The results are shown in figure 1. We may also spectrum average the cross sections. For the electron neutrino case we obtain:

$$< \sigma(\nu_e +^{13} C \longrightarrow e^- +^{13} N_{gs}) >= 1.7 \pm 0.1 \ \times 10^{-41} \ cm^2 \tag{25a}$$

and for the muon neutrino case we obtain:

$$< \sigma(\nu_\mu +^{13} C \longrightarrow +\mu^- +^{13} N_{gs}) >= 1.04 \pm .02 \ \times 10^{-40} \ cm^2. \tag{25b}$$

These numbers are approximately larger than their ^{12}C counterparts by a factor of two.

This result may seem a little suprising because at first it was thought that the ^{13}C cross sections would be substantially higher. However the reason is very clear from Eq.(23). At lower energies, $|M|^2$ is dominated by the terms:

$$3F_A^2 + F_V^2.$$

For many of the processes which are familiar such as $n \leftrightarrow p$ or $^3He \leftrightarrow ^3H$, $F_A^2 \sim F_V^2$. Thus from above approximately 3/4 of the contributions to $|M|^2$ come from the F_A part of the matrix element and 1/4 from the F_V part. For this case $F_A^2 \sim .1\ F_V^2$.

Thus the size of $|M|^2$ is approximately one quarter of what one might expect. Thus the background from the exclusive neutrino reactions in ^{13}C is at around the 2 % level.

INCLUSIVE NEUTRINO REACTIONS IN ^{13}C

Finally we must consider the inclusive neutrino reactions on ^{13}C, $\nu_e +^{13} C \longrightarrow e^- + X$, and $\nu_\mu +^{13} C \longrightarrow \mu^- + X$, which are the counterparts for the reactions $\nu_e + {}^{12}C \longrightarrow e^- + X$ and $\nu_\mu + {}^{12}C \longrightarrow \mu^- + X$ respectively. These latter two reactions have been experimentally observed and so it is necessary to give an estimate for inclusive neutrino reactions in ^{13}C which might conceivably form an important background. We make use of a treatment which has been used successfully for inclusive neutrino reactions[11,12,16] in ^{12}C and ^{127}I.

We consider the above reactions in ^{13}C . The starting point is the matrix element:

$$M_{ki} = \frac{G}{\sqrt{2}} cos\theta \overline{u}_e \gamma^\lambda (1 - \gamma_5) u_\nu < k|J_\lambda^\dagger(0)|^{13}C > \tag{26}$$

where k is a particular final state and:

$$J_\mu^\dagger(0) = V_\mu^\dagger(0) - A_\mu^\dagger(0) \tag{27}$$

We shall only outline the process by which we obtain the cross section as the details have already appeared in the literature[10,11]. We simply note that the cross section may be written as:

$$\sigma_c = \sum_k \frac{m_\nu}{2ME_\nu} \int d^3P_e \, |M_{ki}|^2 \frac{m_e}{E_e(2\pi)^3} \frac{d^3P_k}{2E_k(2\pi)^3} (2\pi)^4 \delta^4(P_k + P_e - P_\nu - P_i). \tag{28}$$

The quantity $|M_{ki}|^2$ is written as:

$$|M_{ki}|^2 = \frac{G^2 \cos^2 \theta_C}{2m_\nu m_\nu} L^{\sigma\lambda} < k|J_\sigma^\dagger(0)|^{13}C >< k|J_\lambda^\dagger(0)|^{13}C >^* . \tag{29}$$

The quantity, $L^{\sigma\lambda}$, is the lepton tensor appropriate to this process and is given by:

$$L^{\sigma\lambda} = p_e^\sigma p_\nu^\lambda - p_e \cdot p_\nu g^{\sigma\lambda} + p_\nu^\sigma p_e^\lambda - \epsilon^{\alpha\sigma\beta\lambda} p_{e\alpha} p_{\nu\beta}. \tag{30}$$

In order to work with average quantities, we assume an average nuclear excitation of δ given by:

$$M_x - M_i = \delta \tag{31}$$

where δ is clearly a function of the incoming neutrino energy and must be determined. We also assume on the basis of our knowledge of individual states that the interaction is largely in the forward direction and that:

$$< E_e > \simeq E_\nu - \delta. \tag{32}$$

and:

$$< \vec{p}_e > \simeq \sqrt{(E_\nu - \delta)^2 - m_e^2}. \tag{33}$$

Although the value of δ, will vary with incoming neutrino energy, above the giant dipole resonance at approximately 15 MeV it should increase slowly in our region of interest. This enables us to easily obtain $< E_e >$ and $< \vec{p}_e >$ over a large part of the range of neutrino energy of interest. We can, using these averages obtain the quantity:

$$\sigma_c = \frac{G^2 cos^2\theta_C}{2ME_\nu} \int d\Omega_e \sum_k < k|J^\dagger_\sigma(0)|^{13}C >< k|J^\dagger_\lambda(0)|^{13}C >^* L^{\sigma\lambda} \times \frac{< |\vec{p_e}| >}{2M - 2E_e + 2E_e \cos\theta_e \frac{<E_e>}{<|\vec{p_e}|>}}. \quad (34)$$

The hadronic part of eq.(17) may be replaced as follows:

$$< k|J^\dagger_\sigma(0)|^{13}C >< k|J^\dagger_\lambda(0)|^{13}C >^* \equiv Q_{\lambda\sigma}(P_i, < q >) \quad (35)$$

which is a tensor. We have previously shown[11,12,16] that this tensor may be reduced to the form which contains only two unknown functions.

$$Q^{\mu\nu(3)} = \alpha g^{\mu\nu} + \frac{\beta}{M^2} P_i^\mu P_i^\nu. \quad (35)$$

The cross section is then found to be:

$$\sigma_c = \frac{G^2 \cos^2(\theta_C)}{4\pi} \frac{< |\vec{p}_e| >< E_e > D}{M(M + E_\nu)} \quad (36)$$

where:

$$D = \beta - 2\alpha \quad (37)$$

and an impulse approximation based calculation[17] gives $D(q^2)$ as:

$$D = a_o - b_o q^2. \quad (38)$$

We assume this simple q^2 dependence for D, which we may also write as $D = a_o + b_o|q^2|$.

Thus our result depends upon two parameters,a_o and b_o. Low error total muon capture results are available for the process, $\mu^- +^{13} C \longrightarrow \nu_\mu + X$. However a little thought shows that this is not the muon capture process which is appropriate to determine the the hadronic part of the matrix element. In the neutrino reaction on ^{13}C there are 7 neutrons on which the neutrinos react. For muon capture the muon reacts with the protons and there are only 6 of those thus we need a nucleus with 7 protons and 6 neutrons. This is obviously the mirror nucleus $^{13}N_{gs}$. However it is not stable and no muon capture is possible. We therefore make use of a trick which we have used earlier[15]. We shall construct a muon capture rate for this nucleus. We therefore note that by proceeding by a calculation very similar to the neutrino reaction case[18] , we find for the total muon capture rate an expression similar to Eq.(36):

$$\Gamma_{TOT} = \frac{C|\Phi(0)|^2 G^2 \cos^2\theta_C < E_\nu >^2 D}{8\pi M_i(M_i + m_\mu)} \quad (39)$$

and one which contains the same D.

As we remarked earlier, the total muon capture rate is needed for a nucleus which has 7 protons,and 6 neutrons and a total $A = 13$ and this nucleus is not stable. However although actual muon capture is not possible we may make use of the fact

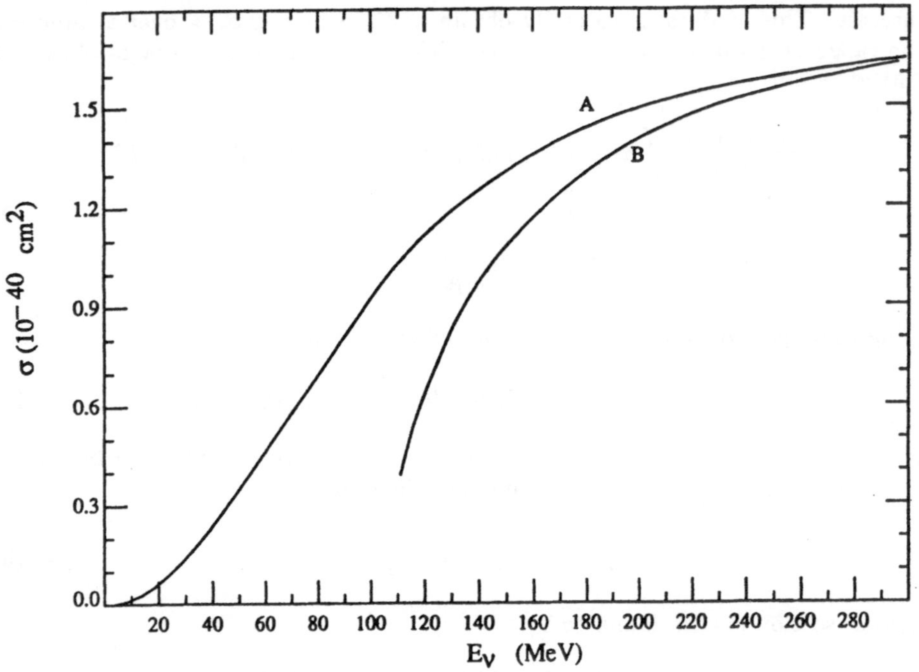

Figure 1 Cross section for the reactions $\nu_e +^{13} C \longrightarrow e^- +^{13} N_{gs}$ and $\nu_\mu +^{13} C \longrightarrow \mu^- +^{13} N_{gs}$ as a function of neutrino energy.

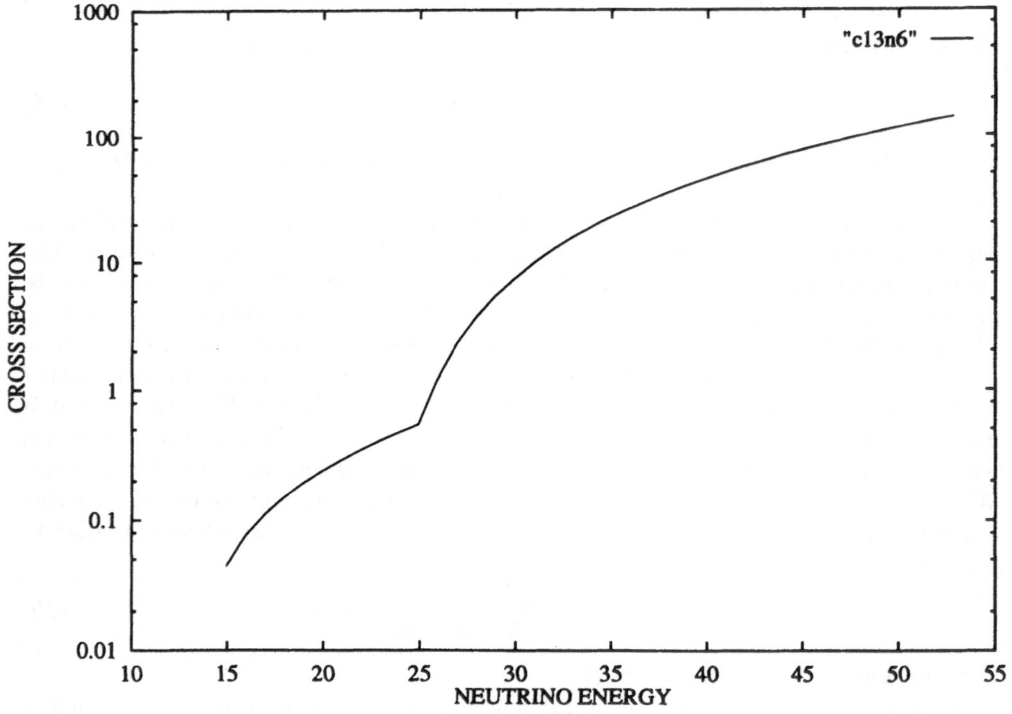

Figure 2 Cross section for the reaction $\nu_e +^{13} C \longrightarrow e^- + X$ as a function of neutrino energy.

that there is an extremely accurate semi-empirical formula for the total muon capture rate in nuclei[19]. This formula is given by:

$$\Gamma = Z_{eff}^4 G_1 [1 + G_2 \frac{A}{2Z} - G_3 \frac{A-2Z}{2Z} - G_4 (\frac{A-Z}{2A} + \frac{A-2Z}{8AZ})] \tag{40}$$

where $G_1 = 261, G_2 = -0.040, G_3 = -0.26$, and $G_4 = 3.24$. This formula, Eq.(40) fits all exisitng to within 20 percent and in most cases is much better. We therefore use Eq.(40) to produce a muon capture rate for $^{13}N_{gs}$. We shall call the muon capture result given by Eq.(39) Γ^{GP} and the result given by Eq.(40),Γ^{MP}. If we take the ratio of these rates for the ^{13}C and $^{13}N_{gs}$ cases,i.e

$$\frac{(\Gamma^{GP})^{^{13}C}}{(\Gamma^{GP})^{^{13}N_{gs}}} = \frac{(\Gamma^{MP})^{^{13}C}}{(\Gamma^{MP})^{^{13}N_{gs}}}, \tag{41}$$

we find that most factors cancel and that we are able to evaluate the remaining ones. We thus obtain:

$$D^{^{13}N_{gs}} = 2.155 \times D^{^{13}C}. \tag{42}$$

Thus from the total muon capture rate for ^{13}C we are able to proceed. This total muon capture rate is known[20] and is given by $\Gamma_{TOT} = 3.57 \times 10^4 sec^{-1}$We note that the error is small of the order of a few percent in this number. We thus obtain:

$$D^{^{13}N_{gs}} = 7.69 \times 10^4 \; sec^{-1}. \tag{43}$$

We have at this point a value for the D which we need at q^2 appropriate for muon capture but we need an additional information to determine D completely. In the case of ^{12}C we relied on inclusive electron scattering data and some impulse approximation results to fully obtain D. Here this is not possible. However an impulse approximation result[17] yields a value for $\tilde{D} = \frac{D(q^2)}{D(0)}$ given by:

$$\tilde{D} = \frac{[1 - (\frac{(A-Z)}{2A})\delta(\vec{q}^2)]}{[1 - \frac{(A-Z)}{2A}\delta(0)} \tag{43a}$$

where:

$$\delta(\vec{q}\,^2) = (\frac{d}{r_o})^3 (1 - \frac{\vec{q}\,^2 d^2}{10}). \tag{43b}$$

and where from Eq.(14) we may write:

$$\tilde{D} = (1 - \frac{b_o}{a_o} q^2) \equiv (1 - b' q^2). \tag{43c}$$

This yields:

$$D^{^{13}N_{gs}} = 6.582[1 + .62 q^2/m_\mu^2]. \tag{43d}$$

We are now able to calculate the cross section. We still need however $\delta(E_\nu)$ as a function of neutrino energy. Above the giant dipole resonance we expect closure to be applicable and so for $E_\nu > 30 MeV$ we set $\delta(E_\nu) = 15\ MeV$. We choose this value from previous experience[11,12,16] with total muon capture rates where good results are obtained at 15 to 20 MeV above the giant dipole resonance with closure. From our experience with the ^{12}C case, below 30 MeV we use a decreasing δ. We have tried several different forms for $\delta(E_\nu)$ and the precise form of the function does not have

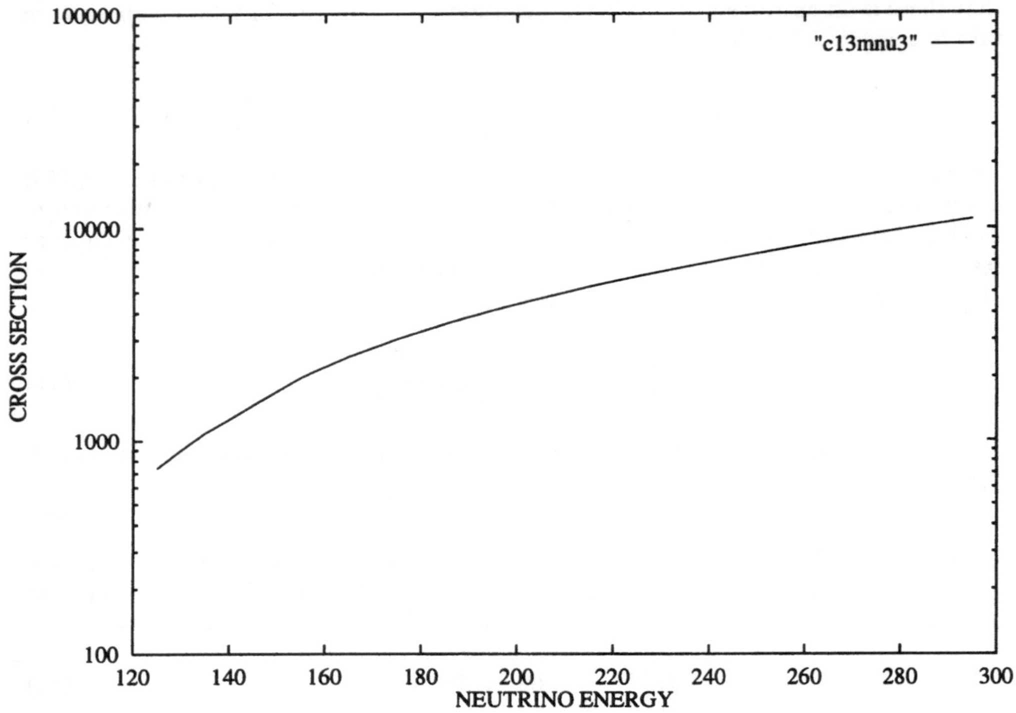

Figure 3 Cross section for the reaction $\nu_\mu +^{13} C \longrightarrow \mu^- + X$ as a function of incident neutrino energy in MeV. The cross section is multiplied by 10^{-42} cm^2.

more than an effect of a few percent on the value of $< \sigma_c >$. We are now able to evaluate the cross section. We do this in figure 2 for the electron neutrino case and in figure 3 for the muon neutrino cross section. . From figure 2 we obtain the Michel spectrum averaged cross section for the inclusive electron neutrino scattering from ^{13}C. The result is:

$$< \sigma(\nu_e +^{13} C \longrightarrow e^- + X) >= 31.08 \times 10^{-42}\ cm^2 \qquad (44a)$$

and for the inclusive muon neutrino cross section on ^{13}C average over the LAMPF spectrum we obtain;

$$< \sigma(\nu_\mu +^{13} C \longrightarrow \mu^- + X) >= 19.4 \times 10^{-40}\ cm^2 \qquad (44b).$$

DISCUSSION

Thus we see that both for the exclusive reactions and inclusive reactions in ^{13}C, the cross sections are roughly twice those of ^{12}C. Because ^{13}C is a relatively small impurity in ^{12}C targets, it seems unlikely that these backgrounds will be important until extremely accurate neutrino measurements are possible. Present measurements carry errors of approximately 25% so that a 2.2% contribution from ^{13}C is not likely to be noticed. At the 10% level however this contribution would be important.

A question which might be asked in the case of the inclusive neutrino reactions in ^{13}C is whether the $D^{^{13}N_{gs}}$ given by Eq.(41) is reasonable. We know that for the ^{127}I case it produced results in line with a very limited experiment. However there is another reason why Eq.(41) might be reasonable. The reasoning leading to Eq.(41)

assumed that the arrangement of protons was the dominating feature in total muon capture rates. If this is true ^{12}C and ^{13}C should both have approximately the same total muon capture rates. These rates have been measured and indeed they are very similar. The rate for ^{14}N has also been measured and it is approximately twice that for ^{13}C. This is exactly what one would expect if the proton arrangements dominate and this is what we obtained in Eq.(42). Thus there are additional arguements supporting the approximation used here.

Finally we should remark that neutrino measurements on ^{12}C are continuing. KARMEN continues to take data both for the reaction discussed here and for other processes. The Los Alamos group intends to move its operations to Fermilab and measure neutrino reactions at higher energies. Thus this is a time of remarkable activity in neutrino physics.

REFERENCES

[1] R.C. Allen,Phys Rev Lett.**64**,1871(1990)

[2] B. Zeitnitz,Prog. in Part. and Nucl. Phys. **13**,445(1965).

[3] T.W. Donnelly,Phys. Lett.**B43**,93(1973).

[4] S.L. Mintz,Phys. Rev. **C 35**,1671(1982).

[5] S.L. Mintz and M. Pourkaviani,Phys. Rev. **C 40**2458(1989).

[6] M. Fukugita,Y. Kohyama and K. Kubodera,Physics Lett **B212** 139(1988).

[7] J. Bernabeu and P. Pascual, Nucl. Phys.**A 324**365(1979).

[8] T.W. Donnelly, D. Hitlin,M. Schwartz, J.D. Walecka, and S.J. Weisner,Physics Lett **B 49**,8(1974).

[9] M. Pourkaviani and S.L. Mintz,J. Phys. **G 15**,569(1990).

[10] E.Kolbe,K. Langanke,S.Krewald and F.-K. Thielmann,Nucl. Phys.**A540**,599(1992).

[11] S.L. Mintz and D.F. King,Phys. Rev. **C 31**1585(1984).

[12] M. Pourkaviani and S.L. Mintz,Nucl. Phys.**A 573**,501(1994).

[13] M. Albert et al.,Measurement for the Reaction $^{12}C(\nu_\mu, \mu^-)X$,LAMPF preprint (1974).

[14] S.L. Mintz and M. Pourkaviani,Nucl. Phys. **A 594**346(1995).

[15] C.W.Kim and H. Primakoff,*Mesons in Nuclei*,eds. M. Rho and D.H. Wilkinson (North-Holland,Amsterdam),68(1979).

[16] S.L. Mintz and M. Pourkaviani,Nucl. Phys. **A 584**665(1995).

[17] C.W. Kim and S.L. Mintz,Phys. Rev. **C 31**,274(1985).

[18] S.L. Mintz and M. Pourkaviani,J. Phys. **G 20**,925(1994).

[19] B. Goulard and H. Primakoff,Phys. Rev. **C 10**,2034(1974).

[20] T.Suzuki et al.,Phys. Rev. **C35**,2212(1987).

Section II
Recent Progress on New and Old Ideas

IS THE COSMOLOGICAL CONSTANT NON-ZERO?

Paul H. Frampton

Department of Physics and Astronomy,
University of North Carolina,
Chapel Hill, NC 27599-3255.

THE ISSUE OF Λ

The present talk is similar to one at COSMO-98 last month. As is well explained in standard reviews of the cosmological constant [1, 2] the theoretical expectation for Λ exceeds its observational value by 120 orders of magnitude.

In 1917, Einstein looked for a static solution of general relativity for cosmology and added a new λ term:

$$R_{\mu\nu} - \frac{1}{2} g_{\mu\nu} R - \lambda g_{\mu\nu} = -8\pi G T_{\mu\nu} \tag{1}$$

A $\lambda > 0$ solution exists with $\rho = \frac{\lambda}{8\pi G}$, radius $r(S_3) = (8\pi\rho G)^{-1/2}$ and mass $M = 2\pi^2 r^3 \rho = \frac{\pi}{4\sqrt{\lambda}G}$.

In the 1920's the universe's expansion became known (more red shifts than blue shifts). In 1929, Hubble enunciated his law that recession velocity is proportional to distance.

Meanwhile Friedmann (1922) discovered the now-standard non-static model with metric:

$$ds^2 = dt^2 - R^2(t) \left[\frac{dr^2}{1 - kr^2} + r^2 (d\theta^2 + sin^2\theta d\phi^2) \right] \tag{2}$$

In 1923, Einstein realized the dilemna. He wrote to his friend Weyl:
"If there is no quasi-static world, then away with the cosmological term".

Setting $\Lambda = 0$ does not increase symmetry. In fact, the issue is one of *vacuum energy* density as follows:

In vacuum:

$$< T_{\mu\nu} = - < \rho > g_{\mu\nu} \tag{3}$$

which changes the λ_{eff} by:

$$\lambda_{eff} = \lambda + 8\pi G < \rho > \tag{4}$$

or equivalently:

$$\rho_V = < \rho > + \frac{\lambda}{8\pi G} = \frac{\lambda_{eff}}{8\pi G} \tag{5}$$

Confluence of Cosmology, Massive Neutrinos, Elementary Particles, and Gravitation
Edited by Kursunoglu *et al.*, Kluwer Academic / Plenum Publishers, New York, 1999.

The observational upper limit on λ comes from:

$$\left(\frac{dR}{dt}\right)^2 = -k + \frac{1}{3}R^2(8\pi G\rho + \lambda) \tag{6}$$

which expresses conservation of energy and leads to the upper bound $|\lambda_{eff}| \leq H_0^2$.

This translates into $|\rho_V| \leq 10^{-29}g/cm^3$. In high-energy units we use $1g \sim 10^{33}eV$ and $(1cm)^{-1} \sim 10^{-4}$eV to rewrite $|\rho_V| \leq [(1/100)eV]^4$

A "natural" value in quantum gravity is:

$$|\rho_V| = (M_{Planck})^4 = (10^{28}eV)^4 \tag{7}$$

which is 10^{120} times too big. This has been called the biggest error ever made in theoretical physics!

Even absent the $(M_{Planck})^4$ term field theory with spontaneous symmetry breaking leads one to expect $< \rho > \quad \gg [(1/100)eV]^4$. As examples, QCD confinement suggests $< \rho > \sim (200MeV)^4$, which is 10^{40} times too big and electroweak spontaneous symmetry breaking would lead to $< \rho > \sim (250GeV)^4$ which is 10^{52} times too big. This is the theoretical issue. I will briefly mention four approaches to its solution.

(1) Supersymmetry, Supergravity, Superstrings

According to global supersymmetry:

$$\{Q_\alpha, Q_\beta^\dagger\}_+ = (\sigma_\mu)_{\alpha\beta}P^\mu \tag{8}$$

and with unbroken supersymmetry:

$$Q_\alpha|0> = Q_\beta^\dagger|0> = 0 \tag{9}$$

which implies a vanishing vacuum value for $< P_\mu >$ and hence zero vacuum energy as required for vanishing Λ.

With global supersymmetry promoted to local supersymmetry the expression for the potential is more complicated than this (one can even have $V < 0$).

When supersymmetry is broken, however, at ≥ 1 TeV one expects again that $|\rho_V| > (1TeV)^4$ which is 10^{54} times too big.

So although unbroken supersymmetry looks highly suggestive, broken supersymmetry does not help. The same is generally true for superstrings.

One new and exciting approach - still in its infancy - involves the compactification of the Type IIB superstring on a manifold $S^5 \times AdS_5$ and give rise to a 4-dimensional $\mathcal{N} = 4$ SU(N) supersymmetric Yang-Mills theory, known to be conformal. Replacing S^5 by an orbifold S^5/Γ can lead to $\mathcal{N} = 0$ non-supersymmetric SU(N) gauge theory and probably (this is presently being checked; see *e.g.* [3]) retain conformal symmetry. If so one may achieve $< \rho >= 0$ without supersymmetry.

(2) Quantum Cosmology

The use of wormholes to derive $\Lambda \rightarrow 0$ has been discredited because of (a) the questionable use of Euclidean gravity, (b) wormholes, if they exist, become macroscopically large and closely-packed, at variance with observation.

(3) Changed Gravity

An example of changing gravity theory [4] is to make $g = det g_{\mu\nu}$ non-dynamical in the generalized action:

$$S = -\frac{1}{16\pi G}\int dx[R + L(g-1)] \tag{10}$$

where L is a Lagrange multiplier. One then finds by variation that $R = -4\Lambda = constant$. Minimizing the action gives $\Lambda = 2\sqrt{6}\pi/\sqrt{V}$ where V is the spacetime volume.

In the path integral

$$Z = \int d\mu(\Lambda) exp(3\pi/G\Lambda) \tag{11}$$

the value $\Lambda \to 0^+$ is exponentially favored.

(4) The Anthropic Principle

If $\Omega_\Lambda \gg 1$ rapid exponential expansion prohibits gravitational condensation to clumps of matter. This requires $\Omega_\Lambda < 400$.

On the other hand if $\Omega_\Lambda \ll 0$ the universe collapses at a finite time, and there is not enough time for life to evolve. For example, if $\Lambda = -(M_{Planck})^4$, R reaches only 0.1mm ($10^{-30}$ of its present value). Taken together these two considerations lead to

$$-1 \leq \Omega_\Lambda \leq 400 \tag{12}$$

– quite a strong constraint.

This shows how important it is to life that Λ is very much closer to zero than to $(M_{Planck})^4$ or even E^4 where E is any vacuum energy scale familiar to High Energy physics.

CBR TEMPERATURE ANISOTROPY

The Cosmic Background Radiation (CBR) was discovered [5] in 1965 by Penzias and Wilson. But detection of its temperature anisotropy waited until 1992 when [6, 7] the Cosmic Background Explorer (COBE) satellite provided impressive experimental support for the Big Bang model. COBE results are consistent with a scale-invariant spectrum of primordial scalar density fluctuations, such as might be generated by quantum fluctuations during inflation [8, 9, 10, 11, 12, 13, 14]. COBE's success inspired many further experiments with higher angular sensitivity than COBE ($\sim 1^o$).

NASA has approved a satellite mission MAP (Microwave Anisotropy Probe) for 2000. ESA has approved the Planck surveyor - even more accurate than MAP - a few years later in 2005.

With these experiments, the location of the first accoustic (Doppler) peak and possible subsequent peaks will be resolved.

The Hot Big Bang model is supported by at least three major triumphs:

- the expansion of the universe
- the cosmic background radiation
- nucleosynthesis calculations

It leaves unanswered two major questions:

- the horizon problem
- the flatness problem

The horizon problem. When the CBR last scattered, the age of the universe was $\sim 100,000y$. The horizon size at that recombination time subtends now an angle $\sim \pi/200$ radians. On the celestial sphere there are 40,000 regions never causally-connected in the unadorned Big Bang model. Yet their CBR temperature is the same to one part in 10^5 - how is this uniformity arranged?

The flatness problem. From the equation (for $\Lambda = 0$)

$$\frac{k}{R^2} = (\Omega - 1)\frac{\dot{R}^2}{R^2} \tag{13}$$

and evaluate for time t and the present $t - t_0$, using $R \sim \sqrt{t} \sim T^{-1}$:

$$(\Omega_t - 1) = 4H_0^2 t^2 \frac{T^2}{T_0^2}(\Omega_0 - 1) \tag{14}$$

Now for high densities:

$$\frac{\dot{R}^2}{R^2} = \frac{8\pi G\rho}{3} \simeq \frac{8\pi GgaT^4}{6} \tag{15}$$

where a is the radiation constant $= 7.56 \times 10^{-9} erg \quad m^{-3} \quad K^{-4}$.

From this we find

$$t(seconds) = (2.42 \times 10^{-6}) g^{-1/2} T(GeV)^{-2} \tag{16}$$

and thence by substitution in Eq. (14)

$$(\Omega_t - 1) = (3.64 \times 10^{-21}) h_0^2 g^{-1} T(GeV)^{-2} (\Omega_0 - 1) \tag{17}$$

This means that if we take, for example, $t = 1 second$ when $T \simeq 1$ MeV, then $|\Omega_t - 1|$ must be $< 10^{-14}$ for Ω_0 to be of order unity as it is now. If we go to earlier cosmic time, the fine tuning of Ω_t becomes even stronger if we want the present universe to be compatible with observation. Why then is Ω_t so extremely close to $\Omega_t = 1$ in the early universe?

Inflation Both the horizon and flatness problems can be solved in the inflationary scenario which has the further prediction (in general) of flatness. That is, if $\Lambda = 0$:

$$\Omega_m = 1 \tag{18}$$

or, in the case of $\Lambda \neq 0$ (which is allowed by inflation):

$$\Omega_m + \Omega_\Lambda = 1 \tag{19}$$

We shall see to what extent this prediction, Eq.(19), is consistent with the present observations.

The goal of the CBR experiments [15, 16, 17, 18] is to measure the temperature autocorrelation function. The fractional perturbation as a function of direction $\underline{\hat{\mathbf{n}}}$ is expanded in spherical harmonics:

$$\frac{\Delta T(\underline{\hat{\mathbf{n}}})}{T} = \sum_{lm} a_{lm} Y_{lm}(\underline{\hat{\mathbf{n}}}) \tag{20}$$

The statistical isotropy and homogeneity ensure that

$$< a_{lm}^{\dagger} a_{l'm'} >= C_l \delta_{ll'} \delta_{mm'} \tag{21}$$

A plot of C_l versus l will reflect oscillations in the baryon-photon fluid at the surface of last scatter. The first Doppler, or accoustic, peak should be at $l_1 = \pi/\Delta\Theta$ where $\Delta\Theta$ is the angle now subtended by the horizon at the time of last scatter: the recombination time at a red-shift of $Z \simeq 1,100$.

The special case $\Lambda = 0$

When $\Lambda = 0$, the Einstein-Friedmann cosmological equation can be solved analytically (not generally true if $\Lambda \neq 0$). We will find $l_1 \sim 1/\sqrt{\Omega_m}$ as follows. Take:

$$ds^2 = dt^2 - R^2 \left[d\Psi^2 + sinh^2 \;\; \Psi \;\; d\Theta^2 + sinh^2 \;\; \Psi \;\; sin^2 \;\; \Theta \;\; d\Phi^2 \right] \tag{22}$$

For a geodesic $ds^2 = 0$ and so:

$$\frac{d\Psi}{dR} = \frac{1}{R} \tag{23}$$

The Einstein equation is

$$\left(\frac{\dot{R}}{R} \right)^2 = \frac{8\pi G\rho}{3} + \frac{1}{R^2} \tag{24}$$

so that

$$\dot{R}^2 R^2 = R^2 + aR \tag{25}$$

with $a = \Omega_0 H_0^2 R_0^3$ and hence

$$\frac{d\Psi}{dR} = \frac{1}{\sqrt{R^2 + aR}} \tag{26}$$

This can be integrated to find

$$\Psi_t = \int_{R_t}^{R_0} \frac{dR}{\sqrt{(R + a/2)^2 - (a/2)^2}} \tag{27}$$

The substitution $R = \frac{1}{2}a(coshV - 1)$ leads to

$$\Psi_t = cosh^{-1}(\frac{2R_0}{a} - 1) - cosh^{-1}(\frac{2R_t}{a} - 1) \tag{28}$$

Using $sinh(cosh^{-1}x) = \sqrt{x^2 - 1}$ gives

$$sinh\Psi_t = \sqrt{\left(\frac{2(1-\Omega_0)}{\Omega_0} + 1 \right)^2 - 1} - \sqrt{\left(\frac{2(1-\Omega_0)R_t}{\Omega_0 R_0} + 1 \right)^2 - 1} \tag{29}$$

The second term of Eq.(29) is negligible as $R_t/R_0 \to 0$ With the metric of Eq.(22) the angle subtended now by the horizon then is

$$\Delta\Theta = \frac{1}{H_t R_t sinh\Psi_t} \tag{30}$$

For $Z_t = 1,100$, the red-shift of recombination one thus finds

$$l_1(\Lambda = 0) \simeq \frac{2\pi Z_t^1/2}{\sqrt{\Omega_m}} \simeq \frac{208.4}{\sqrt{\Omega_m}} \tag{31}$$

This is plotted in Fig. 1 of [19].

The general case $\Lambda \neq 0$

When $\Lambda \neq 0$

$$\dot{R}^2 R^2 = -kR^2 + aR + \Lambda R^4/3 \tag{32}$$

It is useful to define the contributions to the energy density $\Omega_m = 8\pi G\rho/3H_0^2$, $\Omega_\Lambda = \Lambda/3H_0^2$, and $\Lambda_C = -k/H_0^2 R_0^2$. These satisfy

$$\Omega_m + \Omega_\Lambda + \Omega_C = 1 \tag{33}$$

Then

$$l_1 = \pi H_t R_t sinh \Psi_t \tag{34}$$

where

$$\Psi_t = \sqrt{\Omega_C} \int_1^\infty \frac{dw}{\sqrt{\Omega_\Lambda + \Omega_C w^2 + \Omega_m w^3}} \tag{35}$$

After changes of variable one arrives at

$$l_1 = \pi \sqrt{\frac{\Omega_0}{\Omega_C}} \sqrt{\frac{R_0}{R_t}} sinh \left(\sqrt{\Omega_C} \int_1^\infty \frac{dw}{\sqrt{\Omega_\Lambda + \Omega_C w^2 + \Omega_m w^3}} \right) \tag{36}$$

(For positive curvature ($k = +1$) replace *sinh* by *sin*). For the case $k = 0$, the flat universe predicted by inflation, with $\Omega_C = 0$ Eq.(36) reduces to

$$l_1 = \pi \sqrt{\Omega_m} \sqrt{\frac{R_0}{R_t}} \int_1^\infty \frac{dw}{\sqrt{\Omega_\Lambda + +\Omega_m w^3}} \tag{37}$$

These are elliptic integrals, easily do-able by Mathematica. They resemble the formula for the age of the universe:

$$A = \frac{1}{H_0} \int_1^\infty \frac{dw}{w\sqrt{\Omega_\Lambda + \Omega_C w^2 + \Omega_m w^3}} \tag{38}$$

In Fig. 2 of [19] there is a plot of l_1 versus Ω_m for $\Omega_C = 0$. In Fig. 3 are the main result of the iso-l_1 lines on a $\Omega_m - \Omega_\Lambda$ plot for general Ω_C with values of l_1 between 150 and 270 in increments $\Delta l_1 = 10$. The final Fig. 4 of [19] gives a three-dimensional plot of $\Omega_m - \Omega_\Lambda - l_1$.

We can look at the cumulative world data on C_l versus l. Actually even the existence of the first Doppler peak is not certain but one can see evidence for the rise and the fall of C_l. In Fig. 2 of [20] we see such 1998 data and with some licence say that $150 \leq l_1 \leq 270$.

The exciting point is that the data are expected to improve markedly in the next decade. In Fig. 3 of [20] there is an artist's impression of both MAP data (expected 2000) and Planck data(2006); the former should pin down l_1 with a small error and the latter is expected to give accurate values of C_l out to $l = 1000$.

But even the spectacular accuracy of MAP and Planck will specify only one iso-l_1 line in the $\Omega_m - \Omega_\Lambda$ plot and not allow unambiguous determination of Ω_Λ.

Fortunately this ambiguity can be removed by a completely independent set of observations.

III. HIGH-Z SUPERNOVAE IA.

In recent years several supernovae (type IA) have been discovered with high red-shifts $Z > 0.3$ (at least 50 of them). An example of a high red-shift is $Z = 0.83$. How far away is that in cosmic time? For matter-domination

$$\left(\frac{R_0}{R_t}\right) = \left(\frac{t_0}{t}\right)^{2/3} = (Z+1) \tag{39}$$

so the answer is $t = t_0/2.83$. For $t_0 = 14Gy$ this implies $t \simeq 6Gy$. Thus this supervova is older than our Solar System and the distance is over half way back to the Big Bang!

These supernovae were discovered [21, 22] by a 4m telescope then their light-curve monitored by the 10m telescope KEK-II on Mauna Kea, Hawaii *and/or* the Hubble Space Telescope. The light curve is key, because study of nearby supernovae suggests that the breadth of the light curve *i.e.* the fall in luminosity in 15 days following its peak is an excellent indicator of absolute luminosity. Broader (slower) light curves imply brighter luminosity. Clever techniques compare the SN light-curve to a standard template.

It is worth pointing out that although these SN are very far away - over 50% back to the Big Bang they do not penetrate as far back as the CBR discussed earlier which goes 99.998% back to the Big Bang (300,000y out of 14,000,000,000y).

Because of the high Z, just one of these observations, and certainly 50 or more of them, have great influence on the estimation of the deceleration parameter q_0 defined by

$$q_0 = -\frac{\ddot{R}R}{\dot{R}^2} \tag{40}$$

which characterizes departure from the linear Hubble relation $Z = \frac{1}{H_0}d$. In the simplest cosmology ($\Lambda = 0$) one expects that $q_0 = +1/2$, corresponding to a *deceleration* in the expansion rate.

The startling result of the high-Z supernovae observations is that the deceleration parameter comes out *negative* $q_o \simeq -1/2$ implying an *accelerating* expansion rate.

Now if the only sources of vacuum energy driving the expansion are Ω_m and Ω_Λ there is the relationship

$$q_0 = \frac{1}{2}\Omega_m - \Omega_\Lambda \tag{41}$$

So we add a line on the $\Omega_m - \Omega_\Lambda$ plot corresponding to Eq.(41) with $q_0 = -1/2$. Such a line is orthogonal to the iso-l_1 lines from the CBR Doppler peak and the intersection gives the result that values $\Omega_m \simeq 0.3$ and $\Omega_\Lambda \simeq 0.7$ are favored. It is amusing that these values are consistent with Eq.(19) but the data strongly disfavor the values $\Omega_m = 1$ of Eq.(18).

Note that a positive Ω_Λ acts like a negative pressure which accelerates expansion - a normal positive pressure implies that one does work or adds energy to decrease the volume and increase the pressure: a positive cosmological constant implies, on the other hand, that increase of volume goes with increase of energy, only possible if the pressure is negative.

QUINTESSENCE

The non-zero value $\Omega_\Lambda \simeq 0.7$ has two major problems:

- Its value $(1/100 \;\; eV)^4$ is unnaturally small.
- At present Ω_m and Ω_Λ are the same order of magnitude implying that we live in a special era.

Both are addressed by *quintessence*, an inflaton field Φ taylored so that $T_{\mu\nu}(\Phi) = \Lambda(t)g_{\mu\nu}$. The potential $V(\Phi)$ may be

$$V(\Phi) = M^{4+\alpha}\Phi^{-\alpha} \tag{42}$$

or

$$V(\Phi) = M^4(exp(M/\Phi) - 1) \tag{43}$$

where M is a parameter [23].

By arranging that ρ_Φ is a little below ρ_γ at the end of inflation, it can track ρ_γ and then (after matter domination) ρ_m such that $\Omega_{\Lambda(t_0)} \sim \Omega_m$ is claimed [24] not to require fine-tuning. The subject is controversial because, by contrast to [24], [25] claim that slow-roll inflation and quintessence require fine-tuning at the level of $1in10^{50}$.

More generally, it is well worth examining equations of state that differ from the one ($\omega = p/\rho = -1$) implied by constant Λ. Quintessence covers the possibilities $-1 < \omega \leq 0$.

SUMMARY

Clearly more data are needed for both the CBR Doppler peak and the high-Z supernovae. Fortunately both are expected in the forseeable future.

The current analyses favor $\Omega_\Lambda \simeq 0.7$ and $\Omega_m \simeq 0.3$.

Of course, Λ is still 120 orders of magnitude below its natural value, and 52 orders of magnitude below $(250 \;\; GeV)^4$ and that theoretical issue remains.

The non-zero Λ implies that we live in a special cosmic era: Λ was negligible in the past but will dominate the future giving exponential growth $R \sim e^{\Lambda t}, t \to \infty$. This cosmic coincidence is addressed by quintessence.

The principal point of our own work in [19] is that the value of l_1 depends almost completely only on the geometry of geodesics since recombination, and little on the details of the accoustic waves, since our iso-l_1 plot agrees well with the numerical results of White *et al.* [26].

ACKNOWLEDGEMENT

This work was supported in part by the US Department of Energy under Grant No. DE-FG05-85ER41036.

REFERENCES

1. S. Weinberg, Rev. Mod. Phys. **61**, 1 (1989).
2. Y.J. Ng, Int. J. Mod. Phys. **D1**, 145 (1992).
3. P.H. Frampton, hep-th/9812117.

4. P.H. Frampton, Y. J. Ng and H. Van Dam, J. Math. Phys. **33,** 3881 (1992).
5. A.A. Penzias and R.W. Wilson, Ap. J. **142,** 419 (1965).
6. G.F. Smoot *et al.*, Ap. J. Lett. **396,** L1 (1992).
7. K. Sanga *et al.*, Ap. J. **410,** L57 (1993).
8. J.M. Bardeen, P.J. Steinhardt and M.S. Turner, Phys. Rev. **D28,** 679 (1983).
9. A.A. Storobinsky, Phys. Lett. **B117,** 175 (1982).
10. A.H. Guth and S.-Y. Pi, Phys. Rev. Lett. **49,** 1110 (1982).
11. S.W. Hawking, Phys. Lett. **B115,** 295 (1982)
12. A.H. Guth, Phys. Rev. D28, 347 (1981).
13. A.D. Linde, Phys. Lett. **B108,** 389 (1982).
14. A. Albrecht and P.J. Steinhardt, Phys. Rev. Lett. **48,** 1220 (1982).
15. R.L. Davis, H.M. Hodges, G.F. Smoot, P.J. Steinhardt and M.S. Turner, Phys. Rev. Lett. **69,** 1856 (1992).
16. J.R. Bond,R. Crittenden, R.L.Davis, G. Efstathiou and P.J. Steinhardt, Phys. Rev. Lett. **72,** 13 (1994).
17. P.J. Steinhardt, Int. J. Mod. Phys. **A10,** 1091 (1995).
18. M. Kamionkowski and A. Loeb, Phys. Rev. **D56,** 4511 (1997).
19. P.H. Frampton, Y.J. Ng and R. Rohm. Mod. Phys. Lett. **A13,** 2541 (1998).
20. M. Kamionkowski. *astro-ph/9803168.*
21. S.J. Perlmutter *et al.*, (The Supernova Cosmology Project). *astro-ph/9608192.*
22. S.J. Perlmutter *et al.*, (The Supernova Cosmology Project). *astro-ph/9712212.*
23. P.J.E. Peebles and B. Ratra, Ap. J. Lett. **325,** L17 (1988).
24. I. Zlatev, L. Wang and P.J. Steinhardt. *astro-ph/9807002.*
25. C. Kolda and D. Lyth. *hep-ph/9811375.*
26. M. White and D. Scott, Ap.J. **459,** 415 (1996), *astro-ph/9508157.* W. Hu and M. White, Ap. J. **471,** 30 (1996), *astro-ph/9602019*; Phys. Rev. Lett. **77,** 1687 (1996), *astro-ph/9602020.* M.White, Ap. J. **506,** 495 (1998), *astro-ph/9802295.*

CHERN-SIMONS VIOLATION OF LORENTZ AND PTC SYMMETRIES IN ELECTRODYNAMICS

Roman Jackiw

Center for Theoretical Physics
Massachusetts Institute of Technology
Cambridge, MA 02139–4307

The principle of special relativity is very firmly established in the minds of physicists, and it is experimentally confirmed, without known exception. Nevertheless, today's availability of high-precision instruments lets us ask whether this principle is only approximately true, and leads us to seek possible mechanism for its violation. Such an inquiry is not unreasonable, since we know that a relativity principle does **not** apply to the discrete transformations of space and time reversal.

Special relativity arose when the symmetry of Maxwell's electrodynamical field theory, i.e., Lorentz invariance, was elevated to encompass particle mechanics, whose Newtonian, Lorentz noninvariant dynamics had consequently to be modified. Therefore violation of special relativity can be looked for in particle mechanics, in electromagnetism, or in both. I shall restrict my attention to possible nonrelativistic behavior in electromagnetism.

Let me record the conventional equations, both in compact Lorentz covariant, and in explicit vectorial notation. We are dealing with the electric and magnetic fields $(\mathbf{E}, \mathbf{B})$, that are components of a second-rank antisymmetric tensor $F_{\mu\nu} = -F_{\nu\mu}$, or of its dual ${}^*F^{\mu\nu} = \frac{1}{2}\epsilon^{\mu\nu\alpha\beta}F_{\alpha\beta}$:

$$F_{oi} = \tfrac{1}{2}\epsilon^{ijk}\,{}^*F^{jk} = E^i$$
$$\tfrac{1}{2}\epsilon^{ijk}F_{jk} = {}^*F^{oi} = -B^i \ .$$

They satisfy the homogeneous Maxwell equations

$$\boldsymbol{\nabla} \times \mathbf{E} + \frac{1}{c}\frac{\partial \mathbf{B}}{\partial t} = 0 \ , \quad \boldsymbol{\nabla} \cdot \mathbf{B} = 0$$

or

$$\partial_\mu\, {}^*F^{\mu\nu} = 0$$

which permit writing the fields in terms of the potentials $A^\mu = (\phi, \mathbf{A})$, by formulas that are invariant against gauge transforming the potentials:

Confluence of Cosmology, Massive Neutrinos, Elementary Particles, and Gravitation
Edited by Kursunoglu *et al.*, Kluwer Academic / Plenum Publishers, New York, 1999.

$$\phi \to \phi - \frac{1}{c}\frac{\partial}{\partial t}\alpha\,, \quad \mathbf{A} \to \mathbf{A} + \boldsymbol{\nabla}\alpha$$
$$A_\mu \to \mathbf{A}_\mu - \partial_\mu \alpha\,.$$

These formulas are

$$\mathbf{E} = -\boldsymbol{\nabla}\phi - \frac{1}{c}\frac{\partial \mathbf{A}}{\partial t}\,, \quad \mathbf{B} = \boldsymbol{\nabla} \times \mathbf{A}$$
$$F_{\mu\nu} = \partial_\mu A_\nu - \partial_\nu A_\mu\,.$$

The second set of Maxwell's equations, which sees the sources of charge density ρ and current density $\mathbf{j}$, $j^\mu = (c\rho, \mathbf{j})$, reads

$$\boldsymbol{\nabla} \times \mathbf{B} - \frac{1}{c}\frac{\partial \mathbf{E}}{\partial t} = \frac{4\pi}{c}\mathbf{j}\,, \quad \boldsymbol{\nabla}\cdot\mathbf{E} = 4\pi\rho$$

or

$$\partial_\mu F^{\mu\nu} = \frac{4\pi}{c} j^\nu$$

and can be derived from the Lagrange density

$$\mathcal{L} = \frac{1}{8\pi}(\mathbf{E}^2 - \mathbf{B}^2) - \rho\phi + \frac{1}{c}\mathbf{j}\cdot\mathbf{A} = -\frac{1}{16\pi}F_{\mu\nu}F^{\mu\nu} - \frac{1}{c}j_\mu A^\mu$$

where the basic variables are the potentials, and the electromagnetic fields are expressed in terms of them. Consistency of the equations of motion and gauge invariance of the Lagrangian formalism require that the charge density and current satisfy a continuity equation

$$\frac{\partial}{\partial t}\rho + \boldsymbol{\nabla}\cdot\mathbf{j} = 0 = \partial_\mu j^\mu\,.$$

Let us now turn to modifications. In the most obvious departure from the standard formulas, we add a "photon mass term" by supplementing $\mathcal{L}$ with $\frac{\mu^2}{2}A^\mu A_\mu = \frac{\mu^2}{2}\phi^2 - \frac{\mu^2}{2}\mathbf{A}^2$ where μ has dimension of inverse length. In the new equations of motion $-\mu^2 A^\mu$ is added to j^μ, so that when the wave *Ansatz* $e^{ik_\alpha x^\alpha} = e^{i(\omega t - \mathbf{k}\cdot\mathbf{r})}$, $k^\alpha = (\omega/c, \mathbf{k})$, $k \equiv |\mathbf{k}|$ is taken for fields in the source-free case ($j^\mu = 0$), the dispersion law reads

$$k^\alpha k_\alpha = \mu^2\,, \qquad \omega = c\sqrt{k^2 + \mu^2}\,.$$

Of course, this does not violate Lorentz invariance – the mathematical expression of the special relativity principle – but it destroys Einstein's physical reasoning that led him to special relativity: light no longer travels with a universal velocity in all reference frames, and "c" becomes a mysterious limiting velocity that is not attained by any physical particle. Gauge invariance appears to be violated, but today we know that the gauge principle can be obscured by subtle symmetry-breaking mechanisms, for example, the mass μ could arise from a feeble Higgs effect. After solving the modified field equations with prescribed sources, one finds electromagnetic fields that are distorted by the mass term. Comparison to experiment is made with geomagnetic data, leading to the limit $\mu < 3 \times 10^{-24}$ GeV[1] while observations of the galactic magnetic field give $\mu < 3 \times 10^{-36}$ GeV[2] (1 GeV $\sim 10^{13}\text{cm}^{-1}$).

Lorentz invariance and the relativity principle, but not rotational invariance, disappear if the Lagrange density is modified by the addition of a further $\frac{1}{8\pi}\mathbf{B}^2$ term, proportional to ϵ. However, by rescaling $\mathbf{A}$, one sees that this is equivalent to redefining the velocity of light from c to $c_\epsilon \neq c$, in the electromagnetic part of the theory while retaining c as a parameter in the (unspecified) matter kinematics. Hence this modification can be exported into the matter sector and I shall not discuss it further.

It has been studied by S. Coleman and S. Glashow,[3] who use cosmic ray data to bound the magnitude of the addition by 10^{-23}.

I now come to yet another modification, introduced by S. Carroll, G. Field, and me almost a decade ago,[4] which recently came again to attention. To begin, let us note that in addition to $-\frac{1}{2}F_{\mu\nu}F^{\mu\nu} = \mathbf{E}^2 - \mathbf{B}^2$, another Lorentz scalar, quadratic in the field strengths, can be constructed: $-\frac{1}{4}\,{}^*F^{\mu\nu}F_{\mu\nu} = \mathbf{E}\cdot\mathbf{B}$. However, adding this to the electromagnetic Lagrange density does not affect the equations of motion, because that quantity, when expressed in terms of potentials – the dynamical variables in a Lagrangian formulation – involves total derivatives, which do not contribute to equations of motion:

$$\begin{aligned}\tfrac{1}{2}\,{}^*F^{\mu\nu}F_{\mu\nu} &= \tfrac{1}{2}\partial_\mu(\epsilon^{\mu\alpha\beta\gamma}A_\alpha F_{\beta\gamma})\\ -2\mathbf{E}\cdot\mathbf{B} &= \frac{\partial}{\partial t}\Big(\frac{1}{c}\mathbf{A}\cdot\mathbf{B}\Big) + \boldsymbol{\nabla}\cdot(\phi\mathbf{B} - \mathbf{A}\times\mathbf{E})\ .\end{aligned}$$

However, when the $\mathbf{E}\cdot\mathbf{B}$ quantity is multiplied by another space-time–dependent field $\theta(t,\mathbf{r})$, the total derivative feature disappears and such an addition would affect dynamics. Once again using the freedom to modify a Lagrange density by total derivatives, we see that the addition of $\theta\,{}^*FF$ to the Lagrange density is equivalent to adding $-\partial_\mu\theta\epsilon^{\mu\alpha\beta\gamma}A_\alpha F_{\beta\gamma}$. If θ is a dynamical field, then the extended electromagnetism + θ system remains Lorentz invariant. We shall however posit that neither θ nor $\partial_\mu\theta$ are dynamical quantities; rather $\partial^\mu\theta$ is a constant 4-vector $p^\mu = (m,\mathbf{p})$ that picks out a direction in space-time, thereby violating Lorentz invariance. Thus we are led to consider an electromagnetic theory, where the conventional Maxwell Lagrange density is modified by

$$\begin{aligned}\Delta\mathcal{L} &= -\tfrac{1}{4}p_\mu(\epsilon^{\mu\alpha\beta\gamma}A_\alpha F_{\beta\gamma}) = -\tfrac{1}{2}p_\mu\,{}^*F^{\mu\nu}A_\nu\\ &= -\tfrac{1}{2}m\mathbf{A}\cdot\mathbf{B} + \tfrac{1}{2}\mathbf{p}\cdot(\phi\mathbf{B} - \mathbf{A}\times\mathbf{E})\ .\end{aligned}$$

The sourceless Maxwell equations are unchanged (the fields continue to be expressed by potentials). Only the equations with sources are changed, and the change can be viewed as a field-dependent addition to the source current.

$$\partial_\mu F^{\mu\nu} = \frac{4\pi}{c}j^\nu + p_\mu\,{}^*F^{\mu\nu}$$

$$\nabla\times\mathbf{B} - \frac{1}{c}\frac{\partial\mathbf{E}}{\partial t} = \frac{4\pi}{c}\mathbf{j} - m\mathbf{B} + \mathbf{p}\times\mathbf{E}\ , \quad \boldsymbol{\nabla}\cdot\mathbf{E} = 4\pi\rho - \mathbf{p}\cdot\mathbf{B}\ .$$

Note that the field equations are gauge invariant, even though the Lagrange density is not. The quantities m and $\mathbf{p}$ have dimension of inverse length; the latter breaks rotational invariance by selecting a direction in space; m breaks the invariance of the theory against Lorentz boosts. Presumably for the interesting case we should select vanishing $\mathbf{p}$ (rest frame of p^μ, which is taken to be time-like) so that rotational isotropy is maintained. Parity is also broken, since the pseudoscalar $\mathbf{B}$ mixes with vector $\mathbf{E}$, but time inversion and charge conjugation remain intact, hence PTC is broken.[5]

Before describing the consequences of our model, let me recall some facts about the various quantities that we have introduced. ${}^*F^{\mu\nu}F_{\mu\nu}$ is the so-called Chern-Pontryagin density; its non-Abelian generalization plays an important role in the "standard" particle physics model, where it is a measure of the anomalous (quantum mechanical) nonconservation of the axial vector current. It is responsible for the decay of the neutral pion to two photons, and for proton decay. The 4-vector whose divergence gives ${}^*F^{\mu\nu}F_{\mu\nu}$ is called the Chern-Simons density. Both objects are templates for the topologically nontrivial behavior of non-Abelian gauge fields.

In some extensions of the standard model, ${}^*F^{\mu\nu}F_{\mu\nu}$ is coupled to a further dynamical field, like the θ-field mentioned above, which describes a hypothetical particle – the axion – whose role is to ensure CP symmetry of strong interactions. However, no evidence for such a particle has been found thus far.

Note further that if we were living in (2+1)-dimensional space-time, which is on a plane rather than in a three-dimensional volume, the ϵ-tensor would have only three indices and we could introduce the Chern-Simons term into (2+1)-dimensional electrodynamics without the external 4-vector p^μ, i.e., (2+1)-dimensional Lorentz invariance would be preserved by $\Delta\mathcal{L} = m\epsilon^{\alpha\beta\gamma}A_\alpha F_{\beta\gamma}$. Chern-Simons modified electrodynamics plays a role in planar electromagnetic phenomena, as in the quantum Hall effect, and perhaps also in high-T_c superconductivity.

Finally we remark that with vanishing $\mathbf{p}$ and absence of sources, so that $\mathbf{E} = 0$, the remaining modified Maxwell equations read $\boldsymbol{\nabla} \times \mathbf{B} = -m\mathbf{B}$, $\boldsymbol{\nabla} \cdot \mathbf{B} = 0$. These have arisen previously in magnetohydrodynamics. They coincide with the conventional Maxwell equations in the presence of neutral sources and steady currents ($\rho = 0, \boldsymbol{\nabla}\cdot\mathbf{j} = 0$) and are seen to be equivalent to the conventional Ampère's law, when the further condition is imposed that $\mathbf{j}$ is proportional to $\mathbf{B}$.

What is the consequence of our Lorentz invariance violating modification?

Let us examine wave solutions in the absence of sources. ($\rho = \mathbf{j} = 0, j^\mu = 0$). We again make the *Ansatz* that fields behave as exponentials of phases, $e^{i(\omega t - \mathbf{k}\cdot\mathbf{r})} = e^{ik_\alpha x^\alpha}$, $k^\alpha = (\omega/c, \mathbf{k})$, $k \equiv |\mathbf{k}|$, and find the dispersion law

$$(k^\alpha k_\alpha)^2 + (k^\alpha k_\alpha)(p^\beta p_\beta) = (k^\alpha p_\alpha)^2 \, .$$

From this one can show that introducing p^α has the consequence of splitting the photons into two polarization modes, each traveling with different velocities ω/k – forceful evidence of Lorentz and parity violation. This is very easily seen in the rotation invariant case, $\mathbf{p} = 0$. One finds

$$\omega^2 = ck(ck \pm mc) \, .$$

Note that ω can become imaginary for modes with $k < m$. This means there are unstable runaway solutions. These do not contradict energy conservation, because the energy $\mathcal{E}$ is no longer the positive expression of the Maxwell theory, $\frac{1}{2}\int d^3r(\mathbf{E}^2 + \mathbf{B}^2)$. Rather we now have

$$\mathcal{E} = \tfrac{1}{2}\int d^3r\Big[\mathbf{E}^2 + \Big(\mathbf{B} + \frac{m}{2}\mathbf{A}\Big)^2\Big] - \frac{m^2}{8}\int d^3r\mathbf{A}^2 \, .$$

With unstable solutions each of the two terms contributing to $\mathcal{E}$ grows without bound, yet $\mathcal{E}$ stays finite and time-independent owing to a cancellation between the two. (The energy is gauge invariant – in spite of appearances.) However, runaway, exponentially growing modes can be avoided by allowing for noncausal propagation for well-behaved sources (similar to the way runaway solutions are eliminated from the Abraham-Lorentz equation of conventional electrodynamics).

Returning now to our plane wave solutions, we observe that a plane-polarized wave – a superposition to two circularly polarized modes traveling at different velocities – will be rotated when it travels through space. Since p^α is small, we can solve for ω to first order in p^α, and find, (without setting $\mathbf{p}$ to zero)

$$k = \frac{\omega}{c} \mp \tfrac{1}{2}(m - \mathbf{p}\cdot\hat{\mathbf{k}})$$

so the change Δ in the polarization, as the wave travels a distance L is

$$\Delta = -\tfrac{1}{2}(m - \mathbf{p}\cdot\hat{\mathbf{k}})L \, .$$

This is similar to the Faraday effect, where a polarization change is induced by ambient magnetic fields. However our phase change is wavelength independent, while the Faraday effect rotation is proportional to wavelength squared. So the two effects can be distinguished.

When comparing predictions of this theory to experimental data, Carroll, Field and I assumed that rotation invariance holds, we set $\mathbf{p} = 0$, and the entire effect is parameterized by m (time-like p^α). Geomagnetic data can be confronted with the distorted magnetic field that solves the modified equations in the presence of sources. But the data is somewhat difficult to interpret in our context, and the most plausible limit is

$$m \leq 6 \times 10^{-26} \text{GeV} .$$

However, examining the polarization of light from distant galaxies and removing the rotation due to the Faraday effect, yields a much more stringent result

$$m \leq 10^{-42} \text{GeV}$$

(see also M. Goldhaber and V. Trimble.[6]) Since effects of nonzero m can appear only at distances greater than the associated Compton wavelength, which for the above is the distance to the horizon, astrophysical data apparently rules out nonvanishing m.

How about vanishing m and nonvanishing $\mathbf{p}$ (space-like p^μ)?

Our formula for polarization change indicates that here too there should be a non-Faraday rotation. Moreover, one can show that space-like p^α produce no instability. However, Carroll, Field, and I believed that such a violation of rotational symmetry is unlikely.

Thus, we were very surprised when there appeared a *Physical Review Letter* by B. Nodland and J. Ralston[7] alleging that precisely this kind of anisotropy exists. Evidence for this startling assertion was drawn from the same galactic data that we used in our analysis, which gave us the null result.

We were not the only ones surprised. Here is a sampling of news stories about this "discovery" in the popular and semipopular press, following a report of the Nodland-Ralston "result" issued by the American Institute of Physics.

American Institute of Physics
"Is the universe birefringent?", 17 April 1997

New York Times
"'This (don't ask which) side up' may apply to the universe", 18 April 1997
"Theory about the universe has its ups and downs", 25 April 1997

Associated Press
"Space isn't the same in all directions", 18 April 1997

Time
"This side up. New evidence challenges Einstein's universe", 18 April 1997

Tabloid
"Einstein was wrong", 19 April 1997

Science News
"Does the cosmos have a direction?", 26 April 1997
"Cosmic axis begets cosmic controversy", 10 May 1997

Sky and Telescope
"No lopsided universe", 9 May 1997

Galileo (Italy)
News report, 15 May 1997

Physics World (UK)
"Axis of universe debate rumbles on", 15 May 1997

Hilary Price; reprinted with special permission of King Features Syndicate ©1997.

Los Angeles Times
"A zigzag route to the truth", 7 August 1997
Popular Science
"Which way is up?", October 1997
ABQ Journal
"VLA data defends big bang theory", 17 November 1997
Scientific American
"Twist and shout", (web page)

Interest in the result also evoked humorous reactions, in the form of a syndicated cartoon by Hilary Price depicting existential anguish engendered by life in an anisotropic universe, a statement by Lyndon La Rouche that he knew it all the time (interview, 7 May 1997 with A. Papert), and a claim for extraterrestrial life (http://www.enterprisemission.com).

Unfortunately, it appears that Nodland and Ralston made a mistake in their data analysis. Carroll and Field[8] reanalyzed the data, identified their error, and found no anisotropy. Thus our original conclusion that there is no evidence for a Chern-Simons modification to electromagnetism stands, and has been confirmed by several other investigations. The entire matter is nicely reviewed on http://ITP.UCSB.edu/~carroll/aniso.html .

In spite of the negative results, we can nevertheless draw an interesting conclusion. We know that in Nature parity P, time reversal T, and charge conjugation C are violated. While local field theory and Lorentz invariance guarantee that PTC will be conserved, it remains an experimental and interesting question whether PTC is valid in Nature, which perhaps does not make use only of local field theoretic dynamics. But PTC violation in any corner of a grand unified theoretical structure would in general induce PTC violation in electrodynamics, so the stringent limits that we put on the PTC-violating Chern-Simons term, also limit some forms of PTC violation anywhere else in the "final theory".[9]

REFERENCES

1. A. Goldhaber and M. Nieto, *Rev Mod Phys* 43:277 (1971).
2. G. Chibisov, *Usp Fiz Nauk* 119:551 (1976); *Sov Phys Usp* 19:624 (1976).
3. S. Coleman and S. Glashow, *Phys Lett B* 405:249 (1997).
4. S. Carroll, G. Field, and R. Jackiw, *Phys Rev D* 41:1231 (1990).
5. D. Colladay and V.A. Kostelecky, *Phys Rev D* 55:6760 (1997).
6. M. Goldhaber and V. Trimble, *J Astrophys Astr* 17:17 (1996).
7. B. Nodland and J. Ralston, *Phys Rev Lett* 78:3043 (1997).
8. S. Carroll and G. Field, *Phys Rev Lett* 79:234 (1997).
9. R. Jackiw and V.A. Kostelecky (preprint hep-ph/9901358).

SPIN-DEPENDENT FORCES BETWEEN QUARKS IN HADRONS

D. B. Lichtenberg

Physics Department
Indiana University
Bloomington, IN 47405, USA

INTRODUCTION

Most of what I say in this talk is within the framework of the constituent quark model, in which explicit gluon degrees of freedom are integrated out, leaving only a potential between quarks. In this framework, the question, "What per cent of the spin of a hadron is carried by the quarks?" has the simple answer, "100 per cent." The sea of gluons and quark-antiquark pairs surrounding a current quark is considered an integral part of it, and both the quark and the sea together make the constituent quark. When a constituent quark is tweaked gently, it drags its sea around with it, and therefore has more inertia than a current quark. The size of a constituent quark, including its sea, is substantially greater than the size of a current quark, and a constituent quark may not be much smaller than the hadron in which it is bound.

In the constituent quark model there is still a puzzle about the nature of the spin-dependent force between two quarks (or between a quark and an antiquark). Specifically, does the spin-dependent interaction arise from the chromomagnetic interaction of one-gluon exchange,[1] from Goldstone-boson exchange in connection with the breaking of chiral symmetry,[2] or from the effects of instantons[3]?

It is the purpose of my talk to show that the one-gluon-exchange mechanism provides a consistent picture of the spin-splittings in ground-state mesons and baryons, independently of whether those hadrons contain heavy or light quarks. Isgur[4] has recently taken a somewhat similar point of view. However, I cannot rule out the possibility that meson exchange and instantons also contribute to the spin splittings.

Shortly after the invention of QCD, De Rújula et al.[1] discussed consequences of the one-gluon exchange interaction between two (constituent) quarks in a baryon and between a quark and antiquark in a meson. In their work, the spin-dependent splitting of ground-state hadrons arises from the chromomagnetic interaction.

Since then, many authors have written papers in which the mass splittings of the vector and pseudoscalar mesons, and the splittings of the spin-3/2 and spin-1/2 baryons arise from chromomagnetic forces. These forces have three salient features: 1) they are two-body forces, 2) they are short range, and 3) they are attractive in spin-0 states and repulsive in spin-1 states of two quarks in a baryon or a quark-antiquark in a meson.

Confluence of Cosmology, Massive Neutrinos, Elementary Particles, and Gravitation
Edited by Kursunoglu *et al.*, Kluwer Academic / Plenum Publishers, New York, 1999.

Although the chromomagnetic interaction is a two-body operator, Cohen and Lipkin[5] and Richard and Taxil[6] pointed out that in baryons (and more generally in hadrons containing more than two quarks) the spin-dependent interaction *energy* between two quarks is influenced by the remaining "spectator" quarks. Anselmino et al.[7] gave a quantitative measure of this effect in baryons. In cases in which data were available, they found that the spectator quark influenced the interaction energy by less than 12%.

More recently, it was proposed that the spin-dependent interaction between light quarks arises, not from one-gluon exchange, but from Goldstone-boson exchange, the boson being a pseudoscalar meson connected with chiral symmetry breaking. I call this the one-meson-exchange model; see the review by Glozman and Riska[2] for references. A third suggestion is that instantons give the dominant contribution to the the spin-dependent interaction in light quarks. A lattice gauge calculation[8] appears to support this third point of view. I call this the instanton model; references can be found in the review by Shuryak.[3]

FEYNMAN–HELLMANN THEOREM

A useful tool for examining the mass dependence of the hadron masses is the Feynman–Hellmann theorem.[9,10] This theorem says that, given an eigenvalue problem

$$H\psi = E\psi, \tag{1}$$

if the Hamiltonian H depends on a parameter λ, then

$$\frac{\partial E}{\partial \lambda} = \langle \frac{\partial H}{\partial \lambda} \rangle, \tag{2}$$

where $\langle H \rangle = (\psi, H\psi)$. Because the proof of this theorem is so simple, I give it here in a slightly generalized form.[11]

Consider a Hermitian operator $F(E, \lambda)$ satisfying

$$F\psi = 0, \tag{3}$$

where both the eigenvalue E and the wave function ψ depend on the parameter λ. Taking the scalar product and derivative with respect to λ, we get

$$\frac{d}{d\lambda}(\psi, F\psi) = (\frac{\partial \psi}{\partial \lambda}, F\psi) + (\psi, [\frac{\partial F}{\partial E}\frac{\partial E}{\partial \lambda} + \frac{\partial F}{\partial \lambda}]\psi) + (\psi, F\frac{\partial \psi}{\partial \lambda}) = 0. \tag{4}$$

Because of (3) the first term in (4) is zero, and because F is Hermitian, the third term is also zero. Then (4) becomes

$$\frac{d}{d\lambda}(\psi, F\psi) = (\psi, [\frac{\partial F}{\partial E}\frac{\partial E}{\partial \lambda} + \frac{\partial F}{\partial \lambda}]\psi) = 0. \tag{5}$$

Because E is just a number, we can write (5) as

$$\frac{\partial E}{\partial \lambda} = -(\psi, \frac{\partial F}{\partial \lambda}\psi)/(\psi, \frac{\partial F}{\partial E}\psi), \tag{6}$$

which is the generalized Feynman–Hellmann theorem.

If we let $F = H - E$, then

$$\frac{\partial F}{\partial \lambda} = \frac{\partial H}{\partial \lambda}, \quad \frac{\partial F}{\partial E} = -1, \tag{7}$$

and (6) becomes the usual Feynman–Hellmann theorem (2).

APPLICATION TO HADRON MASSES

Consider a Hamiltonian H describing n constituent quarks (and/or antiquarks) moving relativistically in a potential $V(1...n)$, which in general depends on the positions r_i, Pauli spin operators σ_i, and masses m_i $(i = 1, ...n)$ of the quarks. Then H is given by

$$H = \sum_i [(p_i^2 + m_i^2)^{1/2} - m_i] + V, \tag{8}$$

where H satisfies the eigenvalue equation (1), and the eigenvalue E excludes the rest energy by construction. Let the parameter λ be the quark mass m_i. Then, applying the Feynman–Hellmann theorem to (8), we get

$$\frac{\partial E}{\partial m_i} = m_i \langle (p^2 + m_i^2)^{-1/2} \rangle - 1 + \langle \frac{\partial V}{\partial m_i} \rangle. \tag{9}$$

Because p^2 is a positive definite operator, the sum of the first two terms in (9) is negative. Then, if either V is independent of m_i or

$$\langle \frac{\partial V}{\partial m_i} \rangle < 0, \tag{10}$$

then we get from (9)

$$\frac{\partial E}{\partial m_i} < 0, \tag{11}$$

the result we want.

We now consider the potential between quarks in the one-gluon approximation plus a scalar confining term that is independent of quark masses. If we restrict ourselves to ground-state mesons and baryons, it is reasonable to make the further approximation that the quarks have zero orbital angular momenta. In this approximation, the expectation values of the spin-orbit and tensor terms vanish. Then we can write the potential as

$$V = V_0 + V_{\mathcal{M}}, \tag{12}$$

where $V_0 = V_0(r_1...r_n)$ is the static part of the one-gluon potential (a sum of color-Coulomb potentials between quark pairs) plus the confining part of the potential, and

$$V_{\mathcal{M}} = \sum_{i<j} \frac{\sigma_i \cdot \sigma_j}{m_i m_j} f(r_{ij}), \tag{13}$$

is the chromomagnetic (color-hyperfine) part of the potential. Here, r_{ij} is the distance between quarks i and j and $f(r_{ij})$ is a positive-definite short-range potential (a positive constant times a delta-function in the strictly one-gluon approximation).

Taking the derivative of V with respect to m_i, we get

$$\langle\frac{\partial V}{\partial m_i}\rangle = -\sum_{j\neq i}\frac{\langle\sigma_i\cdot\sigma_j\rangle}{m_i^2 m_j}\langle f(r_{ij})\rangle. \tag{14}$$

If all

$$\langle\sigma_i\cdot\sigma_j\rangle > 0, \tag{15}$$

then (10) and (11) hold. But (15) is true if two quarks have spin 1 and false if two quarks have spin 0. It follows that (11) should be true for vector (spin 1) mesons and spin 3/2 baryons but not necessarily for pseudoscalar (spin 0) mesons and spin 1/2 baryons.

We can use the observed vector meson masses to obtain lower limits on quark mass differences. The mass M_{12} of a vector meson containing quarks with masses m_1 and m_2 can be written

$$M_{12} = m_1 + m_2 + E(12), \tag{16}$$

where $E(12)$ is the eigenvalue of the Hamiltonian. We now let the symbol for a particle denote its mass, and we neglect the mass difference between a u and d quark, letting q stand for either. Then the masses of the ρ and K^* mesons can be written

$$\rho = 2q + E(qq), \quad K^* = q + s + E(qs). \tag{17}$$

From (11), we have $E(qq) > E(qs)$, so that (17) yields

$$s - q > K^* - \rho = 124 \text{ MeV}, \tag{18}$$

where the vector meson masses are taken from the tables of the Particle Data group.[12] Similarly, we obtain

$$c - s > D^* - K^* = 1115 \text{ MeV}, \tag{19}$$

$$b - c > B^* - D^* = 3316 \text{ MeV}. \tag{20}$$

By choosing values of the masses q, s, c, and b that satisfy inequalities (18), (19), and (20), Roncaglia et al.[13,14] were able to find sets of quark masses that satisfy the inequalities (18), (19), and (20) and that lead to values of $E(12)$ which decrease monotonically and smoothly as the reduced mass μ increases (μ necessarily increases if m_1 increases and m_2 does not decrease). Figure 1, from Ref. 14, shows how $E(12)$ varies with μ when the quark masses in MeV are given by

$$q = 300, \quad s = 475, \quad c = 1160, \quad b = 4985. \tag{21}$$

We see from Figure 1 that $E(12)$ is monotonically decreasing with increasing μ for the vector mesons. On the other hand, with similar quark mass values, the pseudoscalar meson eigenenergies do not decrease monotonically, as has been shown by Roncaglia et al.[13] Thus, the behavior of the eigenenergies with μ is just what we expect if the spin-dependent interaction is given by the chromomagnetic interaction of one-gluon exchange. The quark masses of (21) were determined[14] from a combined best fit to observed vector meson and spin 3/2 baryon masses. A similar analysis for vector mesons was done by Kwong and Rosner.[15]

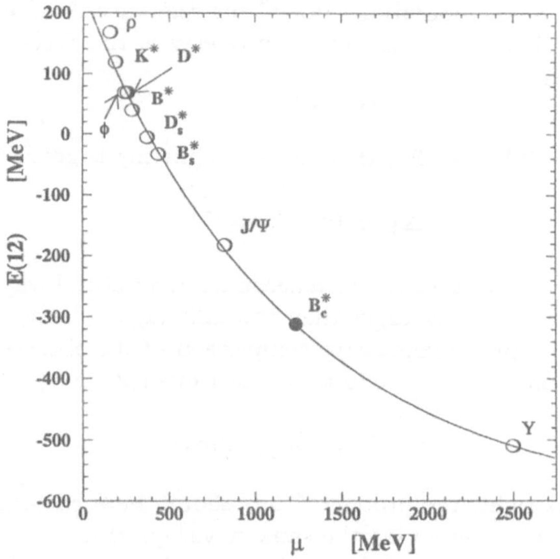

Figure 1. Dependence of the energy eigenvalues $E(12)$ on the reduced mass μ of the two quarks, with the quark masses given by (21). The mass of the B_c^* (solid circle) has not been measured, but is predicted to be 6320 ± 20 MeV.

SPIN SPLITTINGS

I now take up the question of whether the observed spin splittings in ground-state hadrons is consistent with the chromomagnetic interaction arising from one-gluon exchange.

First consider mesons. If the $V_{\mathcal{M}}$ of (13) is taken to be a perturbation (not a very good approximation in light-quark hadrons), then the spin splitting Δ_M between the vector and pseudoscalar mesons containing the same quark flavors is given by

$$\Delta_M = 4\langle f(r)\rangle/(m_1 m_2), \quad r = r_{12}. \tag{22}$$

In order to ascertain how Δ_M varies with the quark masses, we have to estimate how the expectation value $\langle f(r)\rangle$ varies. This problem has been considered nonrelativistically by Frank and O'Donnell[16] for a Coulomb-plus-linear central potential. It has been generalized[17] to any central potential of the form

$$V_0 = V_V + V_S, \tag{23}$$

where

$$V_V \propto -1/r^\beta, \quad V_S \propto r^\beta, \quad 0 < \beta \leq 1. \tag{24}$$

The idea is to assume that V_V transforms like the time component of a Lorentz vector and V_S transforms like a Lorentz scalar. The point is that under these circumstances it has been shown by numerous authors that

$$f(r) = \nabla^2 V_V/6. \tag{25}$$

(See a review of the quark potential model[18] for references.) If V_V is treated like a perturbation, then, because of the scaling property of the Scrödinger equation, it can be shown[16,17] that

$$\langle \nabla^2 V_V \rangle \propto \mu. \tag{26}$$

It follows from (22), (25), and (26) that the spin splitting is given by

$$\Delta_M \propto (m_1 + m_2)^{-1}. \tag{27}$$

Thus, the one-gluon-exchange interaction accounts in a natural way for the fact that spin splitting in light mesons is larger than the splittings in heavy mesons.

In order to to obtain a quantitative comparison of the observed splittings with the splittings of the model, it is useful to define a quantity R by

$$R = (m_i + m_j)/(2m_q). \tag{28}$$

Then, if the model is right, the product $R\Delta_M$ should be a constant for all vector-pseudoscalar meson pairs containing the same flavor quarks.

In Table 1 are given the experimental mass splittings Δ_M from the Particle Data Group[12] and the values of $R\Delta_M$, where $R = (m_i + m_j)/(2m_q)$. If the potential is of the form given in (24) and if the short-range part of the potential can be treated as a perturbation, then the values of $R\Delta_M$ should be the same constant for all the ground-state splittings.

It can be seen from Table 1 that the values of $R\Delta_M$ are not constant, but they are much closer to a constant than the values of Δ_M. In my opinion, the fact that the values of $R\Delta_M$ do not vary by much (the smallest value is 63% of the largest) is a reflection of the underlying validity of the one-gluon-exchange model. If so, the variations in $R\Delta_M$ just show that using a potential in a nonrelativistic Schrödinger equation and treating the short-range part of the potential as a perturbation are not adequate approximations. Note that the values of $R\Delta_M$ are largest for the $\rho - \pi$ and $J/\psi - \eta_c$ splittings. In the case of the pion, the short-range spin-dependent force is sufficiently strong to increase the pion wave function near $r = 0$, thereby decreasing the pion mass more than one would predict from perturbation theory. In the case of the η_c, the color-Coulomb force increases the wave function near $r = 0$, again enhancing the effect of the short-range spin-dependent force.

The effects of the chromomagnetic interaction in ground-state baryons is a little more complicated than in mesons because a baryon contains three quarks. However, if the spin dependence arises from the chromomagnetic force of (13), the effect of the two-body spin-dependent force can be approximately isolated.[7] We distinguish two cases: first, the case in which the baryons have two identical quarks, and second, the case in which all three quarks have different flavors.

If two quarks have the same flavor, we define the spin-dependent splitting Δ_B by $\Delta_B = B^* - B'$, where B^* is the mass of the spin-3/2 baryon and B' is the mass of the spin-1/2 baryon. For example, in the case of baryons made of only u and d quarks, the B^* is the Δ baryon and the B' is the nucleon. In the case in which all three quarks are different, there exists a second spin-1/2 baryon, which we denote

Table 1. Values of the mass differences Δ_M between vector and pseudoscalar mesons containing the same quark content. Also given are the values of $R\Delta_M$, where $R = (m_i + m_j)/(2m_q)$, with the values of the m_i given in (21). Values of Δ_M are from the Particle Data Group[12] . According to the one-gluon-exchange model, the values of $R\Delta_M$ should be approximately constant.

Mesons	Quark content†	Δ_M (MeV)	$R\Delta_M$ (MeV)
$\rho - \pi$	$q\bar{q}$	632	632
$K^* - K$	$q\bar{s}$	398	514
$D^* - D$	$c\bar{q}$	141	456
$D_s^* - D_s$	$c\bar{s}$	144	508
$J/\psi - \eta_c$	$c\bar{c}$	117	640
$B^* - B$	$q\bar{b}$	46	405
$B_s^* - B_s$	$s\bar{b}$	47	428

†The symbol q means either u or d.

by B. The B' has the spin structure of the Σ baryon, while the B has the spin structure of the Λ. If all quarks are different, we define another independent spin splitting $\Delta'_B = (2B^* + B' - 3B)/2$. The quantity R is defined for baryons by (7), just like for mesons, except that the quark masses are those of the two active quarks. The active quarks are those responsible for the spin-dependent force given by a single term in (13); the third quark is the spectator quark for that term. Of course, to obtain the spin splittings in baryons, one must appropriately sum over all three pairs of quarks.

I do not have a derivation that $R\Delta_B$ or $R\Delta'_B$ should be constant, but consider these quantities in analogy with the mesons. In Table 2 are given the experimental values of spin splittings in ground-state baryons from the Particle Data Group.[12] Also given are the two active quarks and the spectator quark (which has a small effect on the spin splitting caused mainly by the active quarks.[7]

bf Table 2. Values of the mass differences Δ_B and Δ'_B between spin-3/2 and spin-1/2 baryons containing the same quark content. See the text for the definitions of the quantities. The table distinguishes between the active quarks, which are mainly responsible for the spin splittings, and the spectator quark, which plays only a minor role. Also given are values of $R\Delta_B$, where $R = (m_i + m_j)/(2m_q)$, and the values of the masses of the active quarks m_i and m_j are given in (21). Values of Δ_B and Δ'_B are from the Particle Data Group.[12]

Baryons	Active quarks†	Spectator quark	Δ_B or Δ'_B (MeV)	$R_B\Delta_B$ or $R_B\Delta'_B$ (MeV)
$\Delta - N$	qq	q	293	293
$\Sigma^* - \Sigma$	qs	q	192	248
$\Xi^* - \Xi$	qs	s	215	278
$\Sigma_c^* - \Sigma_c$	qc	q	65	210
$(\Sigma^* + \Sigma - 3\Lambda)/2$	qq	s	307	307
$(\Sigma_c^* + \Sigma_c - 3\Lambda_c)/2$	qq	c	317	317

†The symbol q means either u or d.

It can be seen from the first, second, and fouth entries of the last column of Table 2 that in baryons the spin-dependent splittings decrease a little faster than $(m_1 + m_2)^{-1}$. In column 4, it can be seen from a comparison of the first, fifth, and sixth entries and from a comparison of the second and third entries that the heavier the spectator quark, the larger the matrix element of the two active quarks. This is in agreement with the conclusion of Ref. 7, and is the result of the heavy spectator quark shrinking the hadron wave function and thereby making the short-range spin-dependent interaction more effective.

DISCUSSION

Because of asymptotic freedom, it is plausible that at sufficiently small separation r between quarks, QCD perturbation theory works well. Furthermore, the chromomagnetic interaction of one-gluon exchange has a short range, and so is more amenable to perturbative treatment than the forces of longer range. Also, QCD dynamics causes hadrons containing only heavy quarks to be smaller in size than hadrons containing only light quarks or both light and heavy quarks. For these three reasons, one-gluon exchange is expected to be the dominant interaction at small distances in heavy-quark hadrons; in particular, the chromomagnetic interaction is expected to be the dominant mechanism of spin splittings in such hadrons.

I have given a plausibility argument that the one-gluon-exchange model is relevant for hadrons containing light, as well as heavy, quarks. It is important to note that nothing I have said rules out the possibility that the one-meson-exchange model and the instanton model also contribute in some measure. For example, Buchmann et al.[19] conclude that the $\Delta - N$ mass splitting arises 2/3 from one-gluon exchange and 1/3 from one-meson exchange.

If the splitting in light hadrons is to be explained by either one-meson exchange or instantons, a way has to be found to "turn off" the one-gluon exchange contribution in light-quark hadrons. I have not seen any convincing way to do this. Negele[8] claims that a lattice calculation shows that the one-gluon-exchange contribution is negligible in light mesons, but how that happens is mysterious to me.

It would be desirable to let experiment decide which of the three models is dominant in light and in heavy hadrons. Buchmann[20] has shown that one-gluon exchange, but not one-meson exchange, contributes to the quadrupole moment of the Ω^- baryon, but this moment has not been measured. Also, an existing calculation shows that the one-gluon-exchange and one-meson-exchange models give different predictions for the properties of some exotic hadrons.[21] However, at the present time, too few exotic hadrons have been identified to distinguish between models. Further progress will require additional theoretical and experimental work.

ACKNOWLEDGMENTS

I have benefited from discussions with Alfons Buchmann and Floarea Stancu, but they do not necessarily agree with the conclusions of this paper. This work was supported in part by the U.S. Department of Energy. Also, part of this work was done when I was a visitor at the Institute of Nuclear Theory of the University of Washington, Seattle, in November, 1998.

REFERENCES

1. A. De Rújula, H. Georgi, and S. L. Glashow, *Phys. Rev. D* 12:147 (1975).
2. L.Ya Glozman and D.O. Riska, *Phys. Rep.* 268:263 (1996).
3. E.V. Shuryak, *Rev. Mod. Phys.* 65:1 (1993).
4. N. Isgur, at Baryons '98 international conference, Bonn, Sept. 22-26, 1998.
5. I. Cohen, H.J. Lipkin, *Phys. Lett.* 106B:119 (1981).
6. J.M. Richard, P. Taxil, *Ann. Phys. (N.Y.)* 150:26 (1983).
7. M. Anselmino, D.B. Lichtenberg, and E. Predazzi, *Z. Phys. C* 48:605 (1990).
8. J.W. Negele, in *Intersections between Particle and Nuclear Physics*, Big Sky, Montana, May, 1997, AIP Conference Proceedings 412, edited by T.W. Donnelly, AIP, Woodbury, New York (1997), p. 3.
9. R. P. Feynman, Phys. Rev. 56:340 (1939).
10. H. Hellmann, Acta Physicochimica URSS I, 6:913 (1935); IV, 2:225 (1936); Einführung in die Quantenchemie (F. Denticke, Leipzig and Vienna, 1937) p. 286.
11. D.B. Lichtenberg, *Phys. Rev. D* 40:4196 (1989).
12. Particle Data Group: C. Caso et al., *Euro. Phys. J. C* 3:1 (1998).
13. R. Roncaglia, A.R. Dzierba, D.B. Lichtenberg, and E. Predazzi, *Phys. Rev. D* 51:1248 (1995).
14. R. Roncaglia, D.B. Lichtenberg, and E. Predazzi, *Phys. Rev. D* 52:1722 (1995).
15. W. Kwong and J. L. Rosner, Phys. Rev. D 44:212 (1991).
16. M. Frank and P.J. O'Donnell, *Phys. Lett. B* 159:174 (1985).
17. D.B. Lichtenberg, *Phys. Lett. B* 193:95 (1987).
18. D.B. Lichtenberg, *Intl. J. Mod. Phys. A* 2:1669 (1987).
19. A.J. Buchmann, E. Hernández, and E. Faessler, *Phys. Rev. C* 55:448 (1997).
20. A.J. Buchmann, *Z. Naturforsch* 52a:877 (1997).
21. Fl. Stancu, *Few Body Phys. Suppl.*, to be published (Proc. of the Workshop on N^* Physics and Non-perturbative QCD, Trento, Italy, May 18-29, 1998).

SUPER KAMIONKANDE DATA ON PROTON LIFETIME AND SUPERGRAVITY MODELS

Pran Nath[a] and R. Arnowitt[b]

[a]Department of Physics
Northeastern University
Boston, MA 02115

[b]Center for Theoretical Physics
Department of Physics
Texas A&M University
College Station, TX 77843-4242

ABSTRACT

A brief review of proton stability in supergravity unified models is given. The results are compared with the most recent lifetime limit measurements on $p \to \bar{\nu}K^+$ from SuperKamiokande experiment. We also discuss the annual modulation signal for Milky Way wimps claimed by the DAMA Collaboration in the framework of supergravity unified models under the constraint of proton stability.

1 INTRODUCTION

SUSY grand unification is very successful in giving the unification of the gauge coupling constants using the high precision LEP data[1, 2]. There is a perhaps a 1-2 σ discrepancy between theory and experiment, but this can be accounted for by including Planck scale corrections which are typically O($\frac{M_{GUT}}{M_{Planck}}$)[3, 4, 5] and hence give corrections of just the right amount to explain the discrepancy[4, 5]. The other aspect of SUSY grand unification is that it predicts proton decay and we discuss here the current situation on proton stability in supergravity unified models in view of the new lower limits on proton decay from Super Kamionkande[6].

$$\tau_p(p \to \bar{\nu}K) > 5.5 \times 10^{32} yr;\ \ 90\% C.L. \tag{1}$$

In this analysis we shall focus on supergravity models with gravity mediated breaking of supersymmetry[7, 8, 9]. We shall consider both the minimal supergavity model (mSUGRA) as well as the non-minimal supergravtiy models with non-universal soft SUSY breaking parameters. The minimal supergravity model with radiative breaking of the electro-weak symmetry is described by four parameters which can be taken to be

$$m_0, m_{\tilde{g}}, A_t, tan\beta;\ \ sign\mu \tag{2}$$

Confluence of Cosmology, Massive Neutrinos, Elementary Particles, and Gravitation
Edited by Kursunoglu *et al.*, Kluwer Academic / Plenum Publishers, New York, 1999.

where m_0 is the scalar mass, $m_{\tilde{g}}$ is the gluino mass, A_t is the trilinear coupling at the electro-weak scale, $tan\beta =< H_2 > / < H_1 >$, where H_2 gives mass to the up quarks and H_1 gives mass to the down quarks and μ is the Higgs mixing papameter which appears in the superpotential as the term $\mu H_1 H_2$. For the non-universal supergravity models[10, 11, 12, 13] we shall consider models with non-universalities in the scalar sector so that

$$m_i^2 = m_0^2(1 + \delta_i) \tag{3}$$

where δ_i parametrize the non-universalities and a reasonable range of non-universalities is $|\delta_i| \leq 1$.

Supersymmetric models in general have B and L violating dimension 4 operators. These can lead to fast proton decay and one needs to suppress them. The simplest way to eliminate them is by the imposition of R parity invariance. Another consequence of R parity invarance is that the lowest supersymmetric particle (LSP) is stable. Thus supesymmetric theories with R partiy invariance naturally provide a candidate for cold dark matter (CDM). Detailed renormalization group analyses in supergravity models show that the LSP in these models is the lightest neutralino χ_1^0. Fits of various cosmological models show that the the neutralino relic density lies in the range

$$0.05 \leq \Omega_{\chi_1^0} h^2 \leq 0.3 \tag{4}$$

In our analysis the relic density is computed using techniques discussed in Ref.[14, 15, 16] and we use the accurate method in its computation[15, 16]. We shall later consider the effects of cold dark matter constraint on the proton lifetime. In the analysis we also impose the $b \rightarrow s + \gamma$ constraint[17, 18, 19].

2 PROTON LIFETIME IN SUGRA MODELS

The dominant proton decay in supergravity theories is expected to arise from the dimension five operator $(qqql)_F$[20, 21, 22]. These operators when dressed by chargino, gluiono, and neutralino exchange gives dimension six operators of chiralities LLLL, $LLRR$, $RRLL$, and $RRRR$. The dominant decay modes involve pseudo-scalar bosons and anti-leptons, i.e.,

$$\bar{\nu}_i K^+, \bar{\nu}_i \pi^+; i = e, \mu, \tau; e^+ K^0, \mu^+ K^0, e^+ \pi^0, \mu^+ \pi^0, .. \tag{5}$$

The relative strength of these modes depends on a number of factors which include the quark mass factors and the Cabbibo-Kobayashi-Maskawa (CKM) factors. Some typical cases are exhibited in the Table below.

Mode	quark factors	CKM factors
$\bar{\nu}_\mu K$	$m_s m_c$	$V_{21}^\dagger V_{21} V_{22}$
eK	$m_d m_u$	$V_{11}^\dagger V_{12}$
$\mu\pi$	$m_s m_u$	$V_{11}^\dagger V_{21}^\dagger$

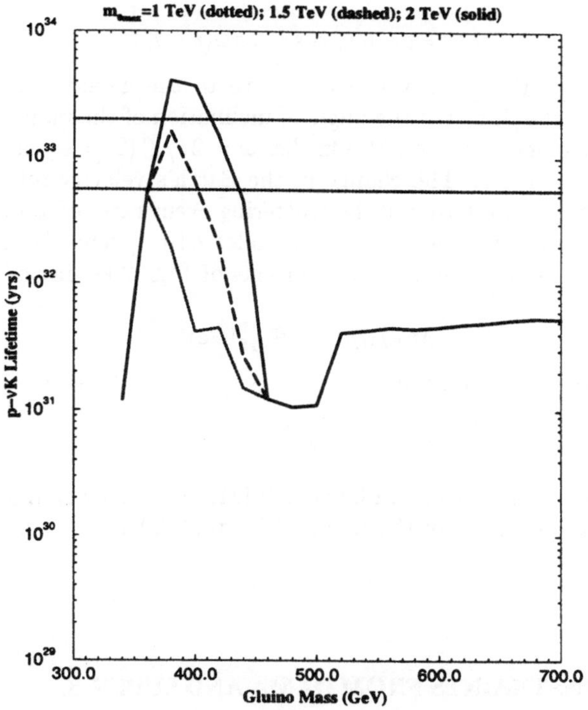

Figure 1. Plot of the maximum lifetime of the proton in mSUGRA as a function of the gluino mass for naturalness limits on m_0 of 1 TeV (solid), 1.5 TeV (dot-dashed), and 2 TeV (dashed) including the dark matter constraint. The lower horizontal dashed line is the current Super K limit on proton lifetime of 5.5×10^{32} years. The upper horizontal dashed line is the maixmum Super K limit on proton lifetime expected from Super K[23].

A detailed analysis of the decay lifetimes for these modes shows a significant amount of model dependence on the SUSY parameters. Thus, for example, the chargino ($\tilde{W}$) contributions to the $\bar{\nu}K^+$ modes give

$$\Gamma(p \to \bar{\nu}K) \sim \frac{1}{M_{H3}^2}(\frac{M_{\tilde{W}} tan\beta}{m_{\tilde{q}}^2})^2 \tag{6}$$

which roughly indicates the regions where the proton lifetime for this mode will be maximized. Typically this will happen for small $tan\beta$, large squark ($\tilde{q}$) masses, and large Higgs triplet ($\tilde{H}_3$) mass. We have analysed the maximum proton lifetime in mSUGRA over the parameter space $m_{\tilde{g}} \leq 1$ TeV, $-7 \leq A_t/m_0 < 7$, $tan\beta \leq 30$, for various naturalness limits on m_0, i.e., $m_0 \leq 1$ TeV, $m_0 \leq 1.5$ TeV, $m_0 \leq 2$ TeV. We also include the dark matter constraints in our analysis. The results are displayed in Fig.1.

3 CONSTRAINTS FROM DAMA

The DAMA Collaboration at Gran Sasso has examined the possibility of the direct detection of Milky Way WIMPS using the annual modulation signal which arises due to the motion of the Earth around the Sun[24, 25]. Thus, v_E, the velocity of the Earth relative to the Galaxy is

$$v_E = v_S + v_0 \cos\gamma \cos\omega(t - t_0) \tag{7}$$

where $v_S = 232km/s$ is the sun's velocity relative to the galaxy, $v_0 = 30km/s$ Earth's orbital velocity around the Sun, γ is the angle of inclination of the plane of the Earth's orbit relative to the galactic plane and $\gamma \cong 60^o$. One has $\omega = 2\pi/T$ ($T = 1$ year) and the maximum velocity occurs at $t_0 =$ June 2. The change in the Earth's velocity relative to the incident WIMPs leads to a yearly modulation of the scattering event rates of about 7%. Using ~ 100 kg of radiopure NaI at Gran Sasso DAMA has collected two sets of data[24, 25]. Set 1[24] consists of 4549 kg-day of data and set 2[25] consists of 14,962 kg-day of data. DAMA gives a WIMP mass of

$$M_{WIMP} = (59^{+17}_{-14})GeV \tag{8}$$

and neutralino-proton cross section of

$$\xi\sigma_{W-p} = (7.0^{+0.4}_{-1.2}) \times 10^{-6}pb \tag{9}$$

where $\xi = \rho_w/\rho_0$. Here ρ_w is the local Milky Way WIMP mass density and $\rho_0 = 0.3\, GeVcm^{-3}$. It is estimated that ρ_w may vary in the range $(0.2 - 0.7)GeVcm^{-3}$, i.e.,

$$0.7 \lesssim \xi \lesssim 2.3 \tag{10}$$

4 COMBINED CONSTRAINTS FROM DAMA AND SUPER K

In the computation of the scattering of neutralinos from protons we follow the techniques discussed in Refs.[26, 27]. We have examined the simultaneous constraints on $\chi_1 - p$ cross sections from DAMA and SUPER K experiments for three different classes of models. These are

- mSUGRA
- Non-minimal GUT Models
- Models with non-universalities

We discuss these in that order.
i. mSUGRA
In the analysis of mSUGRA we find no part of the parameter space where one can generate an annual modulation signal consistent with DAMA and simultaneously satisfy the lower lifetime constraints from Super K for the minimal SU(5) model.
ii. Non-Minimal GUT Models
The minimal SU(5) GUT model makes poor predictions for the quark-lepton mass ratios and one needs extensions to include textures[28]. Inclusion of textures gives an enhancement of a factor of 3-5 for the proton lifetime. Further the Higgs triplet sector can be more complex

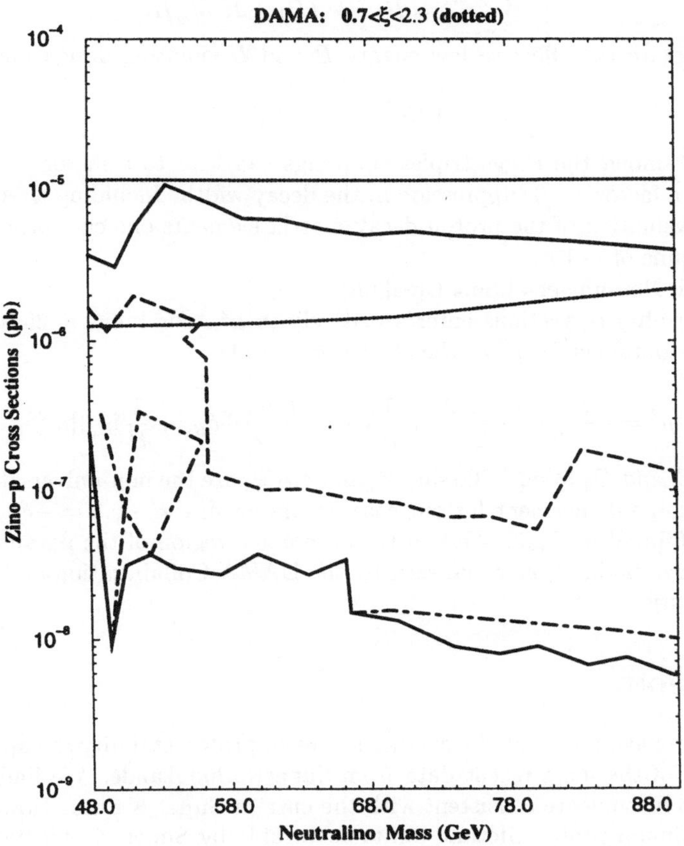

Figure 2. Effects of proton lifetime constraints on neutralino-proton cross section are exhibited for two enhancement factors for the mSUGRA case. The solid curves give the maximum and the minimum of the $\chi_1^0 - p$ cross section when proton lifetime constraint is not imposed. The area enclosed by the dashed curve and the lower solid curve is the area allowed when an enhancement factor of 100 on proton lifetime is included. The area enclosed by the upper and the lower dot-dashed curves on the left and the upper dot-dashed curve and the lower solid curve on the right is the region allowed when the enhancement factor on the proton lifetime is 20. The area enclosed by the dotted lines is the region that allows for the DAMA annual modulation effect (taken from Ref.[29]).

involving many Higgs triplets and anti-triplets (H_i, $\bar{H}_i$; i=1,2,..,n). In the basis where only $\bar{H}_1$ and H_1 couple with matter one can write the superpotential for the Higgs triplets in the form

$$W_3^{triplet} = \bar{H}_1 J + \bar{J} H_1 + \bar{H}_i M_{ij} H_j \tag{11}$$

With this structure the effective low energy B and L violating dimension five operator is given by

$$W_4^{eff} = -\bar{J}(M_{11}^{-1})J \tag{12}$$

A cancellation among the Higgs triplet couplings can lead to a factor ~ 3 suppression in amplitude and a factor ~ 10 suppression in the decay width. Including a factor of 2-3 uncertainty in the evaluation of the proton decay matrix elements one can have an enhancement in p decay lifetime of $\sim 10^2$.

iii. Models with Non-universal Soft Breaking

The non-universality corrections enter sensitively in μ^2. For $\tan\beta \leq 25$ a closed form expression can be obtained for μ^2 at the electroweak scale

$$\mu^2 = \frac{t^2}{1-t^2}[(\frac{1-3D_0}{2} + \frac{1}{t^2}) + (-\frac{1+D_0}{2}\delta_{H_2} + \frac{1}{t^2}\delta_{H_1})]m_0^2 + .. \tag{13}$$

where $t \equiv \tan\beta$, and $D_0 = m_t^2/(200\sin\beta)^2$, and $\delta_{H1,H2}$ are the non-universalities in the Higgs sector. Maximum enhancement for our case occurs for $\delta_{H1} = -0.5 = -\delta_{H2}$. Results of the analysis are exhibited in Fig.3. We find that there is a region of the parameter space which allows the annual modulation signal seen by the DAMA Collaboration under the constraint of proton stability.

5 CONCLUSIONS

In this paper we have reviewed the current status of proton stability in supergravity unified models in view of the most recent data from SuperKamionkande. We find that mSUGRA predictions of p lifetime are consistent with the current Super K data. However, in the near future the maximum proton lifetime limits achievable by Super K will nearly exhaust the mSUGRA parameter space. We also analysed the constraints on mSUGRA arising from the recent claims for the observation of an annual modulation signal by DAMA. We find that the DAMA annual modulation signal cannot be explained under the constraints of proton stability in mSUGRA. However, it is found that in non-minimal GUT models where one has more that one Higgs triplets, and allowing for texture effects, it is possible to enhance the proton lifetime by a factor of 10^2. With this enhancement it is possible to have consistency with the annual modulation signal from DAMA. Analyses were also carried out including the effects of non-universalities and it is found that these effects relieve the tension in the simultaneous satisfaction of the proton stability and dark matter constraints.

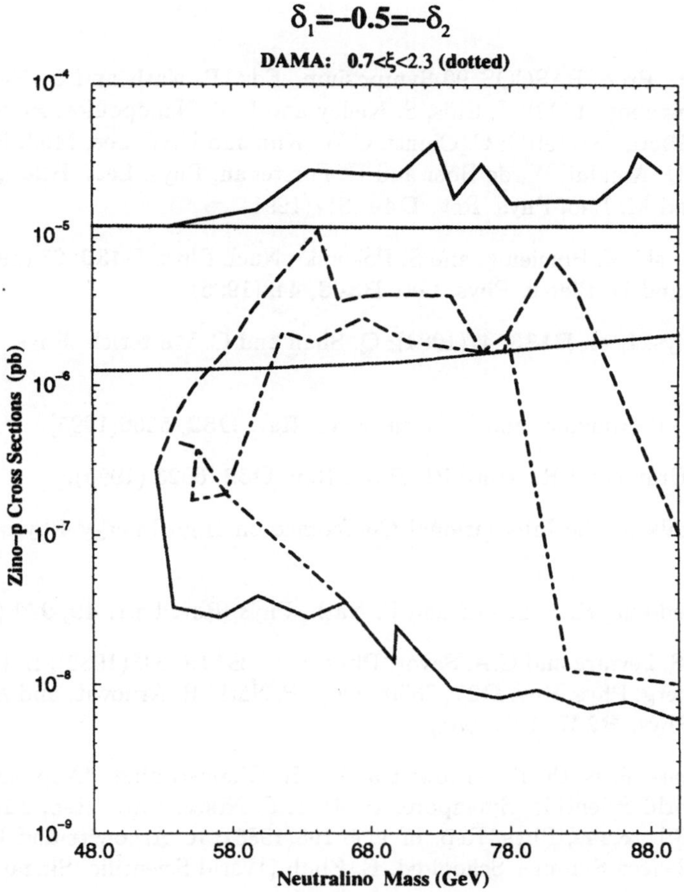

Figure 3. Same as Fig.2 except that $\delta_1 = -0.5$, $\delta_2 = 0.5$, $\delta_3 = 0$, $\delta_4 = 0$ (taken from Ref.[29]).

ACKNOWLEDGMENTS

This work was supported in part by NSF grant number PHY-96020274 and PHY-9722090.

REFERENCES

[1] P. Langacker, Proc. PASCOS 90-Symposium, Eds. P. Nath and S. Reucroft (World Scientific, Singapore 1990); J. Ellis, S. Kelley and D.V. Nanopoulos, Phys. Lett. **B249**, 441(1990); **B260**, 131(1991); C. Ciunti, C.W. Kim and U.W. Lee, Mod. Phys. Lett.**A6**, 1745(1991); U. Amaldi, W. de Boer and H. Furstenau, Phys. Lett. **B260**, 447(1991); P. Langacker and M. Luo, Phys. Rev. **D44**, 817(1991).

[2] P.H. Chankowski, Z. Plucienik, and S. Pokorski, Nucl. Phys. **B439**, 23 (1995);J. Bagger, K. Matchev and D. Pierce, Phys. Lett **B348**, 443(1995).

[3] C.T. Hill, Phys. Lett. **B135**, 47(1984); Q. Shafi and C. Wetterich, Phys. Rev. Lett. **52**, 875(1984).

[4] T. Dasgupta, P. Mamales and P. Nath, Phys. Rev. **D52**, 5366(1995).

[5] D. Ring, S. Urano and R. Arnowitt, Phys. Rev. **D52**, 6623 (1995).

[6] M. Takita, talk at the International Conference on High Energy Physics, Vancouver, July, 1998.

[7] A.H. Chamseddine, R. Arnowitt and P. Nath, Phys. Rev. Lett. **49**, 970 (1982).

[8] R. Barbieri, S. Ferrara and C.A. Savoy, Phys. Lett. **B119**, 343(1982); L. Hall, J. Lykken and S. Weinberg, Phys. Rev. **D27**, 2359(1983); P. Nath, R. Arnowitt and A.H. Chamseddine, Nucl. Phys. **B227**, 121(1983);

[9] For reviews see, P. Nath, R. Arnowitt and A.H. Chamseddine, "Applied N =1 Supergravity" (World Scientific, Singapore, 1984); H.P. Nilles, Phys. Rep. **110**, 1(1984); H. Haber and G.L. Kane, Phys. Rep. bf 117, 195(1985); R. Arnowitt and P. Nath, Proc. of VII J.A. Swieca Summer School ed. E. Eboli (World Scientific, Singapore, 1994).

[10] S. Soni and A. Weldon, Phys. Lett. **B126**, 215(1983); V.S. Kaplunovsky and J. Louis, Phys. Lett. **B306**, 268(1993).

[11] D. Matalliotakis and H. P. Nilles, Nucl. Phys. **B435**, 115(1995); M. Olechowski and S. Pokorski, Phys. Lett. **B344**, 201(1995); N. Polonski and A. Pomerol, Phys. Rev.**D51**, 6532(1995);

[12] P. Nath and R. Arnowitt, Phys. Rev. **D56**, 2820(1997);

[13] R. Arnowitt and P. Nath, Phys. Rev.**D56**, 2833(1997).

[14] J.L. Lopez, D.V. Nanopoulos, and K. Yuan, Phys. Lett. **B267**,219(1991); J. Ellis and L. Roszkowski, Phys. Lett. **B283**, 252(1992); M. Kawasaki and S. Mizuta, Phys. Rev. **D46**, 1634(1992); M. Drees and M.M. Nojiri, Phys. Rev. **D47**, 376(1993); G.L. Kane, C. Kolda,L. Roszkowski, and J.D. Wells, Phys. Rev.**D49**, 6173(1993).

[15] K. Greist and D. Seckel, Phys. Rev. **D43**, 3191(1991); P. Gondolo and G. Gelmini, Nucl. Phys. **B360**, 145(1991).

[16] R. Arnowitt and P. Nath, Phys. Lett. **B299**, 58(1993) and Erratum ibid **B303**, 403(1993); P. Nath and R. Arnowitt, Phys. Rev. Lett. **70**, 3696(1993); M. Drees and A. Yamada, Phys. Rev. **D53**, 1586(1996); H. Baer and M. Brhlik, Phys. Rev.**D53**, 1586(1996); V. Barger and Kao, Phys. Rev. **D57**, 3131(1998).

[17] CLEO Collaboration, see talk at ICHEP98, Vancouver, July 1998 for the latest results.

[18] J.L. Hewett, Phys. Rev. Lett. **70**, 1045(1993); V. Barger, M. Berger, P. Ohmann. and R.J.N. Phillips, Phys. Rev. Lett. **70**, 1368(1993).

[19] M. Diaz, Phys. Lett. **B304**, 278(1993); J. Lopez, D.V. Nanopoulos, and G. Park, Phys. Rev. **D48**, 974(1993); R. Garisto and J.N. Ng, Phys. Lett. **B315**, 372(1993); P. Nath and R. Arnowitt, Phys. Lett. **B336**(1994)395; F. Borzumati, M. Drees, and M.M. Nojiri, Phys. Rev.**D51**, (1995)341; J. Wu, R. Arnowitt and P. Nath, Phys. Rev. **D51**, 1371(1995); V. Barger, M. Berger, P. Ohman and R.J.N. Phillips, Phys. Rev. **D51**, 2438(1995); H. Baer and M. Brhlick, Phys. Rev. **D55**, 3201(1997).

[20] S. Weinberg, Phys. Rev. **D26**, 287(1982); N. Sakai and T. Yanagida, Nucl. Phys.**B197**, 533(1982); S. Dimopoulos, S. Raby and F. Wilczek, Phys.Lett. **112B**, 133(1982); J. Ellis, D.V. Nanopoulos and S. Rudaz, Nucl. Phys. **B202**, 43(1982); B.A. Campbell, J. Ellis and D.V. Nanopoulos, Phys. Lett. **141B**, 299(1984); S. Chadha, G.D. Coughlan, M. Daniel and G.G. Ross, Phys. Lett.**149B**, 47(1984).

[21] R. Arnowitt, A.H. Chamseddine and P. Nath, Phys. Lett. **156B**, 215(1985); P. Nath, R. Arnowitt and A.H. Chamseddine, Phys. Rev. **32D**, 2348(1985); J. Hisano, H. Murayama and T. Yanagida, Nucl. Phys. **B402**, 46(1993); R. Arnowitt and P. Nath, Phys. Rev. **D49**, 1479 (1994).

[22] R. Arnowitt and P. Nath, Phys. Lett. **B437**,344(1998).

[23] Y.Totsuka, Proc. XXIV Conf. on High Energy Physics, Munich, 1988,Eds. R.Kotthaus and J.H. Kuhn (Springer Verlag, Berlin, Heidelberg,1989).

[24] R. Bernabei et.al., Phys. Lett. **B424**, 195(1998).

[25] R. Bernabei et.al., INFN/AE-98/34, (1998).

[26] M.W. Goodman and E. Witten, Phys. Rev. **D31**, 3059(1983); K. Greist, Phys. Rev. **D38**, (1988)2357; **D39**,3802(1989)(E); J. Ellis and R. Flores, Phys. Lett. **B300**,175(1993); R. Barbieri, M. Frigeni and G.F. Giudice, Nucl. Phys. **B313**,725(1989); M. Srednicki and R.Watkins, Phys. Lett. **B225**,140(1989); R. Flores, K. Olive and M. Srednicki, Phys. Lett. **B237**,72(1990).

[27] R. Arnowitt and P. Nath, Mod. Phys. Lett. **A 10**,1257(1995); P. Nath and R. Arnowitt, Phys. Rev. Lett.**74**,4592(1995); R. Arnowitt and P. Nath, Phys. Rev. **D54**,2374(1996); A. Bottino et.al., Astropart. Phys. **1**, 61 (1992); **2**, 77 (1994); V.A. Bednyakov, H. V. Klapdor-Kleingrothaus and S.G. Kovalenko, Phys. Rev. **D50**,7128(1994); L. Bergstrom and P. Gondolo, Astropart. Phys. 5:263-278 (1996); J.D. Vergados, J. Phys. **G22**, 253(1996); P. Nath and R. Arnowitt, Mod. Phys. Lett.A, Vol.13, No. 27, 2239(1998).

[28] P. Nath, Phys. Rev. Lett. **76**, 2218(1996).

[29] R. Arnowitt and P. Nath, "Annual Modulation Signature for the Direct Detection of Milky Way wimps and Supergravity Models", hep-ph/9902237.

DIRECT DETECTION OF DARK MATTER AND GRAND UNIFICATION

R. Arnowitt[a] and Pran Nath[b]

[a]Center for Theoretical Physics
Department of Physics
Texas A&M University
College Station, TX 77843-4242

[b]Department of Physics
Northeastern University
Boston, MA 02115

ABSTRACT

Gravity mediated supergravity grand unified models with R-parity invariance make predictions of what may be expected in future experiments in both accelerator and non-accelerator physics. We review here what parameters of the theory control the relic density of dark matter neutralinos and their direct detection cross section for local (Milky Way) dark matter, and show that this cross section is expected to have a maximum at a neutralino mass of about 60 GeV. Theoretical predictions are compared with the preliminary DAMA data of an annual modulation effect, and it is shown that the theory is consistent with this data for both universal and non-universal soft breaking for $\tan\beta \gtrsim 6-8$. The effect of the DAMA data, were it to be confined, on accelerator predictions at the Tevatron Run II and LHC are discussed.

1 INTRODUCTION

Over the next five years, one can expect a flood of new data in a variety of areas that should help clarify some of the fundamental questions of particle physics. Thus, in accelerator physics, the Tevatron Run II and B factories will begin to collect data shortly. In cosmology, the MAP sattelite (and many balloon and ground based experiments) will obtain precision data on the cosmic microwave background (CMB) radiation, and there will be major galaxy sky surveys (the Sloan Digital Sky Survey and the Two-Degree Field Survey). In addition, numerous non-accelerator experiments will be in operation, e.g. for proton decay (Super Kamiokande and ICARUS) for dark matter (DM) detection (DAMA, CDMS, etc.) and for neutrino physics (SNO, Borexino, NESTOR, AMANDA).

While at this point we do not know what will lie beyond the Standard Model (SM), supersymmetric grand unified (SUSY GUT) models, particularly the gravity mediated models with R-parity invariance [1] are particularly interesting to investigate, as they make predictions in all these areas. This means that a result in one area can influence a prediction in another area, and one generates a more tightly constrained theory. Already accelerator bounds from LEP, Tevatron, and $b \rightarrow s + \gamma$ decay data (CLEO) have begun to narrow the SUSY parameter space that one can use in other fields.

In order to examine how different areas affect each other we will discuss here the following [2]:

1. What parameters of the theory control the amount of neutralino ($\tilde{\chi}_1^0$) relic density left over from the Big Bang (BB) and how the astronomical measurements of the amount of cold dark matter (CDM) constrain these SUSY parameters. (The $\tilde{\chi}_1^0$ are the SUSY candidates for CDM.)

2. What parameters of the theory control the expected detection rate of local Milky Way $\tilde{\chi}_1^0$ particles.

Confluence of Cosmology, Massive Neutrinos, Elementary Particles, and Gravitation
Edited by Kursunoglu *et al.*, Kluwer Academic / Plenum Publishers, New York, 1999.

3. How the preliminary DAMA data [3] on the detection of Milky Way $\tilde{\chi}_1^0$ would limit the parameter space, if this data were confirmed by other DM detector groups.
4. The effects items one through three would have on predictions of the SUSY particle spectrum expected at accelerators (e.g. at the Tevatron and LHC).

2 SUPERGRAVITY MODELS

The coupling of supergravity to chiral matter [1] depends upon three functions of the scalar components ϕ_i of the chiral multiplets (ϕ_i, χ_{iL}): the gauge kinetic function $f_{\alpha\beta}(\phi_i, \phi_i^\dagger)$, which appears in the kinetic energy of the gauge fields $(f_{\alpha\beta}F^{\alpha}_{\mu\nu}F^{\mu\nu\beta})$; the superpotential $W(\phi_i)$, which can be decomposed into a physical sector and a hidden sector part ($W = W_{phys}+W_{hid}$); and the Kahler potential $K(\phi_i, \phi_i^\dagger)$ which appears (among other places) in the kinetic energy of the chiral fields (e.g. $K^i_j\partial_\mu\phi_i\partial^\mu\phi_j^\dagger$, where $K^i_j = \partial K/\partial\phi_i\partial\phi_j^\dagger$). In these models, supersymmetry is assumed to be broken in the hidden sector, the breaking being communicated to the physical sector by gravity, and R-parity is assumed to be preserved.

The functions $f_{\alpha\beta}$, W and K are constrained by gauge invariance and the requirement that, using the renormalization group equations (RGE), the theory be in agreement at low energies with the SM. If one then assumes that the non-renormalizable interactions arising from a power series expansion in ϕ_i are scaled by $\kappa = 1/M_{Pl}$ ($M_{Pl} = (1/8\pi G_N)^{1/2} = 2.4 \times 10^{18}GeV$), one obtains a theory which depends on a number of additional parameters, but with a significant amount of predictability.

The breaking of SUSY in the hidden sector gives rise to a number of soft breaking parameters in the physical sector. We will assume here that this breaking occurs above the GUT scale $M_G \cong 2\times10^{16}GeV$ (e.g., that this is Planck scale physics). For a simple GUT group, this implies to leading order that $f_{\alpha\beta}$ gives rise to a universal gaugino mass $m_{1/2}$ at M_G (with only small corrections of size M_G/M_{Pl} which we neglect here). The Kahler potential determines the nature of the remaining soft breaking parameters. If the hidden sector fields which break SUSY couple universally to the physical fields in K, one obtains the minimal model, mSUGRA, which depends on four additional parameters and one sign. We take these to be m_0 (the universal scalar soft breaking mass at M_G), A_0 (the universal cubic soft breaking at M_G), $tan\beta =< H_2 > / < H_1 >$ at the electroweak scale (where $< H_{2,1} >$ gives rise to (up, down) quark masses), $m_{1/2}$ at M_G and the sign of μ, the Higgs mixing parameter (which appears in the superpotential W as $\mu H_1 H_2$).

If the hidden fields which break SUSY do not couple universally to different matter generations, non-universal m_0 and A_0 occur. In the following, we will assume universal soft breaking for the first two generations to suppress flavor changing neutral currents, and allow non-universality in the Higgs sector and third generation. Thus, at M_G we have

$$m_{H_1^2} = m_0^2(1+\delta_1);\; m_{H_2^2}(1+\delta_2) \tag{1}$$

$$m_{q_L}^2 = m_0^2(1+\delta_3);\; m_{u_R}^2 = m_0^2(1+\delta_4);\; m_{e_R}^2 = m_0^2(1+\delta_4) \tag{2}$$

$$m_{d_R}^2 = m_0^2(1+\delta_6);\; m_{\ell_L}^2 = m_0^2(1+\delta_7) \tag{3}$$

where $q_L = (\tilde{t}_L, \tilde{b}_L)$, $u_R = \tilde{t}_R$, $\ell_L = (\tilde{\nu}_L, \tilde{\tau}_L)$, etc.

In addition, there are three A_0 parameters: A_{0t}, A_{0b}, $A_{0\tau}$. For GUT groups that contain an SU(5) subgroup (e.g. $SU(N), N \geq 5, SO(10), N \geq 10, E_6$) with matter embedded in the 10 and $\bar{5}$ of SU (5) in the usual fashion, one has $\delta_3 = \delta_4 = \delta_5$, $\delta_6 = \delta_7$, and $A_{0b} = A_{0\tau}$.

One of the major quantities that is sensitive to non-universal soft breaking is the μ parameter. The RGE give, e.g. for low and intermediate $\tan\beta$, the result at the electroweak scale [4]:

$$\mu^2 = \frac{t^2}{t^2-1}[(\frac{1-3D_0}{2}+\frac{1}{t^2}) + (\frac{1-D_0}{2}(\delta_3+\delta_4) - \frac{1+D_0}{2}\delta_2 + \frac{1}{t^2}\delta_1)]m_0^2$$

$$+\; other\; universal\; terms;\; t \equiv \tan\beta \tag{4}$$

where $D_0 \cong 1-(m_t/200\sin\beta)^2$ (i.e., $D_0 \leq 0.23$ for the t-quark mass of $m_t = 175GeV$). Thus, for $\delta_3, \delta_4, \delta_1 < 0$ and $\delta_2 > 0$, μ^2 is decreased, while for the reverse signs, μ^2 will be increased.

3 RELIC DENSITY AND DETECTION CROSS SECTION

A measure of the mean amount of matter of type "i" in the universe is given by $\Omega_i = \rho_i/\rho_c$, where ρ_i is the density of matter and $\rho_c = 3H^2/8\pi G_N$ is the "critical density" needed to close the universe. Here H is the Hubble constant, $H = (100/kms^{-1}Mpc^{-1})h$, and $0.5 \leq h \leq 0.75$. Nucleosynthesis in the early universe implies a baryonic (B) component of $\Omega_B \simeq 0.05$. Direct evidence for clustered matter, Ω_m (from large galactic clusters, large scale flows) give $\Omega_m \cong 0.2-0.4$ indicating a large amount of non-baryonic dark matter. Recent measurements of high z Type 1a supernovae have indicated the existance of a vacuum energy Ω_Λ (cosmological constant or "quintessence"), and combining this with the recent CMB data gives [5]

$$\Omega_m = 0.25^{+0.09}_{-0.06};\ \Omega_\Lambda = 0.63^{+0.09}_{-0.12} \tag{5}$$

which is consistent with the direct measurements of Ω_m. Note also that Eq. (5) implies $\Omega_m + \Omega_\Lambda = 0.88 \pm 0.12$ which is consistent with the usual inflationary scenario requirement that $\Omega_{total} = 1$.

The above results still allow for a sizable range in the amount of cold dark matter (CDM) in the universe, and in the following we will assume

$$0.05 \leq \Omega_{CDM}h^2 \leq 0.30 \tag{6}$$

SUGRA models with R-parity invariance automatically predict the existence of CDM, i.e. the relic lightest neutralino, $\tilde{\chi}^0_1$ (which is absolutely stable) left over from the Big Bang. Neutralinos produced by the BB can annihilate into fermion pairs, etc. in the early universe through either s-channel Z° and h (Higgs) poles, or t-channel squark ($\tilde{q}$) or slepton ($\tilde{\ell}$) poles. This leads to two domains:

(i) If $m_{\chi^0_1} \lesssim (50-55)GeV$, rapid annihilation can occur when $2m_{\chi^0_1} \simeq M_Z$ or m_h. In this case, m_0 can be large (to reduce the t-channel contribution) so that too much annihilation does not occur (i.e., the lower bound of Eq. (6) not be violated).

(ii) If $m_{\chi^0_1} \gtrsim 55GeV$, the annihilation proceeds largely through the t-channel poles, and m_0 must be small ($m_0 \lesssim (100-150)GeV$) to lower the $\tilde{q}$ or $\tilde{\ell}$ masses to get sufficient annihilation so that the upper bound of Eq. (6) not be violated.

Rotation curves of gas clouds in the Milky Way around the center of the Galaxy show that the halo of our galaxy contains a large amount of DM which is presumably impinging upon the Earth, thus offering the possibility of direct detection of CDM. Estimates of local DM density in the vicinity of the solar system are

$$0.2\,GeVcm^{-3} \leq \rho_{local} \leq 0.7GeVcm^{-3} \tag{7}$$

with a central value of $0.3GeVcm^{-3}$. (The larger value arises from the possibility that the halo is flattened, while the smaller value assumes that some of the DM are baryonic machos.) One defines $\xi = \rho_{local}/0.3$ allowing for the range

$$0.67 \leq \xi \leq 2.33 \tag{8}$$

Direct detection of this neutralino dark matter is possible by the scattering of the incident $\tilde{\chi}^0_1$ by quarks in a nuclear target. The scattering proceeds mainly via either an s-channel squark pole or a t-channel Z, h and H (heavy Higgs) poles. The cross section thus increases as m_0 decreases (since the $\tilde{q}$ mass decreases). One then has the following behavior for $\sigma_{\chi^0_1-p}$, the $\tilde{\chi}$ - proton cross section. For $m_{\chi^0_1} < 50GeV$, the above relic density analysis implies m_0 is large and hence $\sigma_{\chi^0_1-p}$ will be small. As $m_{\chi^0_1}$ increases up to $55-60GeV$, relic density constraints requires m_0 to decrease and hence $\sigma_{\chi^0_1-p}$ increases. For larger $m_{\chi^0_1}$, the cross section will fall off with increasing $\tilde{\chi}^0_1$ mass. Thus we expect that $\sigma_{\chi^0_1-p}$ has a maximum at $m_{\chi^0_1} \simeq (55-60)GeV$.

In addition to the above dependence on m_0, the dominant spin independent part of $\sigma_{\tilde{\chi}_1^0-p}$ depends on the interference between the Higgsino and gaugino parts of $\tilde{\chi}_1^0$ in the scattering amplitude. This increases generally as μ^2 decreases (increasing $\sigma_{\tilde{\chi}_1^0-p}$). Also, $\sigma_{\tilde{\chi}_1^0-p}$ increases with $\tan\beta$. We will see the above parameter dependences in the following analysis.

4 ANNUAL MODULATION EFFECT

One method to aid in the detection of Milky Way wimps is to make use of the annual modulation effect. As the Earth moves around the Sun, and the Sun moves around the Galaxy, the velocity of the Earth relative to the Galaxy, v_E oscillates:

$$v_E(t) = v_s + v_0 \cos\gamma \cos\omega(t - t_0) \tag{9}$$

Here $v_s = 232km/s$ in the Sun's velocity, $v_0 = 30km/s$ is the velocity of the Earth around the Sun, $\omega = 2\pi/T$, T = one year, $t_0 = 152.2$ day (June 2) and $\gamma \cong 60°$ is the inclination of the Earth's orbit with respect to the disk of the Milky Way. The modulation of the Earth's velocity produces a corresponding modulation in the number of scattering events of about 7%. While this is a small effect, it does act as a veto against backgrounds not possessing the periodic behavior.

The DAMA experiment consists of about 100 kg of radio pure $NaI(T\ell)$ target in the Gran Sasso National Laboratory. Their two published runs, which contains a total of 19,511 kg day of data, has given indication of seeing the modulation effect [3]. They fit the data to a wimp mass and cross section of

$$M_w = (59^{+17}_{-14})GeV; \; \xi\sigma_{w-p} = (7.0^{+0.4}_{-1.2}) \times 10^{-6} pb \tag{10}$$

with a probability of 99.6% that an annual modulation has been observed.

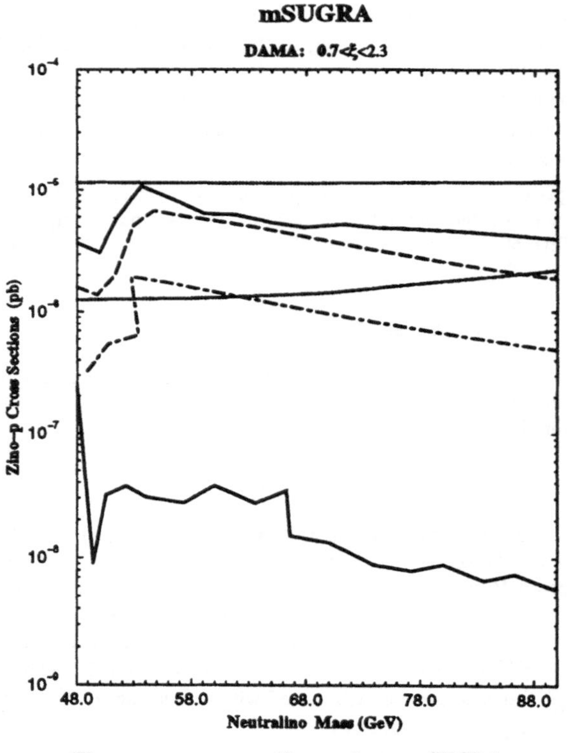

Figure 1. $\sigma_{\tilde{\chi}_1^0-p}$ vs. $m_{\tilde{\chi}_1^0}$. The upper curves are the maximum mSUGRA cross sections for $\tan\beta = 30$ (solid), $\tan\beta = 20$ (dashed), $\tan\beta = 10$ (dash-dot). The lower solid curve is the minimum $\sigma_{\tilde{\chi}_1^0-p}$. The dotted horizontal bands are the DAMA 95% CL combined with the range of ξ of Eq. (8).

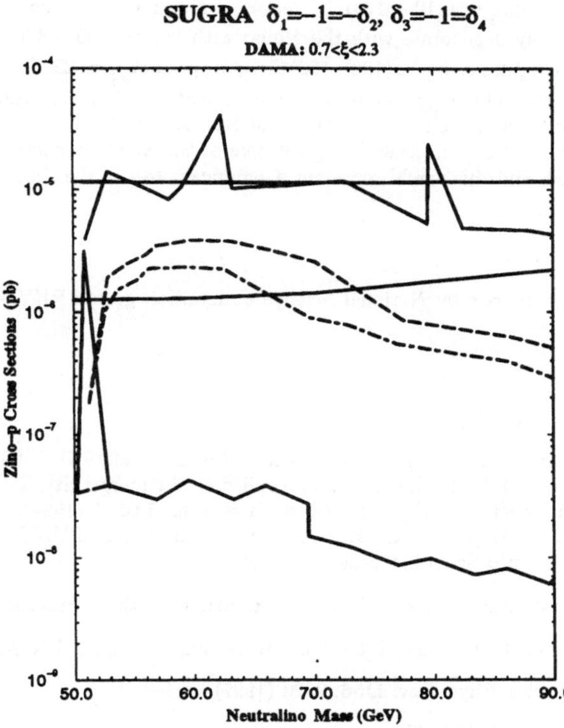

Figure 2. Same as Fig. 1 for non-universal soft breaking with $\delta_1 = \delta_3 = \delta_4 = -1$, $\delta_2 = +1$.

The DAMA experiment remains to be confirmed by another group, and it is likely that other groups will achieve the necessary sensitivity in the relatively near future to test whether the effect is real or spurious. However, it is interesting to examine whether SUGRA models can give results consistent with the DAMA data. We consider first the mSUGRA models. Fig. 1 shows the theoretical $\sigma_{\chi_1^0 - p}$ as a function of $m_{\chi_1^0}$ for $\tan\beta = 10$, 20, and 30. As expected, the curves peak at $m_{\chi_1^0} \approx 55GeV$, and then fall off for larger $m_{\chi_1^0}$, and $\sigma_{\chi_1^0 - p}$ increases with $\tan\beta$. All three curves fall within the allowed DAMA band (dotted curve), and if the DAMA data is indeed confirmed, mSUGRA models are consistent with it for $\tan\beta \gtrsim 8$.

SUGRA models with non-universal soft breaking can even more easily achieve $\chi_1^0 - p$ cross sections of the size necessary to accommodate the DAMA data. Thus, we had that $\sigma_{\chi_1^0 - p}$ increases when μ^2 is decreased, and from Eq. (4), μ^2 will be decreased when $\delta_{3,4,1} < 0$, $\delta_2 > 0$. Fig. 2 shows $\sigma_{\chi_1^0 - p}$ as a function of $m_{\chi_1^0}$ for the case $\delta_1 = \delta_3 = \delta_4 = -1$, $\delta = +1$, and for $\tan\beta = 10$, 20, and 30. The behavior of the curves are similar to Fig. 1, but lie higher, the $\tan\beta = 30$ curve exceeding the DAMA bounds at some points. One finds here that theory would be consistent with data such as the DAMA experiment for $\tan\beta \gtrsim 6$. For the reverse signs of δ_i, the cross sections will be reduced, and so the DM detection cross sections are sensitive to the size and sign of non-universal soft breakings.

5 CONCLUSIONS

SUGRA models with R-parity invariance can treat cosmology and accelerator phenomena in a unified way. Dark matter experiments are now beginning to achieve the sensitivity to influence what one might expect at accelerators. As an example, one might consider the possibility that dark matter experiments determine $m_{\chi_1^0} \simeq 60GeV$ (a suggested by the DAMA data). Then one would expect, approximately that $m_{\chi_1^\pm} \cong m_{\chi_2^0} \simeq (110 - 120)GV$, $m_h \cong (100 - 120)GeV$, the first two generations of squarks are approximately degenerate with the gluino with $m_{\tilde{g}} \simeq (400 - 450)GeV$, the first two generations of sleptons are light, i.e. $m_{\tilde{l}} \simeq (100-150)GeV$, $m_{\chi_{3,4}^0} \cong m_{\chi_2^\pm} \simeq 250GeV$ and $\tan\beta \gtrsim 6-8$. This would mean that h (the light Higgs) would be observable at the Tevatron Run II, and that there is the possibility that the $\chi_1^\pm$ would also be observable at Run II as well [6]. (The gluino and squarks would probably require the LHC.) In general, a good deal of the SUSY particle spectrum would be approximately determined, and this would represent a significant test of the theory.

Acknowledgment

This work was supported in part by National Science Foundation grants PHY-9722090 and PHY-9602074.

References

[1] A. H. Chamseddine, R. Arnowitt, and P. Nath, Phys. Rev. Lett. **49**, 970 (1982). For a review see, P. Nath, R. Arnowitt, and A. H. Chamseddine, Applied N=1 supergravity, Trieste Lectures, 1983 (World Scientific, Singapore, 1984); H. P. Nilles, Phys. Rep. **110**, 1 (1984); H. E. Haber, G. L. Kane, Phys. Rep. **117**, 195 (1984); R. Arnowitt and P. Nath, Proc. of VII J. A. Swieca Summer School, ed. E. Eboli (World Scientific, Singapore, 1994).

[2] For a discussion of proton decay, see P. Nath and R. Arnowitt, these Proceedings.

[3] R. Bernabei et al. Phys. Lett. **B424**, 195 (1998); R. Bernabei et al., INFN/AE-98/34, (1998).

[4] P. Nath and R. Arnowitt, Phys. Rev. **D56**, 2820 (1997).

[5] G. Efstatiou et al., astro-ph/9812226.

[6] E. Barberis, Higgs and SUGRA Searches At The Tevatron, talk at Higgs and Supersymmetry: Search and Discovery, Gainsville, March 9, 1999.

BREAKING LORENTZ SYMMETRY IN QUANTUM FIELD THEORY

Don Colladay

Physics Department
The College of Wooster
Wooster, OH 44691
U.S.A.

INTRODUCTION

Symmetry under the Lorentz group is a fundamental assumption of virtually any fundamental theory used to describe elementary particle physics. For instance, the standard model as well as many extensions including supersymmetry and grand unified theories preserve Lorentz symmetry. Under very mild assumptions, the postulates of a point particle theory that preserves Lorentz invariance lead to the conclusion that CPT is preserved [1].

In this talk, I will discuss the construction of quantum field theories that break Lorentz and CPT symmetry. There are both experimental and theoretical motivations to develop such theories.

Many sensitive experimental tests of Lorentz and CPT symmetry have been performed. For example, high precision tests involving atomic systems [2, 3], clock comparisons [4], and neutral meson oscillations [5] provide stringent tests of Lorentz and CPT symmetry. In the past, each such experiment has bounded phenomenological parameters that lack a clear connection with the microscopic physics of the standard model. It is desirable to have a single theory within the context of conventional quantum field theory and the standard model that could relate various experiments and be used to motivate future investigations.

On the theoretical side, low-energy remnant effects that violate fundamental symmetries may arise in theories underlying the standard model. One example is string theory in which nontrivial structure of the vacuum solutions may induce observable Lorentz and CPT violations [6, 7, 8].

Terms involving standard model fields that violate Lorentz and CPT symmetry are assumed to arise from a general spontaneous symmetry breaking mechanism in which vacuum expectation values for tensor fields are generated in the underlying theory [9].

Confluence of Cosmology, Massive Neutrinos, Elementary Particles, and Gravitation
Edited by Kursunoglu *et al.*, Kluwer Academic / Plenum Publishers, New York, 1999.

Rather than derive the terms from a specific underlying theory, the approach taken here is to examine all possible terms that can arise through spontaneous symmetry breaking that are consistent with the gauge invariance of the standard model and power-counting renormalizability.

The resulting terms lead to modified field equations that can be analyzed within the context of conventional quantum field theory. In this talk, I will develop the modified Feynman rules for a model theory and will explore some possible consequences for quantum electrodynamics.

SPONTANEOUS VIOLATION OF LORENTZ SYMMETRY

Conventional spontaneous symmetry breaking occurs in theories that contain scalar field potentials with nontrivial minima. An example is the conventional Higgs mechanism of the standard model in which Yukawa couplings generate fermion masses through spontaneous symmetry breaking. In theories of this sort, internal symmetries of the original lagrangian such as gauge invariance may be violated, but Lorentz symmetry is always maintained.

Spontaneous Lorentz breaking may occur in a fundamental theory containing a potential for a tensor field that has nontrivial minima. For example, consider the lagrangian describing a fermion ψ and a tensor T of the form

$$\mathcal{L} = \mathcal{L}_0 - \mathcal{L}' \quad , \tag{1}$$

where

$$\mathcal{L}' \supset \frac{\lambda}{M^k} T \cdot \overline{\psi} \Gamma (i\partial)^k \psi + \text{h.c.} + \text{V(T)} \quad . \tag{2}$$

In this expression, λ is a dimensionless coupling constant, M is some heavy mass scale of the underlying theory, Γ denotes a general gamma matrix structure in the Dirac algebra, and $V(T)$ is a potential for the tensor field. (indices are suppressed for notational simplicity) The potential $V(T)$ is assumed to arise in a theory underlying the standard model. Terms contributing to $V(T)$ are precluded from conventional renormalizable four-dimensional field theories, but may arise in the low-energy limit of more general theories such as string theory [6, 7].

If the function $V(T)$ has nontrivial minima, a nonzero expectation value of T is generated in the vacuum. The lagrangian then contains a term of the form

$$\mathcal{L}' \supset \frac{\lambda}{M^k} \langle T \rangle \cdot \overline{\psi} \Gamma (i\partial)^k \chi + \text{h.c.} \quad , \tag{3}$$

that is bilinear in the fermion fields and can violate Lorentz invariance and various discrete symmetries C, P, T, CP, and CPT.

RELATIVISTIC QUANTUM MECHANICS

To develop techniques for treating terms of the form (3) it is useful to examine a subset of all the possible terms. An example applicable to the standard model fermions is furnished by the choice $k = 0$ (no derivatives) and $\Gamma \sim \gamma^\mu$ or $\Gamma \sim \gamma^5 \gamma^\mu$ that violate CPT symmetry. The model lagrangian for a single fermion ψ becomes

$$\mathcal{L} = \frac{i}{2} \overline{\psi} \gamma^\mu \overleftrightarrow{\partial_\mu} \psi - a_\mu \overline{\psi} \gamma^\mu \psi - b_\mu \overline{\psi} \gamma_5 \gamma^\mu \psi - m \overline{\psi} \psi \quad , \tag{4}$$

where the parameters a_μ and b_μ are constant coefficients that denote the tensor expectation values and coupling constants that are present in (3).

Several features of this theory are immediately apparent. First of all, the lagrangian is hermitian and preserves probability. This means that conventional quantum mechanics can be used to describe the evolution of the particle states. The model lagrangian is also invariant under translations and U(1) gauge transformations which leads to conservation of energy, momentum, and charge. The resulting Dirac equation

$$(i\gamma^\mu\partial_\mu - a_\mu\gamma^\mu - b_\mu\gamma_5\gamma^\mu - m)\psi = 0 \quad . \tag{5}$$

obtained using the Euler-Lagrange equations is linear in the field ψ. Equation (5) can be solved exactly using the plane-wave solutions

$$\psi(x) = e^{\pm ip_\mu x^\mu} w(\vec{p}) \quad , \tag{6}$$

where $p^0(\vec{p}) \equiv E(\vec{p})$ is the energy determined by setting the determinant of the matrix acting on $w(\vec{p})$ equal to zero.

The general form of the resulting dispersion relation is complicated, so here the special case $\vec{b} = 0$ is considered. The exact dispersion relations for this case are

$$E_+(\vec{p}) = \left[m^2 + (|\vec{p} - \vec{a}| \pm b_0)^2\right]^{1/2} + a_0 \quad , \tag{7}$$

$$E_-(\vec{p}) = \left[m^2 + (|\vec{p} + \vec{a}| \mp b_0)^2\right]^{1/2} - a_0 \quad . \tag{8}$$

The conventional energy degeneracy of the fermion and antifermion states is broken by a_μ while b_0 splits the degeneracy of the helicities. The corresponding spinor solutions form an orthogonal basis of states.

One interesting feature of the above dispersion relations is the modified relationship that exists between the velocity of a wave packet and its corresponding momentum. For instance, a wave packet formed from a superposition of positive helicity fermions with a four-momentum $p^\mu = (E, \vec{p})$ has a corresponding expectation value of the velocity operator $\vec{v} = i[H, \vec{x}] = \gamma^0\vec{\gamma}$ of

$$\langle\vec{v}\rangle = \langle \frac{(|\vec{p} - \vec{a}| - b^0)}{(E - a^0)} \frac{(\vec{p} - \vec{a})}{|\vec{p} - \vec{a}|} \rangle \quad . \tag{9}$$

Examination of the velocity using a general nonzero b_μ reveals that $|v_j| < 1$, and that the limiting velocity as $\vec{p} \to \infty$ is 1. This implies that effects due to the CPT violating terms are mild enough to preserve causality. This will be verified independently from the perspective of the quantum field theory that will now be developed.

FREE FIELD THEORY

Canonical quantization fails even in the conventional case for fermions, so the approach taken here is to calculate the energy and deduce the quantization conditions from positivity requirements of the energy. The wave function ψ is expanded in terms of its four solutions as

$$\begin{aligned}\psi(x) \;=\; \int \frac{d^3p}{(2\pi)^3} \sum_{\alpha=1}^{2} \Bigg[& \frac{m}{E_u^{(\alpha)}} b_{(\alpha)}(\vec{p}) e^{-ip_u^{(\alpha)}\cdot x} u^{(\alpha)}(\vec{p}) \\ & + \frac{m}{E_v^{(\alpha)}} d^*_{(\alpha)}(\vec{p}) e^{ip_v^{(\alpha)}\cdot x} v^{(\alpha)}(\vec{p}) \Bigg] \quad , \end{aligned} \tag{10}$$

and is promoted in the usual way to an operator acting on a Hilbert space of basis states.

Translational invariance is used to define a conserved energy and momentum as

$$P_\mu = \int d^3x \Theta^0{}_\mu = \int d^3x \tfrac{1}{2} i\overline{\psi}\gamma^0 \overleftrightarrow{\partial}_\mu \psi \quad . \tag{11}$$

The time component P_0 is interpreted as the energy after normal ordering of the operators and is positive definite (for $|a^0| < m$) provided the following anticommutation relations are imposed:

$$\begin{aligned} \{b_{(\alpha)}(\vec{p}), b^\dagger_{(\alpha')}(\vec{p}\,')\} &= (2\pi)^3 \frac{E_u^{(\alpha)}}{m} \delta_{\alpha\alpha'} \delta^3(\vec{p}-\vec{p}\,') \quad , \\ \{d_{(\alpha)}(\vec{p}), d^\dagger_{(\alpha')}(\vec{p}\,')\} &= (2\pi)^3 \frac{E_v^{(\alpha)}}{m} \delta_{\alpha\alpha'} \delta^3(\vec{p}-\vec{p}\,') \quad . \end{aligned} \tag{12}$$

The resulting equal-time anticommutators of the fields become

$$\begin{aligned} \{\psi_\alpha(t,\vec{x}), \psi^\dagger_\beta(t,\vec{x}')\} &= \delta_{\alpha\beta}\delta^3(\vec{x}-\vec{x}') \quad , \\ \{\psi_\alpha(t,\vec{x}), \psi_\beta(t,\vec{x}')\} &= 0 \quad , \\ \{\psi^\dagger_\alpha(t,\vec{x}), \psi^\dagger_\beta(t,\vec{x}')\} &= 0 \quad . \end{aligned} \tag{13}$$

These relations show that the conventional Fermi statistics remain unaltered by the CPT violation.

The conserved charge Q and conserved momentum P_μ are then computed as

$$Q = \int \frac{d^3p}{(2\pi)^3} \sum_{\alpha=1}^{2} \left[\frac{m}{E_u^{(\alpha)}} b^\dagger_{(\alpha)}(\vec{p}) b_{(\alpha)}(\vec{p}) - \frac{m}{E_v^{(\alpha)}} d^\dagger_{(\alpha)}(\vec{p}) d_{(\alpha)}(\vec{p}) \right] \quad , \tag{14}$$

$$\begin{aligned} P_\mu = \int \frac{d^3p}{(2\pi)^3} \sum_{\alpha=1}^{2} \Bigg[& \frac{m}{E_u^{(\alpha)}} p^{(\alpha)}_{u\mu} b^\dagger_{(\alpha)}(\vec{p}) b_{(\alpha)}(\vec{p}) \\ & + \frac{m}{E_v^{(\alpha)}} p^{(\alpha)}_{v\mu} d^\dagger_{(\alpha)}(\vec{p}) d_{(\alpha)}(\vec{p}) \Bigg] \quad . \end{aligned} \tag{15}$$

From these expressions we observe that the charge of the fermion is unperturbed and the energy and momentum satisfy the same energy-momentum relations that were found using relativistic quantum mechanics.

Causality is governed by the anticommutation relations of the fermion fields at unequal times. Explicit integration for the special case of $\vec{b} = 0$ proves that

$$\{\psi_\alpha(x), \overline{\psi}_\beta(x')\} = 0 \quad , \tag{16}$$

for spacelike separations $(x - x')^2 < 0$. This result indicates that physical observables separated by spacelike intervals will in fact commute. This agrees with our previous results regarding the velocity obtained using the relativistic quantum mechanics approach.

EXTENSION TO INTERACTING FIELD THEORY

Next, the issue of extending the free field theory to interacting theory is addressed. Much of the conventional formalism developed for perturbative calculations in conventional interacting field theory carries over directly to the present case. The S matrix and asymptotic in and out states are defined as in the usual case. The LSZ reduction procedure is then used to express the S-matrix elements in terms of Green's functions for the theory. Dyson's formalism is used to express the time-ordered products of the interacting fields in terms of the asymptotic fields. Wick's theorem remains unaffected by the modifications.

The main result is that the usual Feynman rules apply provided that the Feynman propagator is modified as

$$S_F(p) = \frac{i}{p_\mu \gamma^\mu - a_\mu \gamma^\mu - b_\mu \gamma_5 \gamma^\mu - m} \quad , \tag{17}$$

and the exact spinor solutions of the modified free fermion theory are used on external legs. The main reason that conventional techniques work is that the Lorentz violating modifications introduced are linear in the fermion fields.

EXTENSION OF QED AND THE PHOTON

In this section, some implications of Lorentz breaking for photon propagation are studied. The conventional QED lagrangian is given by

$$\mathcal{L}^{\mathrm{QED}}_{\mathrm{electron}} = \tfrac{1}{2} i \overline{\psi} \gamma^\mu \overleftrightarrow{D_\mu} \psi - m_e \overline{\psi} \psi - \tfrac{1}{4} F_{\mu\nu} F^{\mu\nu} \quad , \tag{18}$$

where ψ is the electron field, m_e is its mass, and $F^{\mu\nu}$ is the photon field strength tensor. When all possible Lorentz-violating contributions from spontaneous symmetry breaking are introduced into the standard model that are consistent with gauge invariance and power-counting renormalizability, the resulting modifications to the photon are [9]

$$\mathcal{L}^{\mathrm{CPT-even}}_{\mathrm{photon}} = -\tfrac{1}{4} (k_F)_{\kappa\lambda\mu\nu} F^{\kappa\lambda} F^{\mu\nu} \quad , \tag{19}$$

and

$$\mathcal{L}^{\mathrm{CPT-odd}}_{\mathrm{photon}} = +\tfrac{1}{2} (k_{AF})^\kappa \epsilon_{\kappa\lambda\mu\nu} A^\lambda F^{\mu\nu} \quad . \tag{20}$$

The parameters k_F and k_{AF} are coupling constants related to vacuum expectation values of tensors coupled to gauge bosons. They are classified according to their properties under CPT. The CPT-odd terms have been treated in detail elsewhere [10]. Here the special case of $(k_{AF})^\mu = 0$ (no CPT-odd piece), and $(k_F)_{0j0k} = -\frac{1}{2}\beta_j \beta_k$ is studied.

The resulting modifications to the Maxwell equations are linear just as in the fermion case. Plane waves can therefore be used to solve the modified equations. A solution exists provided p_μ satisfies

$$(p_o)^2 = 0 \quad , \tag{21}$$

$$(p_e)^2 = -\frac{(\vec{\beta} \times \vec{p}_e)^2}{1 + \vec{\beta}^2} \quad , \tag{22}$$

where p_o denotes an ordinary mode and p_e denotes an extraordinary mode of propagation. The ordinary mode propagates as a conventional photon, while the extraordinary mode has modified properties.

For the special case $\vec{\beta} \cdot \vec{p} = 0$ the ordinary mode is polarized with $\vec{A}_o$ along the direction of $\vec{p} \times \vec{\beta}$ while the extraordinary mode $\vec{A}_e$ is polarized along $\vec{\beta}$. Both polarizations are perpendicular to the momentum of the wave $\vec{p}$. The group velocities defined by $\vec{v}_g \equiv \vec{\nabla}_p p^0$ are given by

$$\vec{v}_{g,o} = \hat{p} \quad , \quad \vec{v}_{g,e} = \frac{1}{\sqrt{1+\vec{\beta}^2}} \, \hat{p} \quad . \tag{23}$$

The extraordinary mode travels with a modified velocity that is slightly less than the velocity of the ordinary mode. As a result, an initially plane polarized wave will become elliptically polarized after traveling a distance

$$r \simeq \frac{\pi}{2\left(\sqrt{1+\vec{\beta}^2}-1\right)p^0} \simeq \frac{\pi}{\vec{\beta}^2 p^0} \quad , \tag{24}$$

where the approximation holds for $\vec{\beta}^2 \propto k_F << 1$. The magnetic field behaves similarly. Terms of this form have implications for photon birefringence, in particular they contribute to polarization rotation from distant quasars.

ACKNOWLEDGMENTS

This work was supported in part by the United States Department of Energy under grant number DE-FG02-91ER40661.

REFERENCES

1. See, for example, J. Schwinger, Phys. Rev. **82** (1951) 914.
2. See, for example, R.S. Van Dyck, Jr., P.B. Schwinberg, and H.G. Dehmelt, Phys. Rev. Lett. **59** (1987) 26; G. Gabrielse *et al.*, *ibid.*, **74** (1995) 3544.
3. R. Bluhm, V.A. Kostelecký and N. Russell, Phys. Rev. Lett. **79** (1997) 1432; Phys. Rev. D **57** (1998) 3932; IUHET 388 (1988).
4. See, for example, J.D. Prestage *et al.*, Phys. Rev. Lett. **54** (1985) 2387; S.K. Lamoreaux *et al.*, *ibid.*, **57** (1986) 3125; T.E. Chupp *et al.*, *ibid.*, **63** (1989) 1541.
5. See, for example B. Schwingenheuer *et al.*, Phys. Rev. Lett. **74** (1995) 4376; OPAL Collaboration, R. Ackerstaff *et al.*, Z. Phys. C **76** (1997) 401.
6. V.A. Kostelecký and R. Potting, Nucl. Phys. B **359** (1991) 545; Phys. Lett. B **381** (1996) 389.
7. V.A. Kostelecký, R. Potting, and S. Samuel, in S. Hegarty et al., eds., *Proceedings of the 1991 Joint International Lepton-Photon Symposium and Europhysics Conference on High Energy Physics,* World Scientific, Singapore, 1992;
   V.A. Kostelecký and R. Potting, in D.B. Cline, ed., *Gamma Ray-Neutrino Cosmology and Planck Scale Physics* (World Scientific, Singapore, 1993) (hep-th/9211116).
8. V.A. Kostelecký and R. Potting, Phys. Rev. D **51** (1995) 3923.
9. D. Colladay and V.A. Kosteleck/'y, Phys. Rev. D **55** (1997) 6760; Phys. Rev. D **58** (1998) 116002.
10. S.M. Carroll, G.B. Field, and R. Jackiw, Phys. Rev. D **41** (1990) 1231;
    R. Jackiw, this proceedings.

Section III
Spin and Statistics

SPIN STRUCTURE WITH LEPTON BEAMS

B. W. Filippone

W. K. Kellogg Laboratory
California Institute of Technology
Pasadena, CA 91125

INTRODUCTION

Attempting to understand the origin of the half-integer intrinsic spin of the proton and neutron has been an active area of both experimental and theoretical research for the past twenty years. With the confirmation that the proton and neutron were not elementary particles, physicists were challenged with the task of explaining the nucleon's spin in terms of its constituents. In a simple constituent model one can decompose the nucleon's spin as

$$J_z^N = S_z^q + L_z^q + S_z^g + L_z^g = \frac{1}{2}. \quad (1)$$

where S_z and L_z represent the intrinsic and orbital angular momentum for quarks and gluons. A simple non-relativistic quark model (as described below) gives directly $S_z^q = \frac{1}{2}$ and all the other components = 0.

Because the structure of the nucleon is governed by the strong interaction, the components of the nucleon's spin must be calculable from Quantum ChromoDynamics (QCD). However since the spin is a low energy property, direct calculations with non-perturbative QCD are not possible at present (although some initial estimates are being attempted via lattice QCD). A general theoretical overview of nucleon spin structure was presented by Xiangdong Ji at this conference [1].

This paper summarizes the experimental status of measurements of the nucleon's spin structure as probed via deep inelastic lepton-nucleon scattering with polarized beams and targets. We first present a simple model for the spin structure in terms of up and down constituent quarks, followed by a short introduction to measurements of spin structure via lepton beams and a summary of the early results that confronted the simple model. The experimental program is then reviewed, focusing on the more recent experiments where vastly different techniques are being applied in order to limit possible systematic errors in the measurements. An overview of recent results is followed by a discussion of prospects for future measurements of individual quark contributions (flavor separation), gluon spin contribution, and other aspects of spin structure.

Confluence of Cosmology, Massive Neutrinos, Elementary Particles, and Gravitation
Edited by Kursunoglu *et al.*, Kluwer Academic / Plenum Publishers, New York, 1999.

Simple Model for Proton Spin

A simple non-relativistic wave function for the proton comprising only the valence up and down quarks can be written as

$$|p\uparrow> = \frac{1}{\sqrt{6}}(2|u\uparrow u\uparrow d\downarrow> - |u\uparrow u\downarrow d\uparrow> - |u\downarrow u\uparrow d\uparrow> . \quad (2)$$

Here the up and down quarks give all of the proton's spin. The contribution of the u and d quarks to the proton's spin can be determined by the use of the following matrix element and projection operator:

$$u^{\uparrow} = < p\uparrow |\hat{\mathcal{O}}_{u\uparrow}|p\uparrow> \quad (3)$$
$$\hat{\mathcal{O}}_{u\uparrow} = \frac{1}{4}(1+\hat{\tau}_3)(1+\hat{\sigma}_3). \quad (4)$$

With the above matrix element and a similar one for the down quarks, the quark spin distributions can be defined as

$$\Delta u = u^{\uparrow} - u^{\downarrow} = \frac{1}{2}(< \sigma_3 > + < \sigma_3\tau_3 >) = \frac{4}{3} \quad (5)$$
$$\Delta d = d^{\uparrow} - d^{\downarrow} = \frac{1}{2}(< \sigma_3 > - < \sigma_3\tau_3 >) = -\frac{1}{3} \quad (6)$$

Thus the fraction of the proton's spin carried by quarks in this simple model is

$$\Delta\Sigma = \Delta u + \Delta d = < \sigma_3 > = 2J_z^N = 1 \quad (7)$$

and all of the spin is carried by the quarks. Note however that this simple model overestimates another property of the nucleon, namely the axial-vector weak coupling constant g_A. In fact this model gives

$$\frac{g_A}{g_V} = < \sigma_3\tau_3 > = \Delta u - \Delta d = \frac{5}{3}, \quad (8)$$

compared to the experimentally measured value of $\frac{g_A}{g_V} = 1.267 \pm 0.004$. The difference between the simple non-relativistic model and the data is often attributed to relativistic effects. This "quenching" factor of ~ 0.75 can be applied to the spin carried by quarks to give the following "relativistic" quark model predictions:

$$\Delta\Sigma \approx 0.75 \quad (9)$$
$$\Delta u \approx 1.0 \quad (10)$$
$$\Delta d \approx -0.25 \quad (11)$$
$$\Delta s \approx 0 \quad (12)$$

These predictions will be compared with measurements below.

Lepton Scattering as a Probe of Spin Structure

Deep inelastic scattering (DIS) with charged lepton beams has been an important tool for understanding the electro-magnetic sub-structure of the nucleon. With polarized beams and targets the electro-magnetic *spin* sub-structure of the nucleon becomes

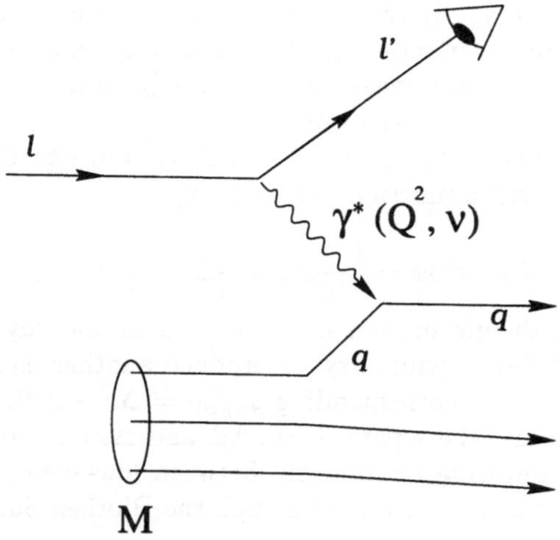

Fig. 1. Schematic diagram of deep-inelastic scattering

accessible. Information from neutral lepton scattering (neutrinos) is complementary to that from charged leptons but generally of lower statistical quality.

The access to nucleon structure via lepton scattering can best be seen within the quark-parton model. An example of a deep-inelastic scattering process is shown in Fig. 1. In this picture a virtual photon (with energy ν and four-momentum transfer squared $-Q^2$) strikes an asymptotically free quark in the nucleon. For unpolarized scattering the quark "momentum" distributions - $q_i(x) = u(x), d(x), s(x), ...$ - are probed in this reaction, where the quark momentum fraction x is given by $x = Q^2/2M\nu$, with $M =$ proton mass. From the cross section for this process, the structure function $F_1(x)$ can be extracted. In the quark-parton model this structure function is related to the quark distributions via

$$F_1(x) = \frac{1}{2}\sum_i e_i^2 q_i(x) \tag{13}$$

where the sum is over quark and anti-quark flavors. With polarized beams and targets the quark spin distributions can be probed. This sensitivity results from the requirement that the quark's spin be anti-parallel to the virtual photon's spin in order for the quark to absorb the virtual photon. With the assumption of nearly massless quarks, angular momentum would not be conserved if the quark absorbs a photon when its spin is parallel to the photon's spin. Thus measurements of the spin-dependent cross section allows the extraction of the spin-dependent structure function $g_1(x)$. Again in the quark-parton model this structure function is related to the quark *spin* distributions via

$$g_1(x) = \frac{1}{2}\sum_i e_i^2 \Delta q_i(x). \tag{14}$$

The structure function is extracted from the measured asymmetry $(A_{||}, A_{\perp})$ of the scattering cross section as the beam or target spin is flipped. These asymmetries are measured with longitudinally polarized beams and longitudinally $(A_{||})$ and transversely $(A_{\perp})$ polarized targets via

$$A_{||} = \frac{N^{\uparrow\downarrow} - N^{\uparrow\uparrow}}{N^{\uparrow\downarrow} + N^{\uparrow\uparrow}}(\frac{1}{fP_bP_T})) + \Delta A_{RC} \tag{15}$$

and a similar expression for $A_\perp$. Here the normalized events per spin state ($N^{\uparrow\downarrow}$), target dilution factor (f), beam polarization (P_B), and target polarization (P_T) contribute to the statistical precision of the measurement. ΔA_{RC} is a calculated correction for higher-order electro-magnetic radiative effects.

The total contribution of quarks to the nucleon spin can then be extracted by integrating the $g_1(x)$ structure function over x giving

$$\int g_1(x)dx = \frac{1}{2}(\frac{4}{9}\Delta u + \frac{1}{9}\Delta d + \frac{1}{9}\Delta s). \tag{16}$$

Then combining this with information from neutron beta-decay giving $\Delta u - \Delta d = g_A/g_V$ and using SU(3) flavor symmetry to produce another equation from hyperon beta-decay (from $\Sigma \rightarrow ne^-\hat{\nu}$ a corresponding $g_A/g_V = \Delta s - \Delta d$), we have three equations in three unknowns and the separate $\Delta u, \Delta d$, and Δs can be derived.

There is also an interesting relationship between low-energy neutron beta-decay and high-energy deep-inelastic scattering through the Bjorken Sum Rule which states

$$\int (g_1^p(x) - g_1^n(x))dx = \frac{1}{2}(\frac{1}{3})\frac{g_A}{g_V}. \tag{17}$$

The validity of this sum rule depends only on isospin symmetry and the application of current algebra.

Summary of Initial Results

Interest in understanding the nucleon's spin structure increased dramatically when the European Muon Collaboration (EMC) made measurements[2,3] of $g_1^p(x)$ over a wider range in x than earlier, pioneering SLAC measurements [4,5]. The EMC measurements permitted the extraction of a reliable integral over x for $g_1^p(x)$, giving

$$\int g_1(x)dx = \frac{1}{2}(\frac{4}{9}\Delta u + \frac{1}{9}\Delta d + \frac{1}{9}\Delta s) = 0.126 \pm 0.018 \tag{18}$$

significantly disagreeing with the prediction from the simple relativistic quark model prediction that gives (using the results from eqs. 9 - 12)

$$\int g_1(x)dx = \frac{1}{2}[\frac{4}{9}(1) + \frac{1}{9}(-\frac{1}{4})] = 0.208. \tag{19}$$

Furthermore with neutron and hyperon beta-decay data combined with the assumption of SU(3) flavor symmetry, the measurements implied

$$\Delta\Sigma \approx 0.1 \tag{20}$$
$$\Delta u \approx 0.8 \tag{21}$$
$$\Delta d \approx -0.5 \tag{22}$$
$$\Delta s \approx -0.2 \tag{23}$$

Thus one is led to the somewhat surprising conclusion that quarks carry very little of the proton's spin (since $\Delta\Sigma \sim 0$) and also that there is significant polarization of strange quarks in the proton.

It is important to note that QCD radiative corrections due to gluon emission give all of the above quantities (eg. structure functions, quark distributions, spin distributions, ...) a Q^2 dependence. This will be discussed later in more detail. While these corrections change many of the numerical expectations presented above, inclusion of these corrections does not significantly alter the main conclusions.

Table 1. Summary of lepton induced spin measurements

Lab	Exp.	Year	Beam	Target
SLAC	E80	75	16 GeV e^-	C_4H_9OH
	E130	80	23 GeV e^-	C_4H_9OH
	E142	92	25 GeV e^-	^3He
	E143	93	29 GeV e^-	NH_3,ND_3
	E154	95	48 GeV e^-	^3He
	E155	97	48 GeV e^-	NH_3, ND_3, LiD
	E155'	97	30 GeV e^-	NH_3, LiD
CERN	EMC	85	100-200 GeV μ^+	NH_3
	SMC	92	100 GeV μ^+	C_4D_9OD
	"	93	190 GeV μ^+	C_4H_9OH
	"	94	190 GeV μ^+	C_4D_9OD
	"	95	190 GeV μ^+	C_4D_9OD
	"	96	190 GeV μ^+	NH_3
DESY	HERMES	95	28 GeV e^+	^3He
		96	28 GeV e^+	H
		97	28 GeV e^+	H
		98	28 GeV e^+	D
		99	28 GeV e^+	D
CERN	COMPASS	00?	190 GeV μ^+	NH_3, LiD

OVERVIEW OF EXPERIMENTS

An extensive program of polarized lepton scattering measurements has developed following the results from the EMC experiment. A summary of the measurements is shown in Table 1, where the beams, targets, and energies are listed for each experiment. The experiments complement each other in their kinematic coverage and in their sensitivity to possible systematic errors associated with the measured quantities. For example for the quantities presented in Eq. 15, the experiments span the range $1 - 300 \times 10^6$ DIS events, $0.1 - 1$ for f, $0.4 - 0.8$ for P_B, and $0.3 - 0.9$ for P_T.

We can summarize the various programs by highlighting the advantages of each:

SLAC: high luminosity, fast beam polarization reversal, low - medium Q^2 coverage;

CERN: two simultaneous oppositely polarized targets, high Q^2 coverage, semi-inclusive hadrons;

DESY: pure atomic targets, semi-inclusive hadrons (with pion and kaon identification after '97), low - medium Q^2 coverage.

OVERVIEW OF RESULTS AND PROSPECTS FOR THE FUTURE

In discussing results, the focus will be on the high precision data that have been acquired over the last 10 years. The proton and neutron spin structure functions g_1^p and g_1^n have been measured by three laboratories and there is generally good agreement between the measurements. This agreement is found only when the momentum transfer dependence (Q^2-dependence) is accounted for due to the different kinematics of the experiments. In fact this momentum transfer dependence allows some crude experimental information to be extracted for the gluon spin contribution ($J_z^g \equiv \Delta G$).

Recently several experiments have used lepton-induced hadron production to gain access to the individual contributions of different quark flavors. These measurements also seek to separate the contributions from the "valence" up and down quarks from the "sea" of quarks and anti-quarks.

A number of experimental programs are underway to provide direct information on the gluon contribution to the nucleon spin. These experiments should begin providing important new information within the next several years.

Lastly there are several other aspects of the nucleon's spin structure that are being addressed in a number of on-going and future experiments. These include measurements of the distribution of quarks polarized perpendicular to the nucleon's spin (in contrast to g_1 which measures the distribution of quarks polarized along the nucleon's spin) and other quark distributions that arise due to relativistic effects and quark momentum transverse to the virtual photon direction. All of these programs are addressed in separate sections below.

Proton and Neutron Spin Structure Functions $g_1(x)$

Precision data on the proton spin structure function $g_1^p(x)$ exists from SLAC experiments E143 [6,7] and E155 [8], as well as from SMC at CERN [9,10,11,12,13] and HERMES at DESY [14]. A comparison of the SLAC and SMC data is shown in Fig. 2. The comparison is made at an extrapolated fixed momentum transfer of 5 GeV^2. The significance of a fixed momentum transfer will be discussed below.

For the neutron, $g_1^n(x)$ can be determined from polarized ^{2}H or ^{3}He targets. For a deuterium target $g_1^n \simeq 2g_1^d - g_1^p$, while for ^{3}He $g_1^n \simeq g_1^{^3He}$ because the wave function for ^{3}He is dominated by the configuration with the protons paired to zero spin. Both targets require small corrections for nuclear effects. The results for $g_1^d(x)$ and $g_1^n(x)$ are shown in Fig. 3 for the SLAC E142 [15,16], E143 [17,7], E154 [18], CERN SMC [19,20,21,13], and DESY HERMES [22] experiments, again at $Q^2 = 5$ GeV^2.

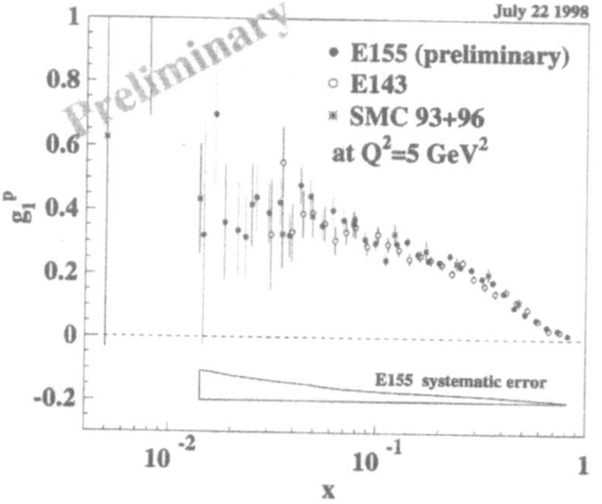

Fig. 2. $g_1^p(x)$ from SLAC E143 and SMC compared to preliminary results from SLAC E155.

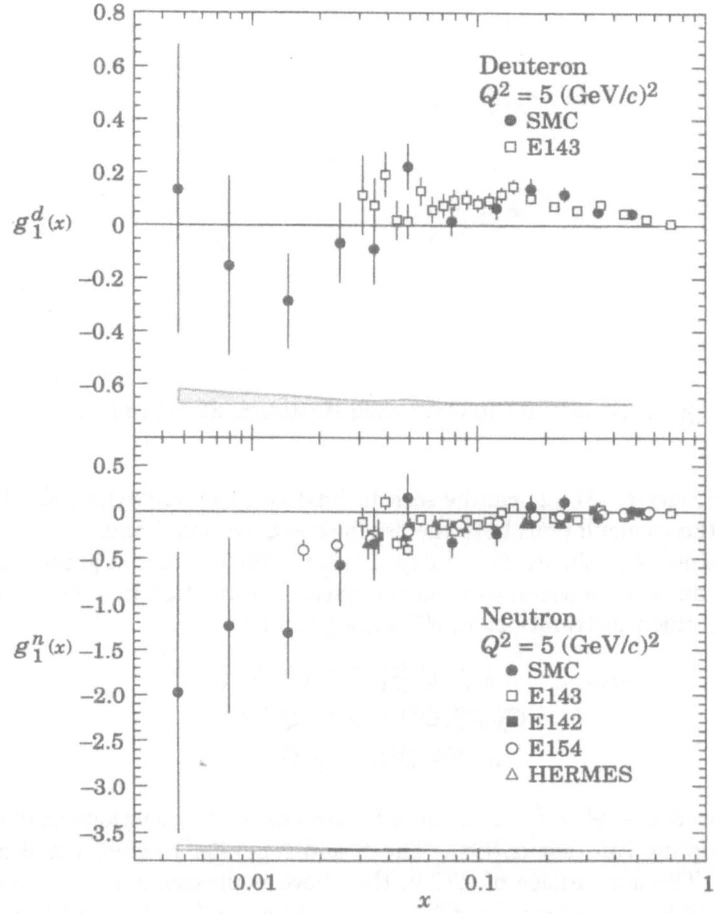

Fig. 3. g_1^d and g_1^n as measured by SLAC, CERN, and DESY.

As mentioned above the structure function is actually a function of two variables - $g_1(x, Q^2)$. This dependence is easily seen in the existing data due to the different average Q^2 of the experiments, namely:

$$\text{HERMES and SLAC E142}: < Q^2 > = \ 2.5 \text{ GeV}^2, \tag{24}$$

$$\text{SLAC E143 and E155}: < Q^2 > = \ 5 \text{ GeV}^2, \tag{25}$$

$$\text{SMC}: < Q^2 > = \ 10 \text{ GeV}^2. \tag{26}$$

This dependence is shown in Fig. 4 where the HERMES and SLAC data are presented at $Q^2 = 2$ GeV^2 and HERMES and SMC data are presented at $Q^2 = 10$ GeV^2. The evolution in Q^2 shown in this figure is calculated with perturbative QCD. The increase with Q^2 seen at lower x is due, at least in part, to the increasing influence of gluons. In fact the measured Q^2 dependence can be used to extract information on the gluon spin contribution $\Delta G(x)$.

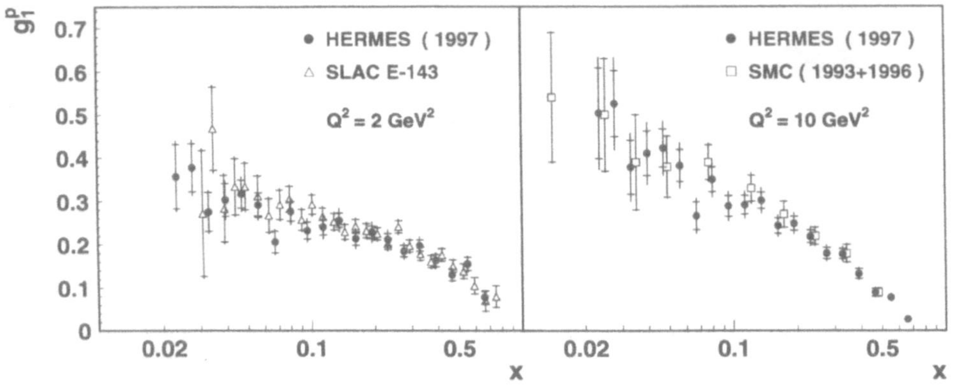

Fig. 4. g_1^p for $Q^2 = 5$ and 10 GeV^2 from HERMES, SLAC E143 and SMC.

The sensitivity to $\Delta G(x)$ can be seen in Next-to-Leading-Order (NLO) QCD calculations. In these calculations both the "radiative corrections" and the variation in Q^2 can be determined, but the explicit x dependence of the quark and gluon distributions cannot; these must be determined from the data. The relation between $g_1(x, Q^2)$ and the quark and gluon distributions in NLO are given by

$$\begin{aligned} g_1(x,Q^2) = \tfrac{1}{2}(\tfrac{2}{9}) \textstyle\int_x^1 \tfrac{dy}{y}[C_S^q(\tfrac{x}{y},Q^2)\Delta\Sigma(y,Q^2) + \\ C_{NS}^q(\tfrac{x}{y},Q^2)\Delta q_{NS}(y,Q^2) + \\ 6C^g(\tfrac{x}{y},Q^2)\Delta G(y,Q^2)] \end{aligned} \tag{27}$$

where $\Delta q_{NS} = \Delta u - \frac{\Delta d}{2} - \frac{\Delta s}{2}$, and the C^i's are coefficient functions that depend on the strong coupling constant $\alpha_s(Q^2)$. The S and NS refer to singlet and non-singlet distributions. The appearance of ΔG in the above is present only in next-to-leading order. In fact in lowest order $C_S^q = C_{NS}^q = \delta(1 - \frac{x}{y})$ and $C^g = 0$, thus to this order eq. 27 reduces to eq. 16.

The evolution in Q^2 of the quark and gluon distributions are given by the so-called DGLAP equations[23]. For the singlet quark distribution $\Delta\Sigma$ this equation is

$$\frac{d}{d\ln Q^2}\Delta\Sigma(x,Q^2) = \frac{\alpha_s(Q^2)}{2\pi}\int_x^1 \frac{dy}{y}[P_{qq}(\frac{x}{y},Q^2)\Delta\Sigma(y,Q^2) + 6P_{qg}(\frac{x}{y},Q^2)\Delta G(y,Q^2)] \tag{28}$$

In addition to the Q^2 dependence of the structure function, the x dependence also plays an important role. As is the case for unpolarized scattering, the x dependence is a basic property of the nucleon and cannot be compared with any rigorous perturbative QCD calculations as it is a result of the non-perturbative confinement. The integral of the structure function $[\int g_1(x)dx]$ which provides the information on the quark spin contribution requires understanding and extrapolating the x dependence outside of the measured x range. For the large x region this is straightforward: since $g_1(x)$ is proportional to the quark distributions it must approach zero as $x \to 1$ as this is the observed behavior of the unpolarized distributions. However the low x region is problematic, as there is no clear dependence expected. At present, simple extrapolations within the NLO QCD calculations are used to include the low x region. This leads to additional systematic uncertainties in the values of the integrals. Additional data at low x would

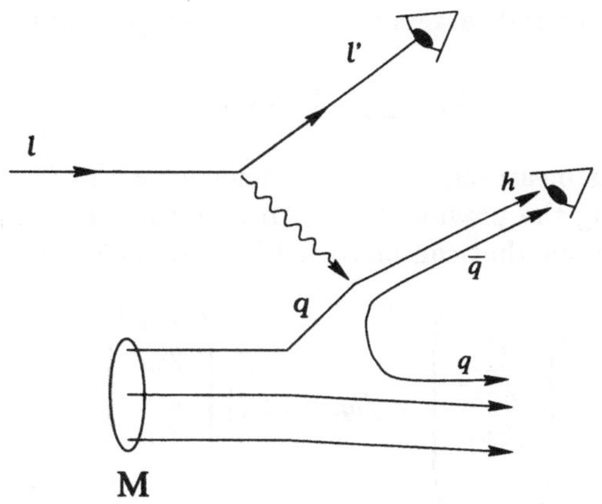

Fig. 5. Schematic diagram of semi-inclusive scattering.

certainly improve this situation. With the present data, global NLO QCD fits have been performed using the measured x and Q^2 dependence of the data. These fits[25,26,24] give the following quark and gluon spin contributions:

$$\Delta\Sigma \approx 0.3 \pm 0.1 \tag{29}$$
$$\Delta G \approx 1.2 \pm 1.0. \tag{30}$$

When combined with the assumption of SU(3) flavor symmetry and data from neutron and hyperon beta-decay the flavor decomposition of the quark contribution is

$$\Delta u \simeq 0.83 \pm 0.03 \tag{31}$$
$$\Delta d \simeq -0.42 \pm 0.03 \tag{32}$$
$$\Delta s \simeq -0.10 \pm 0.06. \tag{33}$$

Serious questions still remain about the applicability of SU(3) flavor symmetry and on the reliability of the ΔG extraction. Alternative approaches to both of these questions are being pursued in other experiments as outlined below.

Flavor Separation

Semi-inclusive asymmetries (detecting a hadron in coincidence with the scattered lepton) can be used to provide new information on the separate quark flavor contributions to the nucleon spin. In addition a separation of the valence and sea quark contributions is possible. This is illustrated in Fig. 5 where, in the quark-parton model, the struck quark "fragments" into an observed hadron.

With appropriate hadron identification, information on the struck quark can be obtained as has been established in a variety of previous unpolarized tests of the quark-parton model. Within the quark-parton model the cross section for semi-inclusive scattering can be written as a product of the probability of striking a given quark q_i and the probability of that quark ending up in a given hadron h:

$$d\sigma^h \sim \sum_i e_i^2 q_i(x) D_i^h. \tag{34}$$

With polarized beams and targets a semi-inclusive spin asymmetry can be formed with

$$A_h \sim \sum_i e_i^2 \Delta q_i(x) D_i^h. \tag{35}$$

Therefore with sufficient measurements of asymmetries with different targets and different hadrons and using the previously measured $q_i(x)$'s and D_i^h's, we have a set of constrained linear equations that can be solved for the Δq_i's, i.e.

$$\begin{pmatrix} A^p_{\pi^+} \\ A^p_{\pi^-} \\ A^n_{K^+} \\ A^n_{K^+} \\ \vdots \end{pmatrix} = f[q_i(x), D_i^h] \begin{pmatrix} \Delta u \\ \Delta d \\ \Delta s \\ \Delta \bar{u} \\ \vdots \end{pmatrix} \tag{36}$$

This technique has recently been applied to extract individual quark distributions in data from SMC [27,28] and HERMES [29] and the results are shown in Fig. 6. The HERMES experiment is continuing to collect data for these reactions and will for the first time include kaon identification which may provide direct information on $\Delta s(x)$, free from the assumption of SU(3) flavor symmetry. Future measurements of $W^\pm$ at the Relativistic Heavy Ion Collider (RHIC) [30] (under construction at Brookhaven National Laboratory) should also provide direct information on the sea quark polarization $\Delta\bar{q}(x)$. This latter topic was addressed by Joel Moss at this conference [31].

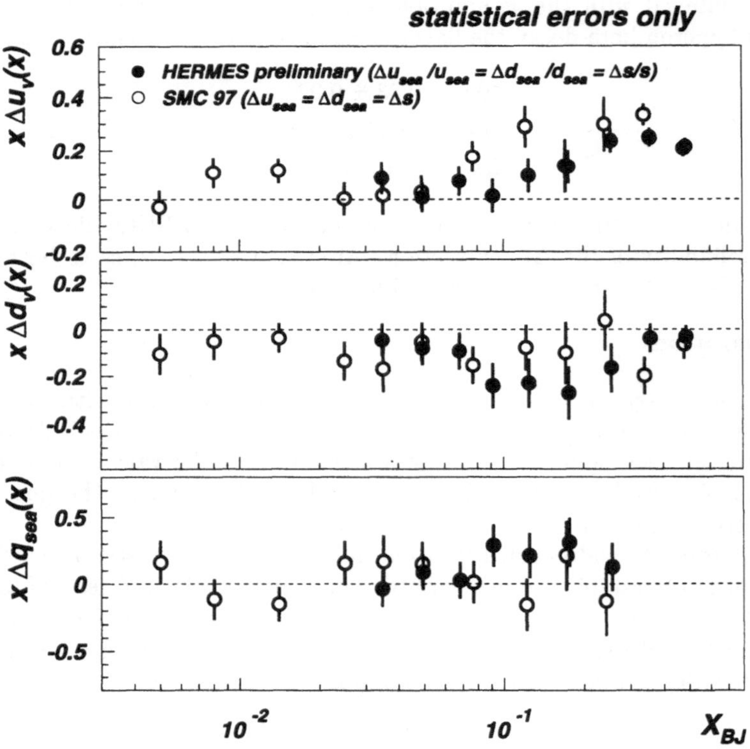

Fig. 6. Valence and sea quark distributions extracted from semi-inclusive scattering by SMC and HERMES.

Prospects for Measuring the Gluon Spin Contribution

Improved information on the gluon contribution to the nucleon spin $\Delta G(x_g)$ is highly desired. Several experimental opportunities are available for the future. Extensions of polarized lepton-nucleon scattering to much higher Q^2 may be possible with the development of polarized proton beams at the HERA collider at DESY. Polarized 28 GeV e^- colliding with polarized 900 GeV protons would allow orders-of-magnitude increases in momentum transfer allowing the Q^2 evolution of the structure function to be precisely determined. At these high Q^2 the influence of gluons should be greatly enhanced allowing significantly better sensitivity to gluons through QCD fits to the structure functions.

Another lepton-induced reaction known as photon-gluon fusion may provide direct access to ΔG. This process is depicted in Fig. 7, where one sees that the sensitivity to gluons is through the production of $q\bar{q}$ pairs. When the $q\bar{q}$ pair are heavy quarks, eg. $c\bar{c}$, there is added sensitivity to gluons. The $c\bar{c}$ can be detected through the production of open charm via D mesons. This approach will be adopted by the COMPASS collaboration at CERN [32]. In addition isolation of the $q\bar{q}$ pair through the production of high transverse momentum pions may be another approach [33] to gluon spin. This process will be studied at COMPASS and possibly at HERMES.

Hadron beams also provide promising access to the gluon spin. Polarized proton-proton collisions at RHIC are discussed in the contribution from Joel Moss to this conference [31].

Other Spin Structure

Other aspects of the nucleon's spin structure are being pursued in future experiments. While $g_1(x)$ probes the probability of quarks having polarization aligned or anti-aligned with the nucleon's spin, another structure function $g_2(x)$ probes quark spins transverse to the nucleon's spin. A number of previous experiments [34,35,36] have measured this structure function. The present data suggests a very small value for this structure function but future experiments at SLAC and HERMES will likely provide improved data on these distributions in the proton and neutron.

With semi-inclusive lepton scattering a number of other structure functions become accessible. A particularly interesting distribution that may be accessible is called

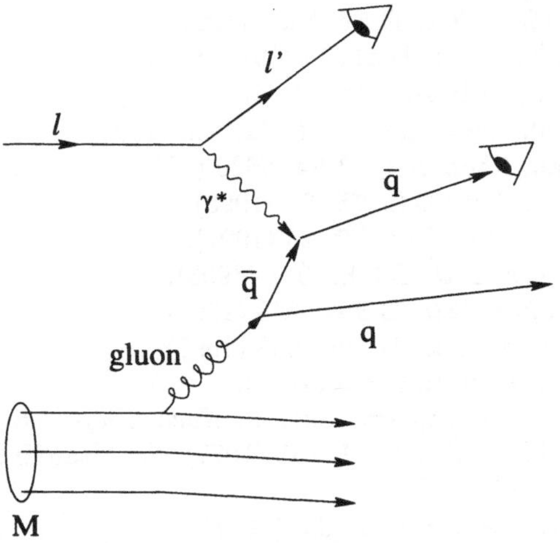

Fig. 7. Schematic diagram of the photon-gluon fusion process.

"transversity"[37,38] - $\delta q(x)$. This distribution reflects the distribution of quarks polarized perpendicular to the nucleon spin for a nucleon polarized transverse to the beam direction. In the absence of relativistic effects this distribution equals $g_1(x)$. Thus it probes directly the relativistic effects in the quark-nucleon wave-function. Future measurements at CERN and DESY may shed light on this third (after the $q(x)$ and $\Delta q(x)$ distributions) quark distribution.

SUMMARY

The next several years should see a significant increase in the amount of data. Precise semi-inclusive data is expected from HERMES. CEBAF [39] has an extensive program of measurements planned at large x and low Q^2. The new CERN proposal COMPASS [32] will extend the high Q^2 data and provide information on semi-inclusive charm production, that may shed light on the contribution of gluons to the nucleon's spin. RHIC[30] (via polarized proton-proton collisions) also has great potential to provide direct information on the gluon contribution. In addition an improved theoretical understanding may be forthcoming with further improvements in the most advanced QCD calculations.

REFERENCES

1. X. Ji, contribution to these proceedings.
2. J. Ashman et al., *Phys. Lett.* B **206**, 364 (1988).
3. J. Ashman et al., *Nucl. Phys.* B **328**, 1 (1989).
4. M. J. Alguard et al., *Phys. Rev. Lett.* **37**, 1261 (1976),*Phys. Rev. Lett.* **41**, 70 (1976).
5. G. Baum et al., *Phys. Rev. Lett.* **51**, 1135 (1983),*Phys. Rev. Lett.* **45**, 2000 (1980).
6. K. Abe et al., *Phys. Rev. Lett.* **74**, 346 (1995).
7. K. Abe et al., *Phys. Rev.* D **58**, 112003 (1998).
8. E. Hughes, private communication for the E155 collaboration.
9. D. Adams et al., *Phys. Lett.* B **329**, 399 (1994).
10. B. Adeva et al., *Phys. Lett.* B **320**, 400 (1994).
11. B. Adeva et al., *Phys. Lett.* B **412**, 414 (1997).
12. D. Adams et al., *Phys. Rev.* D **56**, 5330 (1997).
13. B. Adeva et al., *Phys. Lett.* B **58**, 112001 (1998).
14. A. Airapetian et al., PLB **442**, 484 (1998).
15. P. L. Anthony et al., *Phys. Rev. Lett.* **71**, 959 (1993).
16. P.L. Anthony et al., *Phys. Rev.* D **54**, 6620 (1996).
17. K. Abe et al., *Phys. Rev. Lett.* **75**, 25 (1995).
18. K. Abe et al., *Phys. Rev. Lett.* **79**, 26 (1997).
19. B. Adeva et al., *Phys. Lett.* B **302**, 533 (1993).
20. D. Adams et al., *Phys. Lett.* B **357**, 248 (1995).
21. D. Adams et al., *Phys. Lett.* B **396**, 338 (1997).
22. K. Ackerstaff et al., PLB **404**, 383 (1997).
23. V. N. Gribov and L. N. Lipatov, *Sov. J. Nucl. Phys.* **15**, 138 (1972), Yu. L. Dokahitzer, *Sov. Phys. JETP* **16**, 161 (1977), G. Altarelli and G. Parisi, *Nucl. Phys.* B **126**, 298 (1977).
24. K. Abe et al., *Phys. Lett.* B **405**, 180 (1997).
25. G. Altarelli, R.D. Ball, and S. Forte, *Acta Phys. Polon.* B **29**, 1145 (1998).

26. B. Adeva et al., *Phys. Lett.* B **58**, 112002 (1998).
27. B. Adeva et al., *Phys. Lett.* B **369**, 93 (1996).
28. B. Adeva et al., *Phys. Lett.* B **420**, 180 (1998).
29. C. A. Miller, Proceedings of the ICHEP98.
30. Relativistic Heavy Ion Collider (RHIC) Project (www.rhic.bnl.gov).
31. J. Moss, Contribution to these proceedings.
32. COMPASS Proposal, CERN/SPSLC 96-14 (1996).
33. A. Bravar, D. von Harrach, and A. Kotzinian, *Phys. Lett.* B **421**, 349 (1998).
34. D. Adams et al., *Phys. Lett.* B **336**, 125 (1994).
35. K. Abe et al., *Phys. Rev. Lett.* **76**, 587 (1996).
36. K. Abe et al., *Phys. Lett.* B **404**, 377 (1997).
37. J. Ralston and D. E. Soper, NPB **152**, 109 (1979).
38. R. L. Jaffe, Xuemin Jin, and Jian Tang, *Phys. Rev. Lett.* **80**, 1166 (1998).
39. CEBAF Experiments 89-042, 91-023, 93-009, 94-010, (www.cebaf.gov).

SPIN AND STATISTICS FOR QUANTUM HALL QUASI-PARTICLES

Jon Magne Leinaas

Department of Physics
University of Oslo
P. O. Box 1048 Blindern
N-0316 Oslo, Norway

ABSTRACT

In two space dimensions the possibilities of fractional spin as well as fractional statistics exist. I examine the relation between fractional spin and statistics for Laughlin quasi-particles in a two-dimensional electron system with spherical geometry. The relevance of this for quasi-particles in a planar system is discussed.

1 STATISTICS AND SPIN IN TWO DIMENSIONS

I would like to begin by reminding you of the fact that in two space dimensions there is a richer set of possibilities than in higher dimensions as far as statistics and spin of particles is concerned. Quantum statistics is determined by the symmetry of the wave function under interchange of particle coordinates, and in three and higher dimensions the corresponding symmetry group is the permutation group. However, when particle interchange is viewed as a continuous process under which the coordinates are changed, then the symmetry group in two dimensions is larger, it is the two-dimensional braid group rather than the permutation group [1]. An element of this group does not only specify the permutation of the particles, but also the windings of the particle trajectories under the interchange of the positions. In dimensions higher than two these windings can be disentangled, since only interchanges corresponding to different permutations of the particles are topologically distinct. This is not possible in two dimensions.

For particles on the plane the coordinates can be written as complex variables, $z = x + iy$, and for two particles the symmetry under interchange of the particle positions can be expressed as

$$\psi\left(e^{in\pi}(z_1 - z_2)\right) = e^{in\theta}\psi(z_1 - z_2) \ , \tag{1}$$

where only the relative coordinate has been written out explicitly. In this expression n is the winding number of the particle trajectory in 2-particle space, and θ is the parameter that specifies the statistics. The symmetry follows from the assumption that all configurations which differ only by an interchange of the particle positions are

Confluence of Cosmology, Massive Neutrinos, Elementary Particles, and Gravitation
Edited by Kursunoglu *et al.*, Kluwer Academic / Plenum Publishers, New York, 1999.

physically indistinguishable. The wave function for these configurations should therefore differ at most by a phase factor. Also for more than two (identical) particles the symmetry factors have the form $exp(in\theta)$ and they define a one-dimensional representation of the braid group for the particles. In two dimensions θ is a free parameter, while in higher dimensions it is restricted to the values $\theta = 0$ $(mod\ 2\pi)$ for bosons and $\theta = \pi$ $(mod\ 2\pi)$ for fermions. For values of θ different from these two the particles are said to satisfy intermediate or fractional statistics, and they are referred to as anyons.

Also spin is different in two dimensions. In three dimensions the intrinsic spin of a particle is associated with the rotation group $SO(3)$. It is regarded as the generator of rotations in the rest frame of the particle. As is well known, the unitary representations of the rotation group $SO(3)$ restrict the allowed values of the spin to integer or half-integer multiples of $\hbar$. For particles in two dimensions the rotation group is reduced to $SO(2)$. This is a one-parameter group with unitary representations

$$U(\phi) = e^{i\phi S/\hbar} , \tag{2}$$

where ϕ is the rotation angle. In this case there is no restriction on S, it can take any real value[1].

Thus, statistics as well as spin can be regarded as continuous variables in two dimensions. An obvious question to ask is whether these two variables are linked by some kind of spin-statistics relation. This question has previously been discussed in different ways, and we know from theoretical constructions that many simple explicit models of two-dimensional particles have such a relation. Here I will consider this question in connection with a concrete realization: quasi-particles in the fractional quantum Hall effect. These quasi-particles are believed, on one hand to be real physical realizations of anyons in a quasi two-dimensional electron system, on the other hand to be well described (in some cases) by simple many-electron wave functions. The question of spin and statistics of these quasi-particles can therefore be examined rather directly, and has been done so in the past. One specific study is due to Einarsson et. al. [2], and my talk is inspired by this paper and can be seen as a comment to their result.

2 SPIN-STATISTICS RELATIONS

Since we are considering a non-relativistic system, I would like to stress the point that we cannot expect to find a spin-statistics theorem that on general grounds gives a strict relation between these two particle properties. After all we have a simple counter-example to the standard relation between spin and statistics: spinless fermions described by one-component anti-symmetric wave functions. In the context of non-relativistic many-particle theory there seems to be no problems with such a construction, and this is so for particles in two as well as in three space dimensions. Nevertheless, as soon as one leaves the simple point particle description and makes explicit models where the spin as well as the statistics can be derived from more fundamental fields, the standard spin-statistics relation seems naturally to appear in three-dimensional systems while a linear extension of this relation appear in two dimensions. Let me just mention some examples from two dimensions.

A simple electromagnetic model of an anyon is an electric point charge e with an attached magnetic flux ϕ, that is confined to a small region around the charge.

[1] I am here actually referring to representations of the covering group of $SO(2)$, which are the relevant ones for quantum mechanics.

(The mechanism that binds the flux to the charge is not so important and neither is the detailed profile of the magnetic field surrounding the charge.) In addition to the Coulomb interaction between such charge-flux composites, there will be an Aharonov-Bohm interaction between the charge of one composite and the flux of the other. When two composites are interchanged the latter gives rise to a phase factor that can be identified with the statistics factor. A simple calculation gives for the statistics parameter

$$\theta = -\frac{e\phi}{\hbar c} . \tag{3}$$

There is an electromagnetic spin associated with a charge flux composite, due to the overlap of the electric and magnetic fields. Using the expression for electromagnetic angular momentum reduced to its two-dimensional form, we calculate the spin to be

$$S = -\frac{1}{c}\int d^2 r B \vec{r} \cdot \vec{E} = -\frac{e\phi}{2\pi c} . \tag{4}$$

We note that the statistics parameter and the spin both are determined by the same quantity $e\phi$.

A second example is provided by soliton solutions in the $O(3)$ non-linear σ-model with a topological (Hopf) term [3]. In this case the strength of the topological term determines the spin as well as the statistics of the solitons. A third example is given by the particles described by a scalar field theory with Chern-Simons coupling [4]. The Chern-Simons field gives an explicit realization of fractional statistics in the form of an Aharonov-Bohm effect. It also affects the conserved angular momentum and thereby links the spin to the statistics of the particles.

In the examples referred to above (as well as in some other examples) the relation between spin and statistics has the simple form

$$S = \left[\frac{\theta}{2\pi}\ (mod\ 1)\right] \hbar . \tag{5}$$

It coincides with the standard relation for bosons ($\theta = 0$) and fermions ($\theta = \pi/2$) and extends that linearly to all other values of the statistics parameter θ.

Even if the simple relation (5) is favoured by many anyon models, we do not have a clear specification of the general conditions under which the relation should be satisfied. There do exist, however, some general arguments for a less restrictive form of the spin-statistics relation that are based on the assumption that there exist both anyons and anti-anyons in the system under consideration. Let me briefly give the arguments for this generalized spin-statistics relation, since it is relevant for the quantum Hall quasi-particles.

We then assume that there exist fractional statistics particles of a type we denote by p (with some unspecified statistics parameter θ). There also exist another type of particles $\bar{p}$, that we consider as anti-particles to p. Since we are not considering a relativistic theory, we do not assume charge conjugation symmetry (symmetry between p and $\bar{p}$). The important point is the assumption that a $p - \bar{p}$ pair can be created and annihilated inside the system. This means that all long range effects of a single particle are canceled by the corresponding effects of an anti-particle. This has consequences for statistics as well as for spin.

For a $p - \bar{p}$ pair there are no long-range Aharonov-Bohm effects. That means that the phase factor introduced by transport of another particle of type p around the pair is the trivial factor 1 for a path far away from the two particles. If these two particles also are sufficiently far apart, the phase factor can be written as a product of one factor from each of the particles in the pair. We write this as

$$exp\left(2i\left(\theta_{pp}+\theta_{p\bar{p}}\right)\right)=1\ . \tag{6}$$

We easily see that θ_{pp} is identical to the statistics phase θ of particles p. The other phase $\theta_{p\bar{p}}$ is sometimes referred to as a mutual statistics phase. It describes an Aharonov-Bohm interaction between two non-identical particles p and $\bar{p}$. Clearly we have a similar condition when a particle of type $\bar{p}$ is transported around the pair,

$$exp\left(2i\left(\theta_{\bar{p}\bar{p}}+\theta_{\bar{p}p}\right)\right)=1\ . \tag{7}$$

The two conditions (6) and (7), and the symmetry relation $\theta_{\bar{p}p}=\theta_{p\bar{p}}$, mean that all phases can be expressed in terms of a single phase θ,

$$\begin{aligned}\theta_{\bar{p}\bar{p}}=\theta_{pp} &= \theta\ \ (mod\ \pi)\\ \theta_{\bar{p}p}=\theta_{p\bar{p}} &= -\theta\ \ (mod\ \pi)\ .\end{aligned} \tag{8}$$

A rotation of the $p-\bar{p}$ pair by an angle 2π also has to give rise to a trivial phase factor. We write this as

$$exp\left(2\pi\frac{i}{\hbar}(L_{cm}+L_{rel}+S_p+S_{\bar{p}})\right)=1\ . \tag{9}$$

The orbital angular momentum has here been divided into a center-of-mass part L_{cm} and a part determined by the relative motion, L_{rel}; S_p and $S_{\bar{p}}$ are the intrinsic spins of the two particles. L_{cm} has integer eigenvalues in multiples of $\hbar$, while the spectrum of L_{rel} is shifted due to the nontrivial phase $\theta_{p\bar{p}}$. The eigenvalues are $(n-\theta/\pi)\,\hbar, n=0,\pm1,\pm2....$ With this inserted in (9) we get

$$\frac{1}{2}(S_p+S_{\bar{p}})=\left[\frac{\theta}{2\pi}\ (mod\ \frac{1}{2})\right]\hbar\ . \tag{10}$$

This is the generalized spin statistics relation. It only involves the sum of the spins of the anyon and the anti-anyon. Even if these two spins are equal we note the relation is less restrictive than the relation (5). It does not exclude spinless fermions or bosons with half-integer spin.

3 ANYONS IN THE QUANTUM HALL SYSTEM

The quasi-particles of the quantum Hall system are charged excitations in a 2-dimensional electron gas subject to a strong perpendicular magnetic field. In general the quasi-particles are fractionally charged and obey fractional statistics; they are charged anyons in a strong magnetic field. For special filling fractions of the lowest Landau level, $\nu=1/m$, m odd, there exist simple (trial) wave functions, originally introduced by Laughlin [5], for the ground state of the many-electron system as well as for the quasi-particle excitations. Expressed in complex electron coordinates, the (non-normalized) N-electron ground state has the form

$$\psi_m(z_1,z_2,...,z_N)=\prod_{i<j}(z_i-z_j)^m e^{-\frac{1}{4\ell^2}\sum\limits_{k=1}^{N}|z_k|^2}\ , \tag{11}$$

with $\ell = 1/\sqrt{\frac{\hbar c}{eB}}$ as the magnetic length, and eB taken to be positive. The one quasi-hole state is

$$\psi_Z^{qh}(z_1, z_2, ..., z_N) = \prod_{i=1}^{N} (z_i - Z)\psi_m(z_1, z_2, ..., z_N) \ , \tag{12}$$

with Z as the position of the quasi-hole. Multi-hole wave functions are constructed in a similar way, with several prefactors of the form given in Eq.(12). For the oppositely charged quasi-electron Laughlin has suggested a wavefunction of the form

$$\psi_Z^{qe}(z_1, z_2, ..., z_N) = \prod_{i=1}^{N} (\frac{\partial}{\partial z_i} - Z^*)\psi_m(z_1, z_2, ..., z_N) \ . \tag{13}$$

Supported by general arguments, as well as numerical studies, the ground state and the quasi-hole state are believed to be very well represented by the wave functions (11) and (12) (in a homogeneous system). However there is an asymmetry between the quasi-hole and the quasi-electron, and one should note that there is not a similar strong evidence in favour for the quasi-electron wave function (13)[2].

The form of the quasi-particle wave functions determine the fractional charge as well as their fractional statistics. This was demonstrated by Arovas, Schrieffer and Wilczek who calculated the Berry phases associated with shifts of the quasi-particle coordinates along closed curves [8]. Let me give a brief comment on this in general terms.

The wave functions of configurations with M quasi-holes define a M (complex) dimensional submanifold in the N-electron Hilbert space parameterized by the quasi-hole coordinates. A fractional statistics representation (or anyon representation) [9] of the system can be introduced in terms of wave functions defined on this manifold, $\psi(Z_1, Z_2, ..., Z_M)$. The M-dimensional manifold, on which the wave-functions are defined can be interpreted as the configuration space (alternatively as the phase space) of the (classical) M quasi-hole system. In a low-energy approximation we may consider the system restricted to this space. The kinematics as well as the dynamics of the quasi-hole system are determined from the N-electron system by projection on the complex submanifold. In particular, the kinematics is determined from the geometry of the manifold, and the charge and the statistics appear as geometrically determined parameters.

The scalar product of the N-electron Hilbert space defines, by projection, a complex geometry in the M-dimensional quasi-hole space. It is expressed in terms of the Hermitian matrix

$$\eta_{kl} = \langle D_k\psi | D_l\psi\rangle \ , \tag{14}$$

with

$$D_k = \partial_k + iA_k, \ \ A_k = i\langle\psi|\partial_k\psi\rangle \ . \tag{15}$$

$|\psi\rangle$ denotes the M-quasi-hole state and ∂_k is the partial derivative with respect to a set of real coordinates in the quasi-hole space. A_k is the Berry connection defined by the set of quasi-particle states. The real (and symmetric) part of η_{kl} determines a metric on the M quasi-particle space

$$g_{kl} = Re\langle D_k\psi | D_l\psi\rangle \ , \tag{16}$$

[2]For a recent discussion see Ref. [7].

while the imaginary (and anti-symmetric) part determines a symplectic form, that we identify as the "Berry magnetic field",

$$b_{kl} = 2Im\langle\partial_k\psi|\partial_l\psi\rangle = \partial_k A_l - \partial_l A_k \ . \tag{17}$$

For a single quasi-hole the form of η_{kl} is strongly restricted by translational and rotational invariance (in the limit $N \to \infty$) and by analyticity in the variable Z ,

$$\eta_{kl} = -\frac{b_1}{2}(\delta_{kl} + i\epsilon_{kl}) \, . \tag{18}$$

Here b_1 is a constant that can be expressed in terms of the the real magnetic field, $b_1 = \frac{e^* B}{\hbar c}$, with the coefficient e^* as the effective charge of the quasi-hole. A Berry phase calculation for a loop in the plane determines the flux of b_1 through this loop, and comparison with the real magnetic flux then gives the effective charge e^* [8].

For a two quasi-hole state an expression similar to (18) is valid for η_{kl}, if this now refers to the relative coordinate of the two quasi-holes. In this case b_1 is replaced by a function $b_2(R)$ that depends on the relative distance R. For small R the form of this function is determined by local properties of the quasi-holes. For large R, $b_2(R)$ is expected to approach rapidly the constant $\frac{1}{2}b_1$ when the quasi-holes are well localized objects. The flux of b_2 then has the form

$$\int_{r<R} d^2r \, b_2(R) = \frac{1}{2}\pi R^2 \, b_1 - 2\theta \ , \tag{19}$$

where θ is identified as the statistics parameter of the quasi-holes. Again this parameter can be determined by a Berry phase calculation, that measures the flux of $b_2(R)$ within a given radius.

Berry phase calculations based on the quasi-hole wave function (12) gives $e^* = -e/m$ for the charge and $\theta = -\pi/m$ for the statistics parameter, with e as the electron charge [8]. For the quasi-electron wave function (13) one cannot derive the results so easily [6], but the expected results for the physical quasi-electron is $e^* = e/m$ and $\theta = 2 - \pi/m$, as determined from general reasoning and numerical studies [10].

Whereas charge and statistics can be determined geometrically, in terms of Berry phases associated with closed curves of one and two quasi-particles, the spin cannot be determined quite as easily. However, as pointed out by Einarsson [11] and Li [12] there is a way to derive spin from Berry phases, provided the particles move in a curved space. If the spin can be viewed as a three-dimensional spin constrained to point in the direction orthogonal to the two-dimensional surface, there will be a contribution to the Berry phase when transporting the quasi-particle around a loop that is proportional to the product of the spin value and the solid angle traced out by the spin [13]. This suggests the following form of the Berry magnetic field

$$b_1 = \frac{e^* B}{\hbar c} - \frac{S}{\hbar}\kappa \ , \tag{20}$$

with κ as the Gauss curvature and the coefficient S as the spin. It is not obvious that calculations of Berry phases for quasi-holes will give a separation in two terms of this form, but if they do, the spin can be determined from the Berry phases. This is the assumption made in [2]. In this case a quantum Hall system with the geometry of a sphere is considered. One should note that in this case the magnetic field B as well as the curvature κ are constants. That means that there is no clear distinction between the two contributions to the Berry phase in Eq.(20). However if the charge e^* of the quasi-particle on the sphere is the same as the quasi-particle charge on the plane (which seems reasonable), then the second term can be separated from the first one and the spin can be determined.

4 QUANTUM HALL STATES ON THE SPHERE

In practice, to create a quantum Hall system with the geometry of a sphere can hardly be done. A radially directed magnetic field is then needed, and this means that a magnetic monopole should be found and placed at the center of the sphere. However as a theoretical construction a spherical Hall system can easily be created, and as first shown by Haldane such a geometry may conveniently be used in the study of certain aspects of the quantum Hall effect [14]. Also for numerical calculations it is convenient due to the lack of boundaries [10].

To have a consistent quantum description of the electrons in the monopole field, Dirac's quantization condition has to be satisfied,

$$\frac{e\phi}{4\pi\hbar c} = \frac{1}{2}N_\phi \ , \tag{21}$$

where ϕ is the total flux of the monopole field and N_ϕ is an integer. This means that the total magnetic flux through the sphere is quantized in units of the flux quantum $\phi_0 = \frac{hc}{e}$,

$$\phi = N_\phi \, \phi_0 \ , \tag{22}$$

with N_ϕ as the number of flux quanta.

Laughlin states like (11),(12) and (13) can be constructed on the sphere and can conveniently be expressed in terms of the coordinates $u = \cos(\theta/2)$ and $v = \sin(\theta/2)exp(i\phi)$, with θ and ϕ as the polar coordinates on the sphere. The form of the ground state is (in the Dirac gauge $e\vec{A} = eB \tan\frac{\theta}{2}\,\vec{e}_\phi$)

$$\psi_m = \prod_{i<j}(u_i v_j - u_j v_i)^m \ , \quad N_\phi = m(N-1) \ , \tag{23}$$

and this is non-degenerate, with all particles in the lowest Landau level, provided the number of electrons N is linked to the number of flux quanta N_ϕ as indicated above. If one flux quantum is added, a hole state is created,

$$\psi^{qh}_{UV} = \prod_i (Vu_i - Uv_i)\psi_m \ , \quad N_\phi = m(N-1)+1 \ , \tag{24}$$

with (U,V) as the quasi-hole coordinates, and if one flux quantum is removed, a quasi-electron state is created,

$$\psi^{qe}_{UV} = \prod_i (V^*\frac{\partial}{\partial u_i} - U^*\frac{\partial}{\partial v_i})\psi_m \ , \quad N_\phi = m(N-1)-1 \ , \tag{25}$$

now with (U,V) as the quasi-electron coordinates.

For the quasi-hole state a detailed calculation of the Berry phase has been performed in Ref. [2], with a discussion of the different contributions. I will not repeat that here, let me rather show how the result concerning the spin can be derived directly from rotational invariance, without reference to Berry phases. This derivation is based on the assumption that the quasi-particle can be represented as a particle with charge e^* in the monopole field.

For a single electron moving in a magnetic monopole field, the conserved angular momentum has the form

$$\vec{J} = \vec{r}\times\vec{\pi} + \mu\hat{\vec{r}} \ , \tag{26}$$

with $\vec{\pi}$ as the mechanical momentum,

$$\vec{\pi} = \vec{p} - \frac{e}{c}\vec{A} \,, \tag{27}$$

and

$$\mu = -\frac{e\phi}{4\pi c} \tag{28}$$

as the component of the total angular momentum in the radial direction $\hat{\vec{r}}$. This spin can be identified as the electromagnetic angular momentum due to the overlap of the electric field of the charge with the magnetic monopole field. This radially directed spin is quantized due to the Dirac condition,

$$\mu = \frac{1}{2} N_\phi \hbar \,, \tag{29}$$

and this quantization condition can alternatively be derived directly from the requirement of rotational invariance, *i.e.* from the condition that the operator $\vec{J}$ should generate unitary representations of the rotation group.

Thus, there are two invariants associated with the angular momentum,

$$\vec{J}^2 = j(j+1)\hbar^2 \,, \quad \hat{\vec{r}} \cdot \vec{J} = \mu \,, \tag{30}$$

with the restriction

$$j = |\mu|, |\mu| + 1, \ldots \,. \tag{31}$$

The smallest value of j can be identified as corresponding to the lowest Landau level, and as on the plane, the mechanical part of the angular momentum then has its smallest value. For N electrons the total angular momentum is the sum of the contributions from each electron,

$$\vec{J} = \sum_{i=1}^{N} \vec{J}_i \,. \tag{32}$$

The ground state (23) is rotationally symmetric, with $j = 0$, while the spin of the quasi-hole state (24) is $j = N/2$.

In the anyon representation the quasi-hole is represented as a (single) charged particle in the monopole field. If we assume that it can be treated as a point particle, the angular momentum has the same form as for a single electron,

$$\vec{J} = \vec{r} \times \vec{\pi} + (\mu^* + S)\hat{\vec{r}} \,. \tag{33}$$

In this expression $\vec{r}$ is the quasi-hole coordinate and $\mu^* = -\frac{e^*\phi}{4\pi c}$ is the radially directed electromagnetic spin. S is a possible additional radially directed spin, an intrinsic spin of the quasi-hole. We note that such an additional spin in fact has to be added in order to preserve rotational invariance. If e^* is taken to be identical to the charge e/m of a quasi-hole in a planar system, then $\mu^* = N_\phi/2m$. This is in general not a half-integer, and the condition for rotational invariance is therefore not satisfied with $S = 0$. The value of S can be determined if we identify the anyon coordinates with the coordinates (U, V) of the quasi-hole state (24). The spin component of this state in the (U, V) direction is $N/2$, and this gives the relation

$$\frac{1}{2m} N_\phi + S = \frac{1}{2} N \,. \tag{34}$$

With the number of flux quanta related to the electron number as indicated in Eq.(24) this gives the spin value

$$S_{qh} = \frac{1}{2} - \frac{1}{2m} = \frac{1}{2} + \frac{\theta}{2\pi} \,. \tag{35}$$

where qh now labels the spin of the quasi-hole. This result for the spin is the same as the one determined by Berry phase calculations [12, 2]. We note that the spin-statistics relation given by (35) is not identical to the relation (5) indicated by the anyon models referred to at an earlier stage. There is an additional term 1/2 that looks like a shift between the boson and fermion value of θ. However, one should also note that the contribution from the intrinsic spin of the electrons has not been included here. For fully polarized electrons in the plane this contribution is $-1/2m$. For large electron numbers, this contribution is presumably the same on the sphere. Thus, with all contributions included we get $S_{qh} = \frac{1}{2} - \frac{1}{m} = \frac{1}{2} + \frac{\theta}{\pi}$, and we still do not recover the relation (5). The only exception is for $m = 1$, the case of a fully occupied lowest Landau level. The spin is then $-1/2$, in accordance with the standard spin-statistics relation.

The quasi-electron state (25) can be examined in a similar way. The spin component in the radial direction in this case has the opposite sign and there is also a change in the relation between the number of flux quanta and the electron number. The spin value now is

$$S_{qe} = \frac{1}{2} - \frac{1}{2m} = -\frac{1}{2} + \frac{\theta}{2\pi} \,. \tag{36}$$

The contribution from the intrinsic spin of the electrons in this case is $1/2m$, which gives the total spin $S_{qe} = \frac{1}{2}$. Also here the original spin-statistics relation is not satisfied.

However, the two expressions (35) and (36) show that the generalized spin-statistics relation is satisfied in the form

$$\frac{1}{2}(S_{qh} + S_{qe}) = \frac{\theta}{2\pi} \,. \tag{37}$$

That is the case also when the contribution from the intrinsic spin of the electrons are included, since the contribution to the quasi-electron spin is the same, but with opposite sign as the contribution to the quasi-hole spin.

5 SPIN ON THE SPHERE–SPIN ON THE PLANE

The spin values (35) and (36) are determined for quasi-particles on a sphere. What conclusion can we now draw concerning quasi-particles in a planar system? Is there a local spin associated with the quasi-particles with value identical to the one found on a sphere? The discussion we find in Ref. [2], and also the results found in a paper by Sondhi and Kivelson [15], do not support this conclusion[3]. Thus, if their conclusions are correct, there is no simple relation between the spin of the quasi-particle on the sphere and a spin derived from the angular momentum of the electrons in a planar system. This is somewhat disappointing since the main motivation for putting the quasi-particles on the sphere, I assume, was to be able to visualize the quasi-particle spin, not to create the spin. The usual picture of the quasi-particle excitations is that

[3]Somewhat surprisingly this is not seen as a problem in [2], with the explanation that the spin in the planar system does not have a dynamical significance.

they are strongly localized in space and that they have particle like properties with sharply defined quantum numbers such as charge, mass and possibly spin. If the quasi-particle spin determined on the sphere is not the same as the quasi-particle spin on the plane, that presumably means that it cannot be thought of as a *local* spin associated with the quasi-particle. The spin could in principle be due to a small renormalization of the charge of the quasi-particle when put on a sphere,

$$e^*_{sphere} = e^* \left(1 + \frac{m-1}{mN}\right) , \tag{38}$$

However, the N dependence of the correction term does not seem to fit the picture of the quasi-particle as a strongly localized object.

Let me briefly discuss the question of the quasi-particle spin for a planar system. The normal component of the conserved angular momentum of an electron in a homogeneous magnetic field is

$$J = (\vec{r} \times \vec{\pi})_z + \frac{eB}{2} r^2 , \tag{39}$$

with $\vec{r}$ as a vector in the (x, y)-plane. The first term is the mechanical angular momentum of the circulating electron, whereas the second term can be interpreted as the electromagnetic spin (with an infinite $\vec{r}$-independent term subtracted).

For electrons in the lowest Landau level, the conserved angular momentum can be written in the form

$$J = \hbar \left(- \int d^2 r \rho(r) + \frac{1}{2\ell^2} \int d^2 r r^2 \rho(r) \right) , \tag{40}$$

with ρ as the particle density. The first term, the mechanical angular momentum is proportional to the particle number, since all electrons in the lowest Landau level carry one unit of (mechanical) angular momentum. The second term is the contribution from the electromagnetic angular momentum. It has the opposite sign of the first term and dominates this so that for all angular momentum eigenvalues the spin is non-negative.

The total angular momentum (40) diverges with the size of the system, the first term as the electron number N and the second term as N^2. This is so for the ground state (11) as well as for the quasi-particle states (12) and (13). Clearly, if a local, finite spin should be associated with the quasi-particle, one has in some way to subtract the angular momentum of the ground state. A simple definition of the quasi-particle spin would be

$$S_{qp} = \lim_{R \to \infty} (J_{qp}(R) - J_0(R)) . \tag{41}$$

where $J_{qp}(R)$ is the total angular momentum of the quasi-particle state within a radius R and $J_0(R)$ the angular momentum of the ground state within the same radius. The size of the electron system is here regarded as infinite. Even if these two terms diverge separately for large R, the difference should stay finite and give a well-defined value for the spin.

The first term of the angular momentum (40) gives a contribution to the quasi-particle spin (after the subtraction of the ground state spin) which is determined by the charge of the quasi-particle. The contribution is $\pm 1/m$ with + for the quasi-hole and - for the quasi-electron. The second term is not so easy to determine as the first term, but in the paper by Sondhi and Kivelson [15] (where a similar definition of the quasi-particle spin is used), there is a discussion of the quasi-hole case. In this case the plasma analogy, introduced by Laughlin, can be applied. In the plasma analogy the square modulus of the quasi-hole wave function (12) is interpreted as the partition function of a Coulomb system consisting of N free (unit) charges in a homogeneous neutralizing background, with the presence of an additional fixed charge of value $1/m$ (the quasi-hole). The integrated particle number is then determined as the screening charge of this fixed charge, with the value $-1/m$. Also the second moment of the particle number density, which is relevant for the second term of the angular momentum, can be related to the value of the charge. In fact, assuming that the conditions for "perfect screening" to be satisfied [9], there is a cancellation between the two terms of the the angular momentum so that the quasi-hole spin, as defined above, vanishes. This is the conclusion of Sondhi and Kivelson[4]. With this conclusion it it difficult to see any connection between the physical spin of the quasi-hole state in the plane and the spin determined on the sphere. If the physical spin vanishes for any value of m this in fact rules out any connection between the (physical) spin and the statistics parameter of the quasi-particles.

However, as a final point I would like to pose the question whether the conclusion concerning the spin, which is based on the use of the plasma analogy, is necessarily true, or whether another conclusion may be possible. Clearly, for a full Landau level, with $m = 1$, the quasi-hole spin vanishes since the hole is created simply by removing an electron in a spin 0 state. For $m = 3$ the situation is not quite as obvious and one has to refer to the situation in a one-component plasma with a $1/3$ charge screened by a plasma of integer charges. I am not able to judge the claim that the perfect screening condition is satisfied in this case, but I have noted with interest that in Ref. [16] one refers to a "basic belief" in the underlying assumption when the perfect screening sum rule is derived.

There is of course a way to avoid the reference to the plasma analogy. That is to make a straight forward calculation of the spin (41) of the planar system, and I will cite some preliminary results for Monte-Carlo calculations performed by Heidi Kjønsberg for an electron system consisting of $N = 100$ electrons. The numerical calculations reproduce values for the integrated quasi-hole spin S_{qh}, within a variable radius R around the quasi-hole, which is placed at the center of the circular electron system defined by the Laughlin wave function.

Let me first give some values for the spin evaluated on the sphere, as given by Eq. (35). For $m = 1$ the spin is 0, for $m = 3$ the spin is $1/3$ and for $m = 5$ the spin is $2/5$, all spins expressed in units of $\hbar$.

The numerical results for the the planar system agree well with the value 0 for the $m = 1$ state. However, for $m = 3$ this is not the case. For values of the radius R that lie between the size of the quasi-hole and the size of the full electron system, the results indicate instead a fairly stable value close to $1/3$, that agrees with the value found on the sphere. For $m = 5$ the results are not so clear, due to larger finite size effects and also due to larger statistical fluctuations in the Monte Carlo calculations. Nevertheless, also here the results indicate a spin value different from 0 and possibly consistent with the value $2/5$.

So I would like to finish by referring to the question of the spin of the quasi-hole as an interesting one which deserves a further study. I feel that the situation in a sense would be more satisfying if the spin evaluated on the sphere could be identified as the physical spin of the quasi-particle also for a planar quantum Hall system. But such a

[4]Sondhi and Kivelson also consider corrections to the spin due to the electromagnetic self-interaction of the quasi-hole. Such corrections are important in order to give the correct value of the spin for the physical quasi-hole, but have not been taken into consideration here.

conclusion would raise some new and interesting questions concerning the use of the plasma sum rules for the Laughlin states.

ACKNOWLEDGMENTS

I appreciate the help of Heidi Kjønsberg who have performed the numerical calculations referred to in this paper. I am grateful to Hans Hansson and Anders Karlhede for several helpful comments and would like to thank their group at the Department of Physics, Stockholm University, for hospitality during a stay in February 1999.

REFERENCES

[1] J.M. Leinaas and J. Myrheim, *On the Theory of Identical Particles,* Nuovo Cimento **37B** (1977) 1.

[2] T. Einarsson, S.L. Sondhi, S.M. Girvin and D.P. Arovas, *Fractional Spin for Quantum Hall Effect Quasiparicles,* Nucl. Phys.B **441** (1995) 515.

[3] F. Wilczek and A. Zee, *Linking Numbers, Spin and Statistics of Solitons,* Phys. Rev. Lett. **51** (1983) 2250.

[4] T.H. Hansson, M. Rocek and I. Zahed, *Spin and statistics in massive (2+1)-dimensional QED*, Phys. Lett. B **214** (1988) 475.

[5] R. B. Laughlin, *Anomalous Quantum Hall Effect: An Incompressible Quantum Fluid with Fractionally Charged Excitations*, Phys. Rev. Lett. **50** (1983) 1395.

[6] H. Kjønsberg and J. M. Leinaas, *On the anyon description of the Laughlin hole states*, Int. Jour. Mod. Phys. A **12** (1997) 1975.

[7] H. Kjønsberg and J. M. Leinaas, *Charge and Statistics of Quantum Hall Quasi-Particles: A numerical study of mean values and fluctuations*, preprint Oslo TP 1-99, cond-mat/9901266.

[8] D. Arovas, J. R. Schriffer and F. Wilczek, *Fractional Statistics and the Quantum Hall Effect*, Phys. Rev. Lett. **53** (1984) 722.

[9] R. B. Laughlin, *Elementary Theory: the Incompressible Quantum Fluid*, in *The Quantum Hall Effect*, eds. R. E. Prange and S. M. Girvin, Springer-Verlag, 1990.

[10] M. D. Johnson and G. S. Canright, *Haldane fractional statistics in the fractional quantum Hall effect*, Phys. Rev. B **49** (1994) 2947.

[11] T. Einarsson, *Fractional statistics on compact surfaces*, Mod. Phys. Lett B **5** (1991) 675.

[12] D.P. Li, *The spin of the quasi-particle in the fractional quantum Hall effect*, Phys. Lett. A **169** (1992) 82.

[13] M. B. Berry, *Quantal phase factors accompanying adiabatic changes*, Proc. R. Soc. Lond. A. **392** (1984) 45.

[14] F. D. M. Haldane, *Fractional Quantization of the Hall Effect: A Hierarchy of Incompressible Quantum Fluid States*, Phys. Rev. Lett. **51** (1983) 605.

[15] S.L. Sondhi and S.A. Kivelson, *Long-range interactions and the quantum Hall effect*, Phys. Rev. B. **46** (1992) 13391.

[16] M. Baus and J.-P. Hansen, *Statistical mechanics of simple Coulomb systems*, Phys. Rep. **59** (1980) 1.

ON THE CORRESPONDENCE BETWEEN STRONGLY COUPLED QED_2 AND QCD_2 WITH ANTIFERROMAGNETIC SPIN CHAINS*

F. Berruto, G. Grignani and P. Sodano

Dipartimento di Fisica and Sezione I.N.F.N.
Universita' di Perugia
Via A. Pascoli
I-06123 Perugia, Italy

ABSTRACT

We analyse the correspondence between generalized Heisenberg chains and certain strongly coupled lattice gauge models. We construct the effective Hamiltonians of the strongly coupled lattice multiflavor Schwinger and 't Hooft models and show their equivalence to a suitable $SU(\mathcal{N})$ generalization of the quantum antiferromagnetic Heisenberg model.

INTRODUCTION

One of the analytical approaches to gauge theories with confining spectra is the strong coupling expansion. In the strong coupling limit, confinement is explicit, the confining string is a stable object [1] and some other qualitative features of the spectrum are easily obtained. Since the strong coupling expansion requires a gauge invariant ultraviolet cutoff, it is most conveniently implemented using a lattice regularization. It is well known that many choices of strong coupling theory produce identical continuum physics; in spite of this difficulty, there are strong coupling computations which claim some degree of success [2, 3, 4].

The strong coupling limit of lattice gauge theories with dynamical fermions is related to certain quantum spin systems. There are several similarities between condensed matter systems with lattice fermions and lattice gauge theory systems with staggered fermions, particularly in their strong coupling limit. For example, it is well known that the quantum spin 1/2 Heisenberg antiferromagnet is equivalent to the strong coupling limit of either $U(1)$ or $SU(2)$ lattice gauge theory [5]. For a gauge group $U(\mathcal{N}_c)$ one has a spin-S Heisenberg antiferromagnet with $S = \mathcal{N}_c/2$.

The idea that quantum antiferromagnetic spin chains are related to quantized gauge theories is very appealing. First of all, chiral symmetry breaking in the gauge

⋆ Research supported in part by the I.N.F.N. and M.U.R.S.T.

theory – which, on the lattice, is due to the reduction of the discrete translation symmetry from translation by one site to translation by two sites – corresponds to the Neel ordering of the quantum antiferromagnet [6,7]. Furthermore, at least for some one dimensional gauge theories such as the multiflavor Schwinger models and the 't Hooft model [8], it is possible to compute explicitly the strong coupling spectrum in terms of pertinent excitations of the quantum Heisenberg chain [4].

In the following we analyse some aspects of the correspondence between the Abelian and non-Abelian two-dimensional lattice gauge theories and the antiferromagnetic spin chains. We shall see that [4], for what concerns the spectrum of the gauge theories, the massless mesons can be identified with the spinon excitations [9] of the quantum antiferromagnet, but the massive ones are generated by applying the pertinent fermionic currents on the ground state of the spin chain; the static baryons are generated from the ground state of the spin chain by applying suitable color singlet operators. Our analysis shall also evidence that, except for the one-flavor Schwinger model [4], the chiral symmetry is explicitly broken by staggered fermions and the non-zero vacuum expectation value of a fermion condensate is the only relic on the lattice of the chiral anomaly in the continuum [4].

The Hamiltonian of the antiferromagnetic Heisenberg spin-S chain is

$$H_J = J\sum_{x=1}^{N}(\vec{S}_x \cdot \vec{S}_{x+1} - S^2) \quad . \tag{1}$$

where $J > 0$. Spin operators $\vec{S}_x$ act nontrivially only on the Hilbert space of the xth site. With periodic boundary conditions (1) is invariant under global rotations in the spin space and under translations by one lattice site. Only for $S = 1/2$ the Hamiltonian (1) is completely integrable and the complete spectrum has been derived using the algebraic Bethe ansatz method [9].

There are interesting $U(\mathcal{N})$ generalizations [10] of the spin-1/2 antiferromagnetic Heisenberg chains; for these models the "spins" are the generators of a unitary group. For example an $U(\mathcal{N})$ spin-1/2 quantum antiferromagnetic chain is described by the Hamiltonian

$$H_J^{U(\mathcal{N})} = J\sum_{x=1}^{N} \rho(x)\rho(x+1) + H_J^{SU(\mathcal{N})} \tag{2}$$

where

$$H_J^{SU(\mathcal{N})} = J\sum_{x=1}^{N} S_x^{\alpha} S_{x+1}^{\alpha} \tag{3}$$

is the Hamiltonian of an $SU(\mathcal{N})$ quantum antiferromagnet. In the Eqs.(2) and (3) $S_x^{\alpha} = \psi_{ax}^{\dagger} T_{ab}^{\alpha} \psi_{bx}$ with T^{α} forming a basis of the Lie algebra of $SU(\mathcal{N})$ in the fundamental representation; $\rho(x) = \sum_{a=1}^{\mathcal{N}} \psi_{ax}^{\dagger}\psi_{ax} - \mathcal{N}/2$.

In the following two sections we analyse the correspondence between certain generalized Heisenberg chains and the strongly coupled multiflavor Schwinger and 't Hooft models on the lattice. There we shall show that the effective Hamiltonian of the strongly coupled lattice gauge model is in fact the one describing a generalized Heisenberg model.

THE MULTIFLAVOR SCHWINGER MODELS

The $\mathcal{N}$-flavor Schwinger models have many features in common with four dimensional QCD: at the classical level they have a symmetry group

$$U_L(\mathcal{N}) \otimes U_R(\mathcal{N}) = SU_L(\mathcal{N}) \otimes SU_R(\mathcal{N}) \otimes U_V(1) \otimes U_A(1) \tag{4}$$

that is broken down to $SU_L(\mathcal{N}) \otimes SU_R(\mathcal{N}) \otimes U_V(1)$ by the axial anomaly exactly like in QCD. The massless $\mathcal{N}$-flavor Schwinger models describe no real interactions between their particles as one can infer by writing the model action in a bosonized form. The model exhibits one massive and $\mathcal{N}^2 - 1$ massless pseudoscalar "mesons".

The continuum $SU(\mathcal{N})$-flavor Schwinger models are defined by the action

$$S = \int d^2x \big(\sum_{a=1}^{\mathcal{N}} \overline{\psi}_a (i\gamma_\mu \partial^\mu + \gamma_\mu A^\mu)\psi_a - \frac{1}{4e_c^2} F_{\mu\nu}F^{\mu\nu}\big) \tag{5}$$

where the $\mathcal{N}$ fermions have been introduced in a completely symmetric way. The Dirac fields are an $\mathcal{N}$-plet, *i.e.* transform according to the fundamental representation of the flavor group while the electromagnetic field is an $SU(\mathcal{N})$ singlet. The flavor symmetry of the theory cannot be spontaneously broken due to the Coleman theorem [12]. The particles of the theory belong to $SU(\mathcal{N})$ multiplets. The action is invariant under the symmetry (4).

At the classical level the above symmetry leads to conservation laws for the isovector, vector and axial currents, while at the quantum level the vector and axial currents cannot be simultaneously conserved due to the anomaly phenomenon [11]. With a non-Abelian bosonization [13] it is manifest the relationship between the isovector currents and the bosonic excitations [14].

The Hamiltonian, gauge constraint and non-vanishing (anti-)commutators of the continuum $\mathcal{N}$-flavor Schwinger models are

$$H = \int dx \big[\frac{e^2}{2}E^2(x) + \sum_{a=1}^{\mathcal{N}} \psi_a^\dagger(x)\alpha\,(i\partial_x + eA(x))\,\psi_a(x)\big] \tag{6}$$

$$\partial_x E(x) \; + \sum_{a=1}^{\mathcal{N}} \psi_a^\dagger(x)\psi_a(x) \sim 0 \tag{7}$$

$$[A(x), E(y)] = i\delta(x-y)\;,\;\left\{\psi_a(x), \psi_b^\dagger(y)\right\} = \delta_{ab}\delta(x-y) \quad . \tag{8}$$

On the lattice the Hamiltonian, constraint and (anti-) commutators reducing to (6),(7),(8) in the naive continuum limit are given by

$$H_S = \frac{e_L^2 a}{2}\sum_{x=1}^{N} E_x^2 - \frac{it}{2a}\sum_{x=1}^{N}\sum_{a=1}^{\mathcal{N}} (\psi_{a,x+1}^\dagger e^{iA_x}\psi_{a,x} - \psi_{a,x}^\dagger e^{-iA_x}\psi_{a,x+1}) \tag{9}$$

$$E_x - E_{x-1} + \sum_{a=1}^{\mathcal{N}} \psi_{a,x}^\dagger \psi_{a,x} - \frac{\mathcal{N}}{2} \sim 0 \qquad (10),$$

$$[A_x, E_y] = i\delta_{x,y}\;,\;\left\{\psi_{a,x}, \psi_{b,y}^\dagger\right\} = \delta_{ab}\delta_{xy} \quad .$$

The fermion fields are defined on the sites, $x = 1, \ldots, N$, gauge and the electric fields,

A_x and E_x, on the links $[x; x+1]$, N is an even integer and, when N is finite it is convenient to impose periodic boundary conditions. When N is finite, the continuum limit is the $\mathcal{N}$-flavor Schwinger model on a circle. The coefficient t of the hopping term in (9) plays the role of the lattice light speed. In the naive continuum limit, $e_L = e_c$ and $t = 1$.

The lattice $\mathcal{N}$-flavor Schwinger models are equivalent to a one dimensional quantum Coulomb gas on the lattice with $\mathcal{N}$ kinds of particles. To see this, one can fix the gauge, $A_x = A$ (Coulomb gauge) and in the thermodynamic limit the Schwinger Hamiltonian, rescaled by the factor $e_L^2 a/2$, reads as

$$H = H_0 + \epsilon H_h \tag{11}$$

with

$$H_0 = \sum_{x>y} [\frac{(x-y)^2}{N} - (x-y)]\rho(x)\rho(y) \quad , \tag{12}$$

$$H_h = -i(R - L) \tag{13}$$

and $\epsilon = t/e_L^2 a^2$. In Eq.(13) the right R and left L hopping operators are defined ($L = R^\dagger$) as

$$R = \sum_{x=1}^{N} R_x = \sum_{x=1}^{N}\sum_{a=1}^{\mathcal{N}} R_x^{(a)} = \sum_{x=1}^{N}\sum_{a=1}^{\mathcal{N}} \psi_{a,x+1}^\dagger e^{iA}\psi_{a,x} \quad .$$

On a periodic chain the commutation relation

$$[R, L] = 0$$

is satisfied.

When $\mathcal{N}$ is even the ground state of the Hamiltonian (12) is the state $|g.s. >$ with $\rho(x) = 0$ on every site, *i.e.* with every site half-filled

$$\sum_{a=1}^{\mathcal{N}} \psi_{ax}^\dagger \psi_{ax} |g.s. >= \frac{\mathcal{N}}{2} |g.s. > \quad .$$

It is easy to see that $\rho(x)$ is equal to zero on every site in the ground state by observing that the Coulomb Hamiltonian (12) is a non-negative operator and that the states with zero charge density are zero eigenvalues of (12). $|g.s. >$ is an highly degenerate state; in fact, at each site x the quantum configuration is

$$\prod_{a=1}^{\frac{\mathcal{N}}{2}} \psi_{ax}^\dagger |0> \quad . \tag{14}$$

The state (14) is antisymmetric in the indices $a = 1, \ldots, \frac{\mathcal{N}}{2}$; *i.e.* it takes on any orientation of the vector in the representation of the flavor symmetry group $SU(\mathcal{N})$ with Young tableau of $\frac{\mathcal{N}}{2}$ rows. The energy of $|g.s. >$ is of order 1, since it is non-zero only at the second order in the strong coupling expansion.

When $\mathcal{N}$ is odd the ground states of the Hamiltonian (12) are characterized by the staggered charge distribution

$$\rho(x) = \pm\frac{1}{2}(-1)^x \tag{15}$$

since (15) minimizes the Coulomb Hamiltonian (12); one has $\rho(x) = +1/2$ on the even sublattice and $\rho(x) = -1/2$ on the odd sublattce or viceversa. The electric fields generated by the charge distribution (15) are

$$E_x = \pm\frac{1}{4}(-1)^x \quad . \tag{16}$$

Since

$$H_0|g.s.> = \frac{1}{16}|g.s.> \quad ,$$

the ground state energy is of order e_L^2. The states $|g.s.>$ are highly degenerate since they can take up any orientation in the vector space which carries the representation of the $SU(\mathcal{N})$ group with the Young tableaux of $(\mathcal{N}+1)/2$ rows on one sublattice and $(\mathcal{N}-1)/2$ rows on the other sublattice [7].

For both even or odd $\mathcal{N}$, the ground state degeneracy is resolved at the second order in the strong coupling expansion. First order perturbations to the vacuum energy vanish. The vacuum energy at order ϵ^2 reads

$$E_0^{(2)} = < H_h^\dagger \frac{\Pi}{E_0^{(0)} - H_0} H_h > \quad ; \tag{17}$$

the expectation values are defined on the degenerate subspace of ground states and Π is a projection operator projecting orthogonal to the states of the degenerate subspace. Due to the commutation relation

$$[H_0, H_h] = \frac{N-1}{N} H_h - 2\sum_{x,y} \left[V(x-y) - V(x-y-1)\right] (L_y + R_y)\rho(x)$$

Eq.(17) becomes

$$E_0^{(2)} = -2 < RL > \quad . \tag{18}$$

On the ground state the combination RL can be written in terms of the Heisenberg Hamiltonian of a generalized $SU(\mathcal{N})$ antiferromagnet. By introducing the Schwinger spin operators $S_x^\alpha = \psi_{ax}^\dagger T_{ab}^\alpha \psi_{bx}$ with T^α the generators of the $SU(\mathcal{N})$

group, the $SU(\mathcal{N})$ Heisenberg Hamiltonian reads

$$H_J = \sum_{x=1}^{N} (\vec{S}_x \cdot \vec{S}_{x+1} - \frac{\mathcal{N}}{8} + \frac{1}{2\mathcal{N}} \rho(x)\rho(x+1)) = -\frac{1}{2}\sum_{x=1}^{N} L_x R_x \quad . \tag{19}$$

When $\mathcal{N}$ is even, on the degenerate ground states one has

$$< H_J > = < \sum_{x=1}^{N} (\vec{S}_x \cdot \vec{S}_{x+1} - \frac{\mathcal{N}}{4}) > = < -\frac{1}{2}\sum_{x=1}^{N} L_x R_x > \quad ; \tag{20}$$

when $\mathcal{N}$ is odd one has

$$< H_J > = < \sum_{x=1}^{N} (\vec{S}_x \cdot \vec{S}_{x+1} - \frac{\mathcal{N}^2+1}{8\mathcal{N}}) > = < -\frac{1}{2}\sum_{x=1}^{N} L_x R_x > \quad .te \tag{21}$$

Taking into account that the products of L_x and R_y at different points have vanishing expectation values on the ground states and using Eq.(20) or Eq.(21), Eq.(18) reads

$$E_0^{(2)} = 4 < H_J > \quad .$$

The problem of determining the true ground state, on which to perform the strong coupling expansion, is then reduced to the diagonalization of the $SU(\mathcal{N})$ Heisenberg spin-1/2 Hamiltonian (19). The analysis of the two-flavor Schwinger model has been carried out in [4].

The ground state of the gauge models is very different depending on if $\mathcal{N}$ is even or odd. When $\mathcal{N}$ is even, the ground state $|G.S.>$ of the spin Hamiltonian (19) is non-degenerate and translationally invariant; since it is the ground state of the gauge model in the infinite coupling limit, there is no spontaneous breaking of the chiral symmetry for any $SU(2\mathcal{N})$-flavor lattice Schwinger model. In contrast, when $\mathcal{N}$ is odd, the ground state $|G.S.>$ of the spin Hamiltonian (19) is degenerate of order two and is not translationally invariant; consequently any $SU(2\mathcal{N}+1)$-flavor lattice Schwinger model exhibits spontaneous symmetry breaking of the discrete axial symmetry, given in the continuum by

$$\psi(x) \rightarrow \gamma_5 \psi(x) \quad .$$

By translating of one lattice spacing one ground state, one gets the other.

The $\mathcal{N}$-flavor lattice Schwinger models excitations are generated from $|G.S.>$ by two different mechanisms. There are excitations involving only flavor changes of the fermions without changing the charge density $\rho(x)$; they correspond to spin flips in the $SU(2)$ invariant model and thus are massless excitations. Massive excitations involve fermion transport besides flavor changes and are created by applying to $|G.S.>$ the latticized currents of the Schwinger models which vary the on site value of $\rho(x)$.

LARGE $\mathcal{N}_c$ LATTICE QCD_2

The continuum two-flavor 't Hooft model [8] is defined by the action

$$S = \int d^2x(\overline{\psi}_i^\alpha \gamma_\mu \partial_\mu \psi_i^\alpha + \overline{\psi}_i^\alpha A_{i\ \mu}^j \psi_j^\alpha - \frac{1}{4g^2} F_{\mu\nu\ i}^j F_{\mu\nu\ j}^i) \tag{22}$$

where $\alpha = 1, 2$ is a flavor index, $i, j = 1, \ldots \mathcal{N}_c$ are colour indices and the field strengths read as

$$F_{\mu\nu\ i}^j = \partial_\mu A_{i\ \nu}^j - \partial_\nu A_{i\ \mu}^j + [A_\mu, A_\nu]_i^j \quad .$$

The Hamiltonian of the two-flavor 't Hooft model [8] is

$$H = \int dx \left[\frac{g^2}{2} E^a(x) E^a(x) + \overline{\psi}_i^\alpha \gamma_\mu \partial_\mu \psi_i^\alpha + \overline{\psi}_i^\alpha A_i^j \ {}_\mu \psi_j^\alpha\right] \tag{23}$$

with the electric field operators $E^a(x)$ satisfing

$$\left[E^a(x), E^b(x)\right] = i f^{abc} E^c(x) \delta(x - y) \quad . \tag{24}$$

The lattice Hamiltonian reducing to (23) in the naive continuum limit reads as

$$H = \frac{g^2 a}{2} \sum_{x=1}^{N} E_x^a E_x^a - \frac{it}{2a} [R - L] \quad . \tag{25}$$

The right and left hopping operators are defined ($L = R^\dagger$) as

$$R = \sum_{x=1}^{N} R_x = \sum_{x=1}^{N} \sum_{\alpha=1}^{\mathcal{N}} R_x^{(\alpha)} = \sum_{x=1}^{N} \sum_{\alpha=1}^{\mathcal{N}} \sum_{a,b=1}^{\mathcal{N}_c} \psi_{a,x+1}^{\alpha\dagger} U_{ab}(x) \psi_{b,x}^\alpha \quad ; \tag{26}$$

in Eq. (26) the matrix $U(x)$, associated with the link $[x, x+1]$, is a group element of $U(\mathcal{N}_c)$ in the fundamental representation of $SU(\mathcal{N}_c)$ and carries also a representation of $U(1)$. The generators of the static gauge transformations are

$$\mathcal{G}^A(x) = E_x^A + \psi_{ax}^{\alpha\dagger} T_{ab}^A \psi_{bx}^\alpha \tag{27}$$

with $A = 1, \ldots, \mathcal{N}_c^2 - 1$ and

$$\mathcal{G}^0(x) = E_{x+1}^0 - E_x^0 + \frac{1}{2}\left[\psi_{ax}^{\alpha\dagger}, \psi_{ax}^\alpha\right] \quad . \tag{28}$$

The generators (27) and (28) obey the Lie algebra

$$\left[\mathcal{G}^a(x), \mathcal{G}^b(y)\right] = i f^{abc} \mathcal{G}^c(x) \delta(x - y) \tag{29}$$

and (25) is gauge invariant, *i.e.* $[\mathcal{G}^a(x), H] = 0$.

The Hamiltonian (25), rescaled by the factor $g^2a/2$, reads as

$$H = H_0 + \epsilon H_h \tag{30}$$

with $H_0 = \sum_{x=1}^{N} E_x^a E_x^a$, $H_h = -i(R-L)$ and $\epsilon = t/g^2a^2$. Since H_0 and H_h are both gauge invariant – *i.e.* $[\mathcal{G}^a(x), H_0] = [\mathcal{G}^a(x), H_h] = 0$ – if one finds a gauge invariant eigenstate of H_0, perturbations in H_h still retain gauge invariance. If a state $|g.s. >$ is a singlet of the algebra (24) , *i.e.* $E_x^a|g.s. >= 0$, then $H_0|g.s. >= 0$ and the commutator

$$[H_0, H_h] = C_2^f(\mathcal{N}_c)H_h \tag{31}$$

holds on any linear combination of states $|g.s. >$. In Eq. (31) $C_2^f(\mathcal{N}_c) = (\mathcal{N}_c^2-1)/2\mathcal{N}_c$ is the quadratic Casimir of the fundamental representation of $SU(\mathcal{N}_c)$.

The lowest energy eigenstates of H_o (with $E_0^{(0)} = 0$) are the states which are singlets not only of the algebra (24) but also of (29) , since these states must be gauge invariant colour singlets. The allowed representations of the flavor $SU(2)$ algebra in one site are the empty singlet and those with Young tableaux of $\mathcal{N}_c$ columns and 1 and 2 rows, which are distinguished by the fermion densities $\rho(x) = \mathcal{N}_c(n-1)$, $n = 0, 1, 2$.

The degeneracy of the ground state of H_0 is resolved by diagonalizing the perturbations. The inner product $\langle\ ,\rangle$ is defined in the full Hilbert space of the model, *i.e.* $\langle\ ,\rangle = \prod_x dU(x)\ (\ ,\ )$; dU is the Haar measure on the gauge group manifold and $(\ ,\ )$ is the fermion Fock space inner product. First order perturbations in H_h vanish, since $\int dU\ U_{ab} = 0$. The first non-trivial perturbative order is the second order

$$E_0^{(2)} = \langle g.s.|H_h^\dagger \frac{1}{E_0^{(0)} - H_0} H_h|g.s.\rangle = -\frac{2}{C_2^f(\mathcal{N}_c)}\langle g.s.|LR|g.s.\rangle \quad ; \tag{32}$$

in deriving Eq. (32) the commutator (31) has been used.

Upon introducing the generalized Schwinger spin operators

$$\vec{S}_x = \sum_{a=1}^{\mathcal{N}_c} \sum_{\alpha,\beta=1}^{\mathcal{N}} \psi_{ax}^{\alpha\dagger}(\frac{\vec{\sigma}}{2})_{\alpha\beta}\psi_{ax}^{\beta} \tag{33}$$

with $\vec{\sigma} = (\sigma^1, \sigma^2, \sigma^3)$ the Pauli matrices, using the group integrals $\int dU\ U_{ab}^\dagger\ U_{cd} = \frac{1}{\mathcal{N}_c}\delta_{ad}\delta_{bc}$, Eq. (32)reads as

$$E_0^{(2)} = \frac{8}{\mathcal{N}_c^2 - 1}\langle g.s.|H_J|g.s.\rangle \tag{34}$$

where

$$H_J = \sum_{x=1}^{N}(\vec{S}_x \cdot \vec{S}_{x+1} - S^2) \tag{35}$$

is the spin $S = \mathcal{N}_c/2$ Heisenberg Hamiltonian.

Determining the true ground state, on which to perform the strong coupling expansion, amounts, also for this model, to diagonalize the Heisenberg Hamiltonian (35). The allowed representation of $U(\mathcal{N}_c)$ at each site are now given by the Young tableau of $\mathcal{N}_c$ columns and one row.

If one takes the limit $\mathcal{N}_c \to \infty$, *i.e.* the large spin $S \to \infty$ limit, it is well known [15] that the quantum Hamiltonian (35) becomes a classical one since $[S^\alpha, S^\beta] = i\epsilon^{\alpha\beta\gamma}S^\gamma = O(S) \ll O(S^2)$. The classical ground states of the antiferromagnetic Hamiltonian (35) are the two Neel states

$$|N_1\rangle = |\uparrow\downarrow\uparrow\downarrow\uparrow\downarrow\uparrow\downarrow\ldots\rangle$$

and

$$|N_2\rangle = |\downarrow\uparrow\downarrow\uparrow\downarrow\uparrow\downarrow\uparrow\ldots\rangle$$

Since there is no mixing between $|N_1\rangle$ and $|N_2\rangle$, if not at a perturbative order comparable with N [4], one may choose one of the two as the true ground state and then (34) gives

$$E_0^{(2)} = -4N \quad . \tag{36}$$

In (36)the dependence on $\mathcal{N}_c$ disappears as it should be in order to properly define the $\mathcal{N}_c \to \infty$ limit.

The strongly coupled 't Hooft model on the lattice in the hamiltonian formalism is reduced to a classical Heisenberg model; one may use the exact results of the large S spin chain to solve the spectrum of the gauge theory. The situation parallels the one of the two-flavor Schwinger model, where one exploits the exact solvability of the spin-1/2 chain to study the gauge model spectrum [4]. The spin waves, *i.e.* the small oscillations around the Neel ordered ground state, are the massless pions of the 't Hooft model. Massive mesons are created acting on the Neel state with fermionic currents, giving rise to charge transport. The static baryons are created applying on the Neel state suitable color singlet operators.

CONCLUSIONS

We showed that generalized quantum spin-1/2 $SU(\mathcal{N})$ Heisenberg antiferromagnetic chains correspond to strongly coupled lattice $\mathcal{N}$-flavor Schwinger models and that QCD_2 with large $\mathcal{N}_c$ and 2 fermion flavors is mapped on a spin S Heisenberg antiferromagnet. The spin is determined by the number of colors since $S = \mathcal{N}_c/2$.

It is well known that half-integer spin chains are expected to exhibit gapless excitations, while integer spin chains manifest a gap [16]. It would be an interesting problem to investigate if also QCD_2 exhibits a gapless or gapped spectrum depending on if $\mathcal{N}_c$ is odd or even.

Two dimensional QCD offers an excellent opportunity to study various dynamical questions of gauge theories, since many of its qualitative features are also valid in four dimensions. QCD_2 is exactly solvable in the planar limit $\mathcal{N}_c \to \infty$ [8]. Our analysis shows that the $\mathcal{N}_c \to \infty$ limit of the strongly coupled two flavor QCD_2 corresponds to the $S \to \infty$ limit of the antiferromagnetic Heisenberg chain.

REFERENCES

1. K. G. Wilson, Phys. Rev. **D10**, 2445 (1974).

2. T. Banks, S. Rabi, L. Susskind, J. Kogut, D. R. T. Jones, P. N. Scharbach, D. K. Sinclair, Phys. Rev. **D15**, 1111 (1977).

3. T. Banks, L. Susskind and J. Kogut, Phys. Rev. **D13**, 1043 (1976); A. Carroll, J. Kogut, D. K. Sinclair and L. Susskind, Phys. Rev. **D13**, 2270 (1976);S. Steinhardt, Phys. Rev. **D16**, 1782 (1977).

4. F. Berruto, G. Grignani, G. W. Semenoff and P. Sodano, Phys. Rev. **D57**, 5070 (1998); F. Berruto, G. Grignani, G. W. Semenoff and P. Sodano, Phys. Rev. **D59**, 034504 (1999); F. Berruto, G. Grignani, G. W. Semenoff and P. Sodano, hep-th/9901142; F. Berruto, hep-th/9902036.

5. I. Affleck and J. B. Marston, Phys. Rev. **B37**, 3773 (1988); J. B. Marston, Phys. Rev. Lett. **61**, 1914 (1988); I. Affleck, Z. Zou, T. Hsu and P. W. Anderson, Phys. Rev. **B38**, 745 (1988).

6. J. Smit, Nucl. Phys. **B175**, 307 (1980).

7. G. W. Semenoff, Mod. Phys. Lett. **A7**, 2811 (1992); E. Langmann and G. W. Semenoff, Phys. Lett. **B297**, 175 (1992); M. C. Diamantini, E. Langmann, G. W. Semenoff and P. Sodano, Nucl. Phys. **B405**, 595 (1993).

8. G. 't Hooft, Nucl. Phys. **B75**, 471 (1974).

9. L. D. Faddeev and L. A. Takhtadzhyan, Phys. Lett. **A85**, 375 (1981); L. D. Faddeev and L. A. Takhtadzhyan, Zapiski Nauchnych Seminarov LOMI, **109**, 134 (1981), english translation in J. Sov. Math. **24**, 241 (1984).

10. I. Affleck, Phys. Rev. Lett. **54**, 966 (1985); N. Read and S. Sachdev, Phys. Rev. Lett. **62**, 1694 (1989); N. Read and S. Sachdev, Nucl. Phys. **B316**, 609 (1989); N. Read and S. Sachdev, Phys. Rev. **B42**, 4568 (1990).

11. S. L. Adler and W. A. Bardeen, Phys. Rev. **182**, 1517 (1969); J. S. Bell and R. Jackiw, Nuovo Cimento **A60**, 47 (1969).

12. S. Coleman, Commun. Math. Phys. **31**, 259 (1973).

13. E. Witten, Comm. Math. Phys. **92**, 455 (1984).

14. D. Gepner, Nucl. Phys. **B252**, 481 (1985).

15. See for example A. Auerbach, Interacting Electrons and Quantum Magnetism, Springer-Verlag 1994.

16. F. D. M. Haldane, Phys. Rev. Lett. **50**, 1153 (1983).

SMALL VIOLATIONS OF STATISTICS

O. W. Greenberg[1]

Center for Theoretical Physics
Department of Physics
University of Maryland
College Park, MD 20742-4111

ABSTRACT

There are two motivations to consider statistics that are neither Bose nor Fermi: (1) to extend the framework of quantum theory and of quantum field theory, and (2) to provide a quantitative measure of possible violations of statistics. After reviewing tests of statistics for various particles, and types of statistics that are neither Bose nor Fermi, I discuss quons, particles characterized by the parameter q, which permit a smooth interpolation between Bose and Fermi statistics; $q = 1$ gives bosons, $q = -1$ gives fermions. The new result of this talk is work by Robert C. Hilborn and myself that gives a heuristic argument for an extension of conservation of statistics to quons with trilinear couplings of the form $\bar{f}fb$, where f is fermion-like and b is boson-like. We showed that $q_f^2 = q_b$. In particular, we related the bound on q_γ for photons to the bound on q_e for electrons, allowing the very precise bound for electrons to be carried over to photons. An extension of our argument suggests that all particles are fermions or bosons to high precision.

---

[1]email address, owgreen@physics.umd.edu.

Confluence of Cosmology, Massive Neutrinos, Elementary Particles, and Gravitation
Edited by Kursunoglu *et al.*, Kluwer Academic / Plenum Publishers, New York, 1999.

1 INTRODUCTION

Michael Berry [1] reported on a very interesting new idea to derive the connection of spin and statistics without using relativity in this session. After hearing about this work it is going from the sublime to the ridiculous to consider theories in which particles can have statistics that are neither Bose nor Fermi. Nonetheless, I will do so for two reasons: to stretch the framework of quantum mechanics and of quantum field theory and to provide a formalism that allows a quantitative measure of the accuracy with which a given particle obeys either Bose or Fermi statistics. For an earlier general discussion of violations of statistics see [2].

I first review experiments that test statistics, and then survey the theoretical ways in which violations of statistics can be introduced for identical particles. I discuss quons, a type of particle that can have statistics that interpolate continuously between bosons and fermions, in some detail [3]. At present, the quon theory is the only theory that allows parametrization of small violations of statistics. The new result that I report in this talk is conservation of statistics for quons that relates the q-parameters for particles that couple to each other [4]. For electrons and photons the result is $q_{photon} = q^2_{electron}$, which allows the high-precision bound on possible violations of Fermi statistics for electrons to be carried over to a comparably high-precision bound on violations of Bose statistics for photons. In conclusion, I mention the need for a refined derivation of the above result. I also state a result for the statistics of composite systems of quons that Robert C. Hilborn and I found after the Orbis [5].

2 EXPERIMENTS

Until recently there were no high-precision tests of the Pauli exclusion principle for fermions nor were there such tests for violations of Bose statistics for bosons. The exclusion principle is deeply engrained in our understanding of quantum mechanics and there was no stimulus from either experiment or theory to question it. In the last few years, in part because of the great success of the standard model, long-accepted features of the standard model, such as Lorentz invariance [6] and CPT symmetry [7] have been questioned and, despite the absence of experimental signals of violations, theories have been advanced that allow violations or, if no violations are seen, provide high-precision bounds on such violations for each type of particle. I am going to do the same for violations of statistics.

There are several types of experiments to detect violations of Fermi or Bose statistics if they occur. Here are three types: (i) search for transitions among anomalous states–in either solids or in gases, (ii) search for accumulation of particles in anomalous states, and (iii) search for deviations from the usual statistical properties of bulk matter. R. Amado and H. Primakoff [8] pointed out that there is a superselection rule separating states of identical particles in inequivalent irreducible representations of the symmetric group, and because of this there are no transitions between normal and anomalous states. One has to look for transitions among anomalous states rather than for transitions between normal and anomalous states. If transitions occur between states of the same symmetry type, they occur with the normal rate. Thus, for example, if the electrons in an atom are not in a totally antisymmetric representation so that the K-shell of the atom could have three electrons, then an electron in a higher shell would make the transition to the K-shell at the usual electromagnetic rate.

Atomic spectroscopy is the first place to search for violations of the exclusion principle since that is where Pauli discovered it [9]. One looks for funny lines which do not correspond to lines in the normal theory of atomic spectra. There are such lines, for example in the solar spectrum; however they probably can be accounted for in terms of highly ionized atoms in an environment of high pressure, high density and large magnetic fields. Laboratory spectra are well accounted for by theory and can bound the violation of the exclusion principle for electrons by something like 10^{-6} to 10^{-8} using the parametrization I describe in the next paragraph.

A useful quantitative measure of the violation, v, is that v is the coefficient of the anomalous component of the two-particle density matrix; for fermions, the two-electron density matrix, ρ_2, is

$$\rho_2 = (1 - v_F)\rho_a + v_F \rho_s, \qquad (1)$$

where $\rho_{a(s)}$ is the antisymmetric (symmetric) two-fermion density matrix. Mohapatra and I surveyed a variety of searches for violations of particle statistics in [10].

Next I discuss an insightful experiment by Maurice and Trudy Goldhaber [11] that was designed to answer the question, "Are the electrons emitted in nuclear β-decay quantum mechanically identical to the electrons in atoms?" We know that the β-decay electrons have the same spin, charge and mass as electrons in atoms; however the Goldhabers realized that if the β-decay electrons were not quantum mechanically identical to those in atoms, then the β-decay electrons would not see the K-shell of a heavy atom as filled and would fall into the K-shell and emit an x-ray. They looked for such x-rays by letting β-decay electrons from a natural source fall on a block of lead. No such x-rays were found. The Goldhabers were able to confirm that electrons from the two sources are indeed quantum mechanically identical. At the same time, they found that any violation of the exclusion principle for electrons must be less than 5%.

E. Ramberg and G. Snow [12] developed this experiment into one which yields a high-precision bound on violations of the exclusion principle. Their idea was to replace the natural β source, which provides relatively few electrons, by an electric current, in which case Avogadro's number is on our side. The possible violation of the exclusion principle is that a given collection of electrons can, with different probabilities, be in different permutation symmetry states. The probability to be in the "normal" totally antisymmetric state presumably would be close to one, the next largest probability would occur for the state with its Young tableau having one row with two boxes, etc. The idea of the experiment is that each collection of electrons has a possibility of being in an anomalous permutation state. If the density matrix for a conduction electron together with the electrons in an atom has a projection onto such an anomalous state, then the conduction electron will not see the K-shell of that atom as filled. Then a transition into the K-shell with x-ray emission is allowed. Each conduction electron which comes sufficiently close to a given atom has an independent chance to make such an x-ray-emitting transition, and thus the probability of seeing such an x-ray is proportional to the number of conduction electrons which traverse the sample and the number of atoms which the electrons visit, as well as the probability that a collection of electrons can be in the anomalous state. Ramberg and Snow chose to run 30 amperes through a thin copper strip for about a month. They estimated the energy of the x-rays which would be emitted due to the transition to the K-shell. No excess of x-rays above background was found in this energy region. Ramberg and Snow set the limit

$$v_F \leq 1.7 \times 10^{-26}. \tag{2}$$

This is high precision, indeed! K. Deilamian, J.D. Gillaspy and D.E. Kelleher [13] searched for transitions atoms of helium in which the two electrons are in a symmetric state under permutations. They used precision calculations of the levels of such atoms made by G.W.F. Drake [14]. They found the limit $v_F \leq 2 \times 10^{-7}$. M. De Angelis, et al [15] and, independently, R.C. Hilborn and C.L. Yuca [16] searched for forbidden bands in the O_2 spectrum and found the bounds $v_B \leq 5 \times 10^{-7}$ and $v_B \leq 5 \times 10^{-7}$, respectively, on violations of Bose statistics for the oxygen nuclei. Modugno, Ingusicio, and Tino [17] found that the probability of finding the two ^{16}O nuclei (spin 0) in carbon dioxide in a permutation antisymmetric state is less than 5×10^{-9}. Preliminary results on an experiment to bound violations of Bose statistics for photons give $v_B \leq 10 \times 10^{-7}$ [18].

3. WAYS TO VIOLATE STATISTICS

It is difficult to violate the statistics of identical particles. The Hamiltonian must be totally symmetric in the dynamical variables of the identical particles; H cannot change the permutation symmetry type of the wave function. In particular, one cannot dial in a small violating term using $H = H_S + \epsilon H_V$, since then the Hamiltonian would not be totally symmetric. Also one cannot, for example, have red electrons and blue electrons even if there were only red electrons in our neighborhood. This would lead to a doubling of the cross section $\sigma(\gamma X \to e^+ e^- X)$, since photons couple universally.

3.1 Gentile's Intermediate Statistics

The first attempt to go beyond Bose and Fermi statistics seems to have been made by G. Gentile [19] who suggested an "intermediate statistics" in which at most n identical particles could occupy a given quantum state. In intermediate statistics, Fermi statistics is recovered for $n = 1$ and Bose statistics is recovered for $n \to \infty$; thus intermediate statistics interpolates between Fermi and Bose statistics. However Gentile's statistics is not a proper quantum statistics because the condition of having at most n particles in a given quantum state is not invariant under change of basis. For example for intermediate statistics with $n = 2$ the state $|\psi\rangle = |k, k, k\rangle$ does not exist; however the state $|\chi\rangle = \sum_{l_1,l_2,l_3} U_{k,l_1} U_{k,l_2} U_{k,l_3} |l_1, l_2, l_3\rangle$ obtained from $|\psi\rangle$ by the unitary change of single-particle basis $|k\rangle' = \sum_l U_{k,l}|l\rangle$ does exist.

By contrast, parafermi statistics of order n (to be discussed just below) is invariant under change of basis. Parafermi statistics of order n not only allows at most n identical particles in the same state, but also allows at most n identical particles in a symmetric state. In the example just described, neither $|\psi\rangle$ nor $|\chi\rangle$ exist for parafermi statistics of order two.

3.2 Green's Parastatistics

H.S. Green [20] proposed the first proper quantum mechanical generalization of Bose and Fermi statistics. Green noticed that the commutator of the number operator with the annihilation and creation operators is the same for both bosons and fermions

$$[n_k, a_l^\dagger]_- = \delta_{kl} a_l^\dagger. \tag{3}$$

The number operator can be written

$$n_k = (1/2)[a_k^\dagger, a_k]_\pm + \text{const}, \tag{4}$$

where the anticommutator (commutator) is for the Bose (Fermi) case. If these expressions are inserted in the number operator-creation operator commutation relation, the resulting relation is *trilinear* in the annihilation and creation operators. Polarizing the number operator to get the transition operator n_{kl} that annihilates a free particle in state k and creates one in state l leads to Green's trilinear commutation relation for his parabose and parafermi statistics,

$$[[a_k^\dagger, a_l]_\pm, a_m^\dagger]_- = 2\delta_{lm} a_k^\dagger. \tag{5}$$

Since these rules are trilinear, the usual vacuum condition,

$$a_k|0\rangle = 0, \tag{6}$$

does not suffice to allow calculation of matrix elements of the a's and $a^\dagger$'s; a condition on one-particle states must be added,

$$a_k a_l^\dagger|0\rangle = \delta_{kl}|0\rangle. \tag{7}$$

Green found an infinite set of solutions of his commutation rules, one for each integer, using an ansatz in terms of Bose and Fermi operators. Let

$$a_k^\dagger = \sum_{p=1}^{n} b_k^{(\alpha)\dagger}, \quad a_k = \sum_{p=1}^{n} b_k^{(\alpha)}, \tag{8}$$

and let the $b_k^{(\alpha)}$ and $b_k^{(\beta)\dagger}$ be Bose (Fermi) operators for $\alpha = \beta$ but anticommute (commute) for $\alpha \neq \beta$ for the "parabose" ("parafermi") cases. This ansatz clearly satisfies Green's relation. The integer p is the order of the parastatistics. The physical interpretation of p is that for parabosons p is the maximum number of particles that can occupy an antisymmetric state, while for parafermions p is the maximum number of particles that can occupy a symmetric state (in particular, the maximum number that can occupy the same state). The case $p = 1$ corresponds to the usual Bose or Fermi statistics. Later Messiah and I [21] proved that Green's ansatz gives all Fock-like solutions of Green's commutation rules. Local observables have a form analogous to the usual ones; for example, the local current for a spin-1/2 theory is $j_\mu = (1/2)[\bar{\psi}(x), \psi(x)]_-$. From Green's ansatz, it is clear that the squares of all norms of states are positive, since sums of Bose or Fermi operators give positive norms. Thus parastatistics gives a set of orthodox theories. Parastatistics is one of the possibilities found by Doplicher, Haag and Roberts [22] in a general study of particle statistics using algebraic field theory methods. Haag's recent book [23] gives a good review of this work.

This is all well and good; however the violations of statistics provided by parastatistics are gross. Parafermi statistics of order two has up to two particles in each quantum state. High-precision experiments are not necessary to rule this out for the all particles we think are fermions.

3.3 The Ignatiev-Kuzmin Model and "Parons"

Interest in possible small violations of the exclusion principle was revived by a paper of Ignatiev and Kuzmin [24] in 1987. They constructed a model of one oscillator with three possible states: a vacuum state, a one-particle state and, with small probability, a two-particle state. They gave trilinear commutation relations for their oscillator. Mohapatra and I showed that the Ignatiev-Kuzmin oscillator could be represented by a modified form of the order-two Green ansatz [25]. We suspected that a field theory generalization of this model having an infinite number of oscillators would not have local observables and set about trying to prove this. To our surprize, we found that we could construct local observables and gave trilinear relations that guarantee the locality of the current [25]. We also checked the positivity of the norms with states of three or fewer particles. At this stage, we were carried away with enthusiasm, named these particles "parons" since their algebra is a deformation of the parastatistics algebra, and thought we had found a local theory with small violation of the exclusion principle. We did not know that Govorkov [26] had shown in generality that any deformation of the Green commutation relations necessarily has states with negative squared norms in the Fock-like representation. For our model the first such negative-probability state occurs for four particles in the representation of $\mathcal{S}_4$ with three boxes in the first row and one in the second. We were able to understand Govorkov's result qualitatively as follows [27]: Since parastatistics of order p is related by a Klein transformation to a model with exact $SO(2)$ or $SU(2)$ internal symmetry, a deformation of parastatistics that interpolates between Fermi and parafermi statistics of order two would be equivalent to interpolating between the trivial group whose only element is the identity and a theory with $SO(2)$ or $SU(2)$ internal symmetry. This is impossible, since there is no such interpolating group.

3.4 Apparent Violations of Statistics Due to Compositeness

Before getting to "quons," the final type of statistics I will discuss, I want to interpolate some comments about apparent violations of statistics due to compositeness. Consider two ^3He nuclei, each of which is a fermion. If these two nuclei are brought in close proximity, the exclusion principle will force each of them into excited states, plausibly with small amplitudes for the excited states. Let the creation operator for the nucleus at location A be

$$b_A^\dagger = \sqrt{1-\lambda_A^2}\,b_0^\dagger + \lambda_A b_1^\dagger + \cdots, |\lambda_A| << 1, \tag{9}$$

and the creation operator for the nucleus at location B be

$$b_B^\dagger = \sqrt{1-\lambda_B^2}\,b_0^\dagger + \lambda_B b_1^\dagger + \cdots, |\lambda_B| << 1. \tag{10}$$

Since these nuclei are fermions, the creation operators obey fermi statistics,

$$[b_i^\dagger, b_j^\dagger]_+ = 0 \tag{11}$$

Then,

$$b_A^\dagger b_B^\dagger|0\rangle = [\sqrt{1-\lambda_A^2}\lambda_B - \lambda_A\sqrt{1-\lambda_B^2}]b_0^\dagger b_1^\dagger|0\rangle, \tag{12}$$

$$||b_A^\dagger b_B^\dagger|0\rangle||^2 \approx (\lambda_A - \lambda_B)^2 << 1, \tag{13}$$

so with small probability, the two could even occupy the same location, because each could be excited into higher states with different amplitudes. This is not an intrinsic violation of the exclusion principle but rather only an apparent violation due to compositeness.

4 QUONS

4.1 Quon Algebra and Fock Representation

Now I come to my last topic, quons [3]. The quon algebra is

$$a_k a_l^\dagger - q a_l^\dagger a_k = \delta_{kl}. \tag{14}$$

For the Fock-like representation I impose the vacuum condition

$$a_k|0\rangle = 0. \tag{15}$$

These two conditions determine all vacuum matrix elements of polynomials in the creation and annihilation operators. In the case of free quons all non-vanishing vacuum matrix elements must have the same number of annihilators and creators. For such a matrix element with all annihilators to the left and creators to the right, the matrix element is a sum of products of "contractions" of the form $\langle 0|aa^\dagger|0\rangle$ just as in the case of bosons and fermions. The only difference is that the terms are multiplied by integer powers of q. The power can be given as a graphical rule: Put o's for each annihilator and $\times$'s for each creator in the order in which they occur in the matrix element on the x-axis. Draw lines above the x-axis connecting the pairs that are contracted. The minimum number of times these lines cross is the power of q for that term in the matrix element. Thus a modified Wick's theorem holds for quon operators.

The physical significance of q for small violations of Fermi statistics is that $q = 2v_F - 1$, where the parameter v_F appears in Eq.(1). For small violations of Bose statistics, the two-particle density matrix is

$$\rho_2 = (1 - v_B)\rho_s + v_B\rho_a, \tag{16}$$

where $\rho_{s(a)}$ is the symmetric (antisymmetric) two-boson density matrix. Then $q = 1 - 2v_B$.

For q in the open interval $(-1, 1)$ all representations of the symmetric group occur. As $q \to 1$ the symmetric representations are more heavily weighted and at $q = 1$ only the totally symmetric representation remains; correspondingly, as $q \to -1$ the antisymmetric representations are more heavily weighted and at $q = -1$ only the totally antisymmetric representation remains. Thus for a general n-quon state there are $n!$ linearly independent states for $-1 < q < 1$, but there is only one state for $q = \pm 1$. I emphasize something that many people find very strange: *there is no commutation relation between two creation or between two annihilation operators,* except for $q = \pm 1$, which, of course, correspond to Bose and Fermi statistics. Indeed, the fact that the general n-particle state with different quantum numbers for all the particles has $n!$ linearly independent states proves that there is no such commutation relation between any number of creation (or annihilation) operators. An even stronger statement holds: There is no two-sided ideal containing a term with only creation operators. Note that here quons differ from the "quantum plane" in which

$$xy = qyx \tag{17}$$

holds.

Quons are an operator realization of the "infinite statistics" that were found as a possible statistics by Doplicher, Haag and Roberts [22] in their general classification of particle statistics. The simplest case, $q = 0$ [28], suggested to me by Hegstrom [29], was discussed earlier in the context of operator algebras by Cuntz [30]. It seems likely that the Fock-like representations of quons for $|q| < 1$ are homotopic to each other and, in particular, to the $q = 0$ case, which is particularly simple. Thus it is convenient, as I will now do, to illustrate qualitative properties of quons for this simple case. All bilinear observables can be constructed from the number operator, $n_k \equiv n_{kk}$, or the transition operator, n_{kl}, that obey

$$[n_k, a_l^\dagger]_- = \delta_{kl} a_l^\dagger, \quad [n_{kl}, a_m^\dagger]_- = \delta_{lm} a_k^\dagger. \tag{18}$$

Although the formulas for n_k and n_{kl} in the general case are complicated, the corresponding formulas for $q = 0$ are simple [28]. Once Eq.(18) holds, the Hamiltonian and other observables can be constructed in the usual way; for example for free particles

$$H = \sum_k \epsilon_k n_k, \quad \text{etc.} \tag{19}$$

The obvious thing is to try

$$n_k = a_k^\dagger a_k. \tag{20}$$

Then

$$[n_k, a_l^\dagger]_- = \delta_{kl} a_k^\dagger - a_l^\dagger a_k^\dagger a_k. \tag{21}$$

The first term in Eq.(21) is $\delta_{kl} a_k^\dagger$ as desired; however the second term is extra and must be canceled. This can be done by adding the term $\sum_t a_t^\dagger a_k^\dagger a_k a_t$ to the term in Eq.(20). This cancels the extra term, but adds a new extra term, that must be canceled by another term. This procedure yields an infinite series for the number operator and for the transition operator,

$$n_{kl} = a_k^\dagger a_l + \sum_t a_t^\dagger a_k^\dagger a_l a_t + \sum_{t_1, t_2} a_{t_2}^\dagger a_{t_1}^\dagger a_k^\dagger a_l a_{t_1} a_{t_2} + \ldots \tag{22}$$

As in the Bose case, this infinite series for the transition or number operator defines an unbounded operator whose domain includes states made by polynomials in the creation operators acting on the vacuum. (As far as I know, this is the first case in which the number operator, Hamiltonian, etc. for a free field are of infinite degree. Presumably this is due to the fact that quons are a deformation of an algebra and are related to quantum groups.) For nonrelativistic theories, the x-space form of the transition operator is [32]

$$\rho_1(\mathbf{x}; \mathbf{y}) = \psi^\dagger(\mathbf{x})\psi(\mathbf{y}) + \int d^3z \psi^\dagger(\mathbf{z})\psi^\dagger(\mathbf{x})\psi(\mathbf{y})\psi(\mathbf{z})$$

$$+ \int d^3z_1 d^3z_2 \psi(\mathbf{z_2})\psi^\dagger(\mathbf{z_1})\psi^\dagger(\mathbf{x})\psi(\mathbf{y})\psi(\mathbf{z_1})\psi(\mathbf{z_2}) + \cdots, \tag{23}$$

which obeys the nonrelativistic locality requirement

$$[\rho_1(\mathbf{x};\mathbf{y}), \psi^\dagger(\mathbf{w})]_- = \delta(\mathbf{y}-\mathbf{w})\psi^\dagger(\mathbf{x}), \quad \text{and} \quad \rho(\mathbf{x};\mathbf{y})|0\rangle = 0. \tag{24}$$

The apparent nonlocality of this formula associated with the space integrals has no physical significance. To support this last statement, consider

$$[Qj_\mu(x), Qj_\nu(y)]_- = 0, \quad x \sim y, \tag{25}$$

where $Q = \int d^3x j^0(x)$. Equation (25) seems to have nonlocality because of the space integral in the Q factors; however, if

$$[j_\mu(x), j_\nu(y)]_- = 0, \quad x \sim y, \tag{26}$$

then Eq.(25) holds, despite the apparent nonlocality. What is relevant is the commutation relation, not the representation in terms of a space integral. (The apparent nonlocality of quantum electrodynamics in the Coulomb gauge is another such example.)

In a similar way,

$$[\rho_2(\mathbf{x},\mathbf{y};\mathbf{y}',\mathbf{x}'), \psi^\dagger(\mathbf{z})]_- = \delta(\mathbf{x}'-\mathbf{z})\psi^\dagger(\mathbf{x})\rho_1(\mathbf{y},\mathbf{y}') + \delta(\mathbf{y}'-\mathbf{z})\psi^\dagger(\mathbf{y})\rho_1(\mathbf{x},\mathbf{x}'). \tag{27}$$

Then the Hamiltonian of a nonrelativistic theory with two-body interactions has the form

$$H = (2m)^{-1}\int d^3x \nabla_x \cdot \nabla_{x'}\rho_1(\mathbf{x},\mathbf{x}')|_{\mathbf{x}=\mathbf{x}'} + \frac{1}{2}\int d^3x d^3y V(|\mathbf{x}-\mathbf{y}|)\rho_2(\mathbf{x},\mathbf{y};\mathbf{y},\mathbf{x}). \tag{28}$$

$$[H, \psi^\dagger(\mathbf{z}_1)\ldots\psi^\dagger(\mathbf{z}_n)]_- = [-(2m)^{-1}\sum_{j=1}^{n}\nabla^2_{\mathbf{z}_i} + \sum_{i<j}V(|\mathbf{z}_i-\mathbf{z}_j|)]\psi^\dagger(\mathbf{z}_1)\ldots\psi^\dagger(\mathbf{z}_n)$$
$$+\sum_{j=1}^{n}\int d^3x V(|\mathbf{x}-\mathbf{z}_j|)\psi^\dagger(\mathbf{z}_1)\cdots\psi^\dagger(\mathbf{z}_n)\rho_1(\mathbf{x},\mathbf{x}'). \tag{29}$$

Since the last term on the right-hand-side of Eq.(29) vanishes when the equation is applied to the vacuum, this equation shows that the usual Schrödinger equation holds for the n-particle system. Thus the usual quantum mechanics is valid, with the sole exception that any permutation symmetry is allowed for the many-particle system. This construction justifies calculating the energy levels of (anomalous) atoms with electrons in states that violate the exclusion principle using the normal Hamiltonian, but allowing anomalous permutation symmetry for the electrons [14].

4.2 Positivity of Squares of Norms

I have not yet addressed the question of positivity of the squares of norms that caused grief in the paron model. Several authors have given proofs of positivity [33, 34, 35, 36]. The proof of Zagier provides an explicit formula for the determinant of the $n! \times n!$ matrix of scalar products among the states of n particles in different quantum states. Since this determinant is one for $q = 0$, the norms will be positive unless the determinant has zeros on the real axis. Zagier's formula

$$det\ M_n(q) = \Pi_{k=1}^{n-1}(1 - q^{k(k+1)})^{(n-k)n!/k(k+1)}, \tag{30}$$

has zeros only on the unit circle, so the desired positivity follows. Although quons satisfy the requirements of nonrelativistic locality, the quon field does not obey the relativistic requirement, namely spacelike commutativity of observables. Since quons interpolate smoothly between fermions, which must have odd half-integer spin, and bosons, which must have integer spin, the spin-statistics theorem, which can be proved, at least for free fields, from locality would be violated if locality were to hold for quon fields. It is amusing that, nonetheless, the free quon field obeys the TCP theorem and Wick's theorem holds for quon fields [3].

4.3 Speicher's ansatz

Speicher [35] has given an ansatz for the Fock-like representation of quons analogous to Green's ansatz for parastatistics. Speicher represents the quon annihilation operator as

$$a_k = \lim_{N\to\infty} N^{-1/2} \sum_{\alpha=1}^{N} b_k^{(\alpha)}, \tag{31}$$

where the $b_k^{(\alpha)}$ are Bose oscillators for each α, but with relative commutation relations given by

$$b_k^{(\alpha)} b_l^{(\beta)\dagger} = s^{(\alpha,\beta)} b_l^{(\beta)\dagger} b_k^{(\alpha)}, \alpha \neq \beta, \text{ where } s^{(\alpha,\beta)} = \pm 1. \tag{32}$$

Equation(31) is taken as the weak limit, $N \to \infty$, in the vacuum expectation state of the Fock space representation of the $b_k^{(\alpha)}$. In this respect, Speicher's ansatz differs from Green's, which is an operator identity. Further to get the Fock-like representation of the quon algebra, Speicher chooses a probabilistic condition for the signs $s^{(\alpha,\beta)}$,

$$\text{prob}(s^{(\alpha,\beta)} = 1) = (1+q)/2, \tag{33}$$

$$\text{prob}(s^{(\alpha,\beta)} = -1) = (1-q)/2. \tag{34}$$

Since a sum of Bose operators acting on a Fock vacuum always gives a positive-definite norm, the positivity property is obvious with Speicher's construction.

Speicher's ansatz leads to the conjecture that there is an infinite-valued hidden degree of freedom underlying q-deformations analogous to the hidden degree of freedom underlying parastatistics.

If one asks "How well do we know that a given particle obeys Bose or Fermi statistics?," we need a quantitative way to answer the question. That requires a formulation in which either Bose or Fermi statistics is violated by a small amount. As stated earlier, we cannot just add to the Hamiltonian a small term that violates Bose or Fermi statistics; such a term would not be invariant under permutations of the identical particles and thus would clash with the particles being identical. As mentioned above parastatistics, which does violate Bose or Fermi statistics, gives gross violations. The only way presently available to allow small violations of statistics is the quon theory just described.

Unfortunately, the quon theory is not completely satisfactory. The observables in quon theory do not commute at spacelike separation. If they did, particle statistics could change continuously from Bose to Fermi without changing the spin. Since spacelike commutativity of observables leads to the spin-statistics theorem, this would be a direct contradiction. Kinematic Lorentz invariance can be maintained, but without spacelike commutativity or anticommutativity of the fields the theory may not be consistent.

For nonrelativistic theories, however, quons are consistent. The nonrelativistic version of locality is

$$[\rho(\mathbf{x}), \psi(\mathbf{y})] = -\delta(\mathbf{x}-\mathbf{y})\psi(\mathbf{y}) \tag{35}$$

for an observable $\rho(\mathbf{x})$ and a field $\psi(\mathbf{y})$ and this does hold for quon theories. It is the antiparticles that prevent locality in relativistic quon theories.

5. CONSERVATION OF STATISTICS

5.1 Conservation of Statistics for Bosons and Fermions

The first conservation of statistics theorem states that terms in the Hamiltonian density must have an even number of Fermi fields and that composites of fermions and bosons are bosons, unless they contain an odd number of fermions, in which case they are fermions [37, 38].

5.2 Conservation of Statistics for Parabosons and Parafermions

The extension to parabosons and parafermions is more complicated [21]; however, the main constraint is that for each order p at least two para particles must enter into every reaction.

Reference [39] argues that the condition that the energy of widely separated subsystems be additive requires that all terms in the Hamiltonian be "effective Bose operators" in that sense that

$$[\mathcal{H}(\mathbf{x}), \phi(\mathbf{y})]_- \to 0, |\mathbf{x}-\mathbf{y}| \to \infty. \tag{36}$$

For example, $\mathcal{H}$ should not have a term such as $\phi(x)\psi(x)$, where ϕ is Bose and ψ is Fermi, because then the contributions to the energy of widely separated subsystems would alternate in sign. Such terms are also prohibited by rotational symmetry. This discussion was given in the context of external sources.

It is well known that external fermionic sources must be multiplied by a Grassmann number in order to be a valid term in a Hamiltonian. This is necessary, because additivity of the energy of widely separated systems requires that all terms in the Hamiltonian must be effective Bose operators. I constructed the quon analog of Grassmann numbers [39] in order to allow external quon sources. Because this issue was overlooked, the bound on violations of Bose statistics for photons claimed in [40] is invalid.

For a fully quantized field theory, one can replace Eq.(36) by the asymptotic causality condition, asymptotic local commutativity,

$$[\mathcal{H}(\mathbf{x}), \mathcal{H}(\mathbf{y})]_- = 0, |\mathbf{x}-\mathbf{y}| \to \infty \tag{37}$$

or by the stronger causality condition, local commutativity,

$$[\mathcal{H}(\mathbf{x}), \mathcal{H}(\mathbf{y})]_- = 0, \mathbf{x} \neq \mathbf{y}. \tag{38}$$

Studying this condition for quons in electrodynamics is complicated, since the terms in the interaction density will be cubic. It is simpler to use the description of the electron current or transition operator as an external source represented by a quonic Grassmann number.

5.3 Conservation of Statistics for Quons

Here we give a heuristic argument for conservation of statistics for quons based on a simpler requirement in the context of quonic Grassmann external sources [4]. The commutation relation of the quonic photon operator is

$$a(k)a^\dagger(l) - q_\gamma a^\dagger(l)a(k) = \delta(k-l), \tag{39}$$

where q_γ is the q-parameter for the photon quon field. We call the quonic Grassmann numbers for the electron transitions to which the photon quon operators couple $c(k)$. The Grassmann numbers that serve as the external source for coupling to the quon field for the photon must obey

$$c(k)c(l)^\star - q_\gamma c(l)^\star c(k) = 0, \tag{40}$$

and the relative commutation relations must be

$$a(k)c(l)^\star - q_\gamma c(l)^\star a(k) = 0, \tag{41}$$

etc. Since the electron current for emission or absorption of a photon with transition of the electron from one atomic state to another is bilinear in the creation and annihilation operators for the electron, a more detailed description of the photon emission would treat the photon as coupled to the electron current, rather than to an external source. We impose the requirement that the leading terms in the commutation relation for the quonic Grassmann numbers of the source that couples to the photon should be mimicked by terms bilinear in the electron operators. The electron operators obey the relation

$$b(k)b^\dagger(l) - q_e b^\dagger(l)b(k) = \delta(k-l), \tag{42}$$

where q_e is the q-parameter for the electron quon field.

To find the connection between q_e and q_γ we make the following associations,

$$c(k) \Rightarrow b^\dagger(p)b(k+p), \quad c^*(l) \Rightarrow b^\dagger(l+r)b(r) \tag{43}$$

We now replace the c's in Eq.(40) with the products of operators given in Eq.(43) and obtain

$$[b^\dagger(p)b(k+p)][b^\dagger(l+r)b(r)] - q_\gamma[b^\dagger(l+r)b(r)][b^\dagger(p)b(k+p)] = 0. \tag{44}$$

This means that the source $c(k)$ is replaced by a product of b's that destroys net momentum k; the source $c^*(l)$ is replaced by a product of b's that creates net momentum l. We want to rearrange the operators in the first term of Eq.(44) to match the second term, because this corresponds to the standard normal ordering for the transition operators. For the products $bb^\dagger$ we use Eq.(42). For the products bb, as mentioned above, there is no operator relation; however *on states in the Fock-like representation* there is an approximate relation,

$$b(k+p)b(r) = q_e b(r)b(k+p) + \text{ terms of order } 1 - q_e^2. \tag{45}$$

In other words, in the limit $q_e \to -1$, we retrieve the usual anticommutators for the electron operators. (The analogous relation for an operator that is approximately bosonic would be that the operators commute in the limit $q_{bosonic} \to 1$.) We also use the adjoint relation

$$b^\dagger(p)b^\dagger(l+r) = q_e b^\dagger(l+r)b^\dagger(p) + \text{terms of order } 1 - q_e^2 \tag{46}$$

and, finally,

$$q_e b^\dagger(p)b(r) = b(r)b^\dagger(p) - \delta(r-p). \tag{47}$$

We require only that the quartic terms that correspond to the quonic Grassmann relation Eq.(40) cancel, so we drop terms in which either $k+p = l+r$ or $r = p$. We also drop terms of order $1 - q_e^2$. In this approximation, we find that Eq.(44) becomes

$$(q_e^2 - q_\gamma)[b^\dagger(l+r)b(r)][b^\dagger(p)b(k+p)] \approx 0, \tag{48}$$

and conclude that

$$q_e^2 \approx q_\gamma. \tag{49}$$

This relates the bound on violations of Fermi statistics for electrons to the bound on violations of Bose statistics for photons and allows the extremely precise bound on possible violations of Fermi statistics for electrons to be carried over to photons. Eq.(49) is the quon analog of the conservation of statistics relation that the square of the phase for transposition of a pair of fermions equals the phase for transposition of a pair of bosons.

Arguments analogous to those just given, based on the source-quonic photon relation, Eq.(41), lead to

$$q_{e\gamma}^2 \approx q_\gamma, \tag{50}$$

where $q_{e\gamma}$ occurs in the relative commutation relation

$$a(k)b^\dagger(l) = q_{e\gamma} b^\dagger(l)a(k). \tag{51}$$

Since the normal commutation relation between Bose and Fermi fields is for them to commute [41], this shows that $q_{e\gamma}$ is close to one.

6. HIGH-PRECISION BOUNDS

Since the Ramberg-Snow bound on Fermi statistics for electrons is

$$v_e \leq 1.7 \times 10^{-26} \iff q_e \leq -1 + 3.4 \times 10^{-26}, \tag{52}$$

the bound on Bose statistics for photons is

$$q_\gamma \geq 1 - 6.8 \times 10^{-26} \iff v_\gamma \leq 3.4 \times 10^{-26}. \tag{53}$$

This bound for photons is much stronger than could be gotten by a direct experiment. Nonetheless D. DeMille and N. Derr are performing an experiment that promises to give the best *direct* bound on Bose statistics for photons [18]. It is essential to test every basic property in as direct a way as possible. Thus experiments that yield direct bounds on photon statistics, such as the one being carried out by DeMille and Derr, are important.

Teplitz, Mohapatra and Baron have suggested a method to set a very low limit on violation of the Pauli exclusion principle for neutrons [42].

The argument just given that the q_e value for electrons implies $q_\gamma \approx q_e^2$ for photons can be run in the opposite direction to find ${q_\phi}^2 \approx q_\gamma$ for each charged field ϕ that couples bilinearly to photons. Isospin and other symmetry arguments then imply that almost all particles obey Bose or Fermi statistics to a precision comparable to the precision with which electrons obey Fermi statistics.

7 CONCLUSION

In concluding, we note that further work should be carried out to justify the approximations made in deriving Eq.(49) and also to derive the relations among the q-parameters that follow from couplings that do not have the form $\bar{f}fb$. We plan to return to this topic in a later paper. After the Orbis, Hilborn and I derived a generalization of the Wigner–Ehrenfest-Oppenheimer rule of the statistics of bound states in terms of the quon statistics of their constituents, $q_{composite} = q_{constituent}^{n^2}$, where n is the number of constituents in the bound state [5].

Acknowledgements

I thank the Aspen Center for Physics for a visit during which part of this work was carried out. The unique atmosphere of the Center encourages concentrated work without the distractions of one's home university. This work was supported in part by the National Science Foundation. The work on conservation of statistics for quons was done in collaboration with Robert C. Hilborn.

References

[1] M.V. Berry, talk in this session and M.V. Berry and J.M. Robbins, *Proc. R. Soc. Lond. A* 453:1771(1997).

[2] O.W. Greenberg, D.M. Greenberger and T.V. Greenbergest, in *Quantum Coherence and Reality*, eds. J.S. Anandan and J.L. Safko, (World Scientific, Singapore, 1994), p. 301.

[3] O.W. Greenberg, *Phys. Rev. D* 43:4111(1991).

[4] O.W. Greenberg and R.C. Hilborn, Univ. of Maryland Physics Paper 99-005, hep-th/9808106, to appear in *Foundations of Physics* 29:March, 1999 (special issue in honor of D.M. Greenberger).

[5] O.W. Greenberg and R.C. Hilborn, Univ. of Maryland Physics Paper 99-089, quant-ph/9903020.

[6] S. Coleman and S.L. Glashow, HUTP-98-A082, Dec 1998, 33pp, hep-ph/9812418.

[7] R. Jackiw and V.A. Kostelecky, IUHET-400, Jan 1999. 4pp, hep-ph/9901358; and V.A. Kostelecky, IUHET-397, Oct 1998. 13pp, hep-ph/9810365.

[8] R. Amado and H. Primakoff, *Phys. Rev. C* 22:1388(1980).

[9] W. Pauli, *Zeitschr. f. Phys.* 31:765(1925).

[10] O.W. Greenberg and R.N. Mohapatra, *Phys. Rev. D* 39:2032(1989).

[11] M. Goldhaber and G.S. Goldhaber, *Phys. Rev.* 73:1472(1948).

[12] E. Ramberg and G. Snow, *Phys. Lett. B* 238:438(1990).

[13] K. Deilamian, J.D. Gillaspy and D.E. Kelleher, *Phys. Rev. Lett.* 74:4787(1995).

[14] G.W.F. Drake, *Phys. Rev. A* 39:897(1989).

[15] M. de Angelis, G. Gagliardi, L. Gianfrani, and G. Tino, *Phys. Rev. Lett.* 76:2840(1996).

[16] R.C. Hilborn and C.L. Yuca, *Phys. Rev. Lett.* 76:2844(1996).

[17] G. Modugno, M. Inguscio, and G.M. Tino, *Phys. Rev. Lett.* 81:4790(1998).

[18] D. DeMille and N. Derr, in preparation, (1999).

[19] G. Gentile, *Nuovo Cimento* 17:493(1940).

[20] H.S. Green, *Phys. Rev.* 90:270(1953).

[21] O.W. Greenberg and A.M.L. Messiah, *Phys. Rev. B* 138:1155(1965).

[22] S. Doplicher, R. Haag and J. Roberts, *Commun. Math. Phys.* 23:199(1971) and *ibid* 35:49(1974).

[23] R. Haag, *Local Quantum Physics* (Springer-Verlag, Berlin, 1992).

[24] A.Yu. Ignatiev and V.A. Kuzmin, *Yad. Fiz.* 46:786(1987), [*Sov. J. Nucl. Phys.* 46:444(1987)].

[25] O.W. Greenberg and R.N. Mohapatra, *Phys. Rev. Lett.* 59:2507(1987).

[26] A.B. Govorkov, *Teor. Mat. Fis.* 54:361(1983) [*Sov. J. Theor. Math. Phys.* 54:234(1983)]; and *Phys. Lett. A* 137:7(1989).

[27] O.W. Greenberg and R.N. Mohapatra, *Phys. Rev. Lett.* 62:712(1989).

[28] O.W. Greenberg, *Phys. Rev. Lett.* 64:705(1990).

[29] Roger Hegstrom, private communication.

[30] J. Cuntz, *Commun. Math. Phys.* 57:173(1977).

[31] S. Stanciu, *Commun. Math. Phys.* 147:211(1992).

[32] O.W. Greenberg, *Physica A* 180:419(1992).

[33] D. Zagier, *Commun. Math. Phys.* 147:199(1992).

[34] M. Bożejko and R. Speicher, *Commun. Math. Phys.* 137:519(1991).

[35] R. Speicher, *Lett. Math. Phys.* 27:97(1993).

[36] D.I. Fivel, *Phys. Rev. Lett.* 65:3361(1990); erratum, *ibid* 69:2020(1992).

[37] E.P. Wigner, *Math. und Naturwiss. Anzeiger der Ungar. Ak. der Wiss.* 46:576(1929).

[38] P. Ehrenfest and J.R. Oppenheimer, *Phys. Rev.* 37:333(1931).

[39] O.W. Greenberg, *Phys. Lett. A* 209:137(1995).

[40] D.I. Fivel, *Phys. Rev. A*. 43:4913(1991).

[41] H. Araki, *J. Math. Phys.* 2:267(1961).

[42] V.L. Teplitz, R.N. Mohapatra and E. Baron, University of Maryland Preprint 99-014, (1998).

Section IV
Strings

QUANTIZED MEMBRANES

Chiara R. Nappi*

School of Natural Sciences
Institute for Advanced Study
Olden Lane
Princeton, NJ 08540

INTRODUCTION

String theorists today have good reasons to believe that there is a fundamental theory in eleven dimensions, known as M-theory, which generates all known string theories. The basic objects in M-theory are the supergraviton, the twobrane and the fivebrane. The purpose of this lecture is to discuss the quantization of the M-theory fivebrane [1, 2].

The physical degrees of freedom of the fivebrane consist of an N=(2,0), D=6 tensor supermultiplet. Aside from scalars and fermions, this multiplet contains a chiral two-form B_{MN}, *i.e.* a two-form with a three-form field strength H_{LMN} which is self-dual. While the worldvolume actions of membranes whose physical degrees of freedom consist of scalars and fermions are known, the action for the self-dual two-form is problematic. This is generally true for chiral p-forms in field theory, starting from the well-known case of the two-dimensional chiral boson. The difficulty is writing down covariant actions which incorporate the self-duality (chirality) constraints.

Recently many attempts [3, 4, 5] have been made to write the action for the M-theory fivebrane. Some of the attempts have led to non-covariant Lagrangians [6, 7], while others have succeeded in writing down a covariant Lagrangian but have required the introduction of auxiliary fields [8]. It has also been shown that these various approaches are equivalent [9].

While the search for a covariant Lagrangian is an interesting problem in itself, a covariant Lagrangian is not strictly needed in order to quantize the theory. Indeed in [1] we computed the partition function of the fivebrane on a six-dimensional torus and showed that it is modular invariant, *i.e.* invariant under the $SL(6, \mathcal{Z})$ mapping class group of T^6.

This result is somehow unexpected, since the wisdom was that the situation for the partition function of the M-theory fivebrane would be similar to that of the chiral two-dimensional boson. As we review in the next section, there is no modular invariant partition function for a single chiral field in two dimensions. A posteriori, one would think that this is the reason why one cannot write a covariant Lagrangian for such a field. Such a Lagrangian, if it were to exist, could then be quantized on a

⋆ Research supported in part by the Ambrose Monell Foundation.

Confluence of Cosmology, Massive Neutrinos, Elementary Particles, and Gravitation
Edited by Kursunoglu *et al.*, Kluwer Academic / Plenum Publishers, New York, 1999.

Riemann surface of genus g and would yield results that depend only on the metric of the Riemann surface in a modular invariant way. Instead, a two-dimensional chiral scalar on a Riemann surface of genus g has 2^{2g} candidate partition functions.

The reason why the chiral two-form in six dimensions avoids the problems of the chiral boson in two dimensions and manages to be modular invariant is that we are compactifying on T^6, which can be viewed as the product $T^2 \times T^4$. From the point of view of T^2, the three degrees of freedom of the two-form potential (which is the (3,1) representation of Spin(4)$\cong SU(2) \times SU(2)$, the little group in six dimensions) behave like three massive scalars in two dimensions, therefore mimicking the situation of three non-chiral bosons. Therefore, at least in the case of compactification on the torus, there is an $SL(6, \mathbb{Z})$ modular invariant partition function for the M-theory fivebrane chiral two-form, *i.e.* a quantum theory with symmetry analogous to the modular invariance of consistent interacting strings.

The details of our calculation have been reported in [1]. Here we will try to avoid technicalities and instead will stress the ideas behind the calculation. It is useful to review the analogous calculation for the chiral boson in two dimensions, both to clarify our procedure and to point out the differences between the two cases.

THE TWO-DIMENSIONAL CHIRAL BOSON

A non-chiral massless two-dimensional boson has Lagrangian

$$\mathcal{L} = \frac{1}{2}\partial_\alpha\phi\partial^\alpha\phi = \frac{1}{2}\partial_+\phi\partial_-\phi \tag{1}$$

where $\partial_+ = \partial_t + \partial_x$ and $\partial_- = \partial_t - \partial_x$. The equations of motions are $\partial_+\partial_-\phi = 0$ and imply that the solution is a sum of a right-moving field and a left-moving field $\phi = \phi_+ + \phi_-$, with $\phi_+ = \phi(x+t)$ and $\phi_- = \phi(x-t)$. The Hamiltonian of a non-chiral boson splits in a sum of two Hamiltonians, one for the left-movers and one for the right-movers

$$H = \frac{1}{2}((\partial_t\phi)^2 + (\partial_x\phi)^2) = \frac{1}{2}((\partial_t\phi_+)^2 + (\partial_x\phi_+)^2 + (\partial_t\phi_-)^2 + (\partial_x\phi_-)^2) = L_0 + \tilde{L}_0 \tag{2}$$

where $L_0 = \frac{1}{2}((\partial_t\phi_+)^2 + (\partial_x\phi_+)^2)$ and $\tilde{L}_0 = \frac{1}{2}((\partial_t\phi_-)^2 + (\partial_x\phi_-)^2)$.

The chiral boson satisfies the self-dual condition $\partial_\mu\phi = \epsilon_{\mu\nu}\partial^\nu\phi$, *i.e.* $\partial_t\phi = \partial_x\phi$ or equivalently $\partial_-\phi = 0$. This means that a chiral boson in two dimensions propagates only in one direction. No Lorentz covariant Lagrangian will reproduce this constraint. (For a more in-depth discussion, however, see [10]) .

Since we do not have a Lagrangian, we cannot use the path integral formalism to compute the partition function, but we can use the Hamiltonian approach. On a twisted two-torus with radii R_1 and R_6 and twist angle α, we can introduce the modular parameter $\tau = \alpha + i\frac{R_6}{R_1}$. If we think of the torus as a cylinder of circumference 2π and length $2\pi\mathrm{Im}\tau$ with end twisted by an angle $2\pi\mathrm{Re}\tau$ and then sewn together, then the partition function is given in terms of the Hamiltonian and the momentum [11, 12] by

$$Z(\tau) = tre^{-2\pi\mathrm{Im}\tau H + i2\pi\mathrm{Re}\tau P} \tag{3}$$

where the momentum P is given by $P = L_0 - \tilde{L}_0$.

To compute this partition function, we need the normal mode expansion of L_0 and $\tilde{L}_0$

$$L_0 = \frac{p_L^2}{2} + \sum_{n=1}^{\infty} n{a^\dagger}_n a_n - \frac{1}{24} \tag{4}$$

$$\tilde{L}_0 = \frac{p_R^2}{2} + \sum_{n=1}^{\infty} n\tilde{a}_n^\dagger \tilde{a}_n - \frac{1}{24}, \tag{5}$$

where $-\frac{1}{24}$ is the normal ordering constant. The zero modes p_L and p_R are the left and right momenta defined as $p_L = \frac{n}{R} + mR$ and $p_R = \frac{n}{R} - mR$, respectively. Since we are compactifying on a torus, m and n are integers.

If we impose the self-duality constraint by eliminating the right-moving modes, namely by setting $\tilde{L}_0 = 0$, then the partition function (3) will reduce to

$$\begin{aligned} Z(\tau) = & tr e^{i2\pi\tau L_0} = \\ & \sum_{p_L} (e^{i2n\tau})^{\frac{{p_L}^2}{2}} (e^{i2\pi\tau})^{-\frac{1}{24}} tr e^{i2\pi\tau \sum_n n{a^\dagger}_n a_n} \end{aligned} \tag{6}$$

We can now use the standard Fock space argument

$$tr \omega^{\sum_p p a_p^\dagger a_p} = \prod_p \sum_{k=o}^{\infty} \langle k | \omega^{p a_p^\dagger a_p} | k \rangle = \prod_p \frac{1}{1-\omega^p} \tag{7}$$

to do the sum over the oscillators and get

$$Z(\tau) = \frac{\theta_3(\tau)}{\eta(\tau)} \tag{8}$$

Above, $\theta_3(\tau) = \sum_n (e^{i2\pi\tau})^{\frac{n^2}{2}}$ is the Jacobi theta function and comes from the zero modes in (6). We are working here at the special radius $R = 1$ and therefore summing on integer values of p_L. The Dedekind eta function in the denominator $\eta(\tau) = (e^{i2\pi\tau})^{\frac{1}{24}} \prod_p (1 - e^{i2\pi\tau p})$ comes instead from the sum over the oscillators.

Had we left in the right-moving modes as well, we would have obtained the partition function for a non-chiral boson. In the limit of continuous momenta the answer would have been [11, 12]

$$Z(\tau) = \frac{(4\pi^2 \mathrm{Im}(\tau))^{-\frac{1}{2}}}{\eta(\tau)\bar{\eta}(\bar{\tau})} \tag{9}$$

where $\bar{\eta}$ comes from the sum over the right-moving oscillators. The difference between the partition function of the chiral boson (8) and that of the non-chiral boson (9) is that the former is not modular invariant. This can be checked [11, 12] by looking how (8) and (9) transform under $\tau \to -\frac{1}{\tau}$ and $\tau \to \tau + 1$, the two generators of $SL(2, \mathcal{Z})$, the modular group of the two-torus. The anomalies under these transformations cancel between the numerator and the denominator in (9), but not in (8). Modular invariance is important in string theory since it ensures perturbative unitarity.

THE CHIRAL TWO-FORM ON THE SIX-TORUS

In order to compute the partition function for the chiral two-form on the six-torus we try to follow a procedure similar to the one just outlined for the two-dimensional scalar on the two-torus. The six-dimensional chiral two-form B_{MN} has a field-strength

$$H_{MNL} = \partial_L B_{MN} + \partial_M B_{NL} + \partial_N B_{LM}$$

which is self-dual, *i.e.* it satisfies

$$H_{LMN}(\vec{\theta}, \theta^6) = \frac{1}{6\sqrt{-G}} G_{LL'} G_{MM'} G_{NN'} \epsilon^{L'M'N'RST} H_{RST}(\vec{\theta}, \theta^6) \tag{10}$$

where G_{MN} is the six-dimensional metric and the indices L, M and N vary from 1 to 6. Using (10) one can eliminate [13] the components H_{6mn} in terms of the other components H_{lmn} with $l, m, n = 1...5$ as follows

$$H_{6mn} = -\frac{G^{6l}}{G^{66}} H_{lmn} + \frac{1}{6\sqrt{-G}G^{66}} \epsilon_{mnlrs} H^{lrs}, \tag{11}$$

where now indices are raised with the 5d-metric G_5^{mn} and $\epsilon_{12345} \equiv G_5 \epsilon^{12345} = G_5$. H_{lmn} is a totally anti-symmetric three-form and has ten components.

One can start from the unconstrained Lagrangian for a non-chiral two-form $\mathbf{L} = \int d^6\theta (-\frac{\sqrt{-G}}{24}) H_{LMN} H^{MNL}$ and derive the canonical Hamiltonian and momenta. By using (11) one can then eliminate H_{6mn} and write the Hamiltonian and the momenta in a fully five-dimensional covariant way:

$$\mathcal{H} = \frac{1}{12} \int_0^{2\pi} d\theta^1 ... d\theta^5 \sqrt{G_5} G_5{}^{ll'} G_5{}^{mm'} G_5{}^{nn'} H_{lmn}(\vec{\theta}, \theta^6)\, H_{l'm'n'}(\vec{\theta}, \theta^6) \tag{12}$$

$$P_l = -\frac{1}{24} \int_0^{2\pi} d\theta^1 ... d\theta^5 \epsilon^{rsumn} H_{umn}(\vec{\theta}, \theta^6)\, H_{lrs}(\vec{\theta}, \theta^6) \tag{13}$$

with $1 \leq l, m, n, r, s, u \leq 5$. These expressions are the equivalent of the Hamiltonian and the momentum obtained in the chiral boson case after imposing the constraint $\tilde{L}_0 = 0$. Here too we will use these expressions to compute the partition function.

A general metric on T^6 is a function of 21 parameters and can be represented by the line element

$$\begin{aligned} ds^2 =& R_1{}^2 (d\theta^1 - \alpha d\theta^6)^2 + R_6{}^2 (d\theta^6)^2 \\ &+ \sum_{i,j=2...5} g_{ij} (d\theta^i - \beta^i d\theta^1 - \gamma^i d\theta^6)(d\theta^j - \beta^j d\theta^1 - \gamma^j d\theta^6) \end{aligned} \tag{14}$$

where $0 \leq \theta^I \leq 2\pi$, $1 \leq I \leq 6$ and we single out directions 1 and 6. The latter is our time direction. The 21 parameters are as follows: R_1 and R_6 are the radii for directions 1 and 6, g_{ij} is a 4d metric, β^i, γ^j are the angles between directions 1 and i, and between 6 and j, respectively, and α is related to the angle between 1 and 6, as in the two-dimensional case. The six-dimensional metric can be read off from the line element above.

Generalizing from string theory [11], the partition function on the twisted six-torus with metric (14) is given in terms of the Hamiltonian and momenta by

$$Z(R_1, R_6, g_{ij}, \alpha, \beta^i, \gamma^i) = tr(e^{-2\pi R_6 \mathcal{H} + i2\pi \frac{G^{l6}}{G^{66}} P_l}) \tag{15}$$

This is the analogue of (3) in the two-dimensional case.

The expression in the exponent in (15) can be written in a more compact form in terms of the dependent field strength H_{6mn} given in (11)

$$\begin{aligned} -2\pi R_6 \mathcal{H} + i2\pi \frac{G^{l6}}{G^{66}} P_l =& \frac{i\pi}{12} \int_0^{2\pi} d^5\theta H_{lrs} \epsilon^{lrsmn} H_{6mn} \\ =& \frac{i\pi}{2} \int_0^{2\pi} d^5\theta \sqrt{-G} H^{6mn} H_{6mn} \end{aligned} \tag{16}$$

where $H^{6mn} = \frac{1}{2\sqrt{-G}} \epsilon^{mnlrs} H_{lrs}$ from (10).

It is interesting to realize that the expression $-2\pi R_6 \mathcal{H} + i2\pi \frac{G^{l6}}{G^{66}} P_l$ in (16) is exactly the Hamiltonian derivable from the non-covariant Lagrangian that is shown in [6, 13] to give rise to the self-duality equation (10). Indeed, the equation of motion for H_{lmn} derived from the self-duality condition (10) is

$$\frac{1}{3} \epsilon^{lptrs} \partial_6 H_{trs} = \epsilon^{lpijk} \partial_i H_{6jk}. \tag{17}$$

Equation (17) can be obtained by varying the Lagrangian $L_1 + L_2 + L_3$ where

$$L_1 = \frac{1}{6} \sqrt{-G} H_{trs} H^{trs} \tag{18}$$

$$L_2 = \frac{1}{12} \epsilon^{mntrs} H_{trs} \partial_6 B_{mn} \tag{19}$$

$$L_3 = \frac{1}{12} \frac{G^{6t}}{G^{66}} H_{trs} \epsilon^{rslmn} H_{lmn} \tag{20}$$

The usual canonical procedure gives the Hamiltonian $\tilde{\mathcal{H}} = -(L_1 + L_3)$. One can explicitly check that this Hamiltonian is exactly the exponent in (15)

$$i\pi \int d^5\theta \tilde{\mathcal{H}} = -2\pi R_6 \mathcal{H} + i2\pi \frac{G^{l6}}{G^{66}} P_l.$$

It is somehow puzzling that a Lagrangian written to reproduce the self-duality condition of a chiral two-form emerges from a prescription for computing its partition function on a twisted torus. This observation might contain some interesting hint about Lagrangians for general chiral p-forms and probably deserves further investigation.

THE CALCULATION OF THE PARTITION FUNCTION

The trace in the partition function in (15) is over all independent Fock space operators which appear in the normal mode expansion of B_{MN}. To compute the zero mode part of the partition function, *i.e.* the equivalent of the Jacobi theta function in the two-dimensional chiral boson case, one should start from the zero modes in the normal mode expansion of B_{MN} and find the zero modes for the 'Hamiltonian' in (15). However, in writing the normal mode expansion one assumes that B_{MN} is a free field and therefore needs some prescription to take into account the chirality constraint.

Our approach in computing the zero modes is to sum over the integer values of the ten components of H_{lmn}. The analogy in string theory would be to sum over the winding modes. In parallel with the zero mode calculation for the two-dimensional chiral boson, this sum is given [14] by the Riemann theta function $\vartheta\begin{bmatrix}\vec{0}\\ \vec{0}\end{bmatrix}(\vec{0},\Omega)$, where the 10x10 symmetric non-singular complex matrix Ω can be reconstructed from (15) using (12) and (13) . As derived in [1]

$$\begin{aligned}Z_{\text{zero modes}} = \sum_{n_7,\ldots,n_{10}\in Z^4} & \exp\{-\pi\frac{R_6R_1}{6}\sqrt{g}g^{ii'}g^{jj'}g^{kk'}H_{ijk}H_{i'j'k'} \\ & -\frac{\pi}{4|\tau|^2}\frac{R_6}{R_1}\sqrt{g}(g^{jj'}g^{kk'}-g^{jk'}g^{j'k})H_{ijk}H_{i'j'k'}\gamma^i\gamma^{i'}\} \\ & \cdot\sum_{n_1,\ldots,n_6\in Z^6} e^{-\pi(n+x)\cdot A\cdot(n+x)}\end{aligned} \tag{21}$$

where $A_{11}=\mathcal{A}^{2323}$, $A_{16}=\mathcal{A}^{2345},\ldots$, $x_1=x_{23}$, $x_2=x_{24},\ldots$ and $H_{123}=n_1$, $H_{124}=n_2$, $H_{125}=n_3$, $H_{134}=n_4$, $H_{135}=n_5$, $H_{145}=n_6$.

We have defined

$$\mathcal{A}^{jkj'k'}=\frac{R_6}{R_1}\sqrt{g}(g^{jj'}g^{kk'}-g^{jk'}g^{kj'})+i\alpha\epsilon^{jkj'k'}. \tag{22}$$

$$x_{jk}\equiv\beta^iH_{ijk}+\frac{i}{4}\gamma^i\mathcal{A}^{-1}{}_{jkj'k'}\epsilon^{j'k'gh}H_{igh} \tag{23}$$

and

$$\tau=\alpha+i\frac{R_6}{R_1}. \tag{24}$$

The form (21) is particularly useful for the zero mode calculation since it allows us to use a generalization of the Poisson summation formula [11]

$$\sum_{n\in Z^p} e^{-\pi(n+x)\cdot A\cdot(n+x))}=(\det A)^{-\frac{1}{2}}\sum_{n\in Z^p} e^{-\pi n\cdot A^{-1}\cdot n}e^{2\pi i n\cdot x}. \tag{25}$$

Indeed, applying (25) we discover that under

$$R_1\to R_1|\tau|,\ R_6\to R_6|\tau|^{-1},\ \alpha\to-|\tau|^{-2}\alpha,\ \beta^i\to\gamma^i,\ \gamma^i\to-\beta^i,\ g_{ij}\to g_{ij}. \tag{26}$$

the zero modes of the partition function transform as

$$Z_{\text{zero modes}}(R_1|\tau|,R_6|\tau|^{-1},g_{ij},-\alpha|\tau|^2,\gamma^i,-\beta^i)=(\det A)^{\frac{1}{2}}Z_{\text{zero modes}}(R_1,R_6,g_{ij},\alpha,\beta^i,\gamma^i) \tag{27}$$

where $\det A=|\tau|^6$.

The transformation (26) is the generalization of the $SL(2, \mathcal{Z})$ generator $\tau \to -\tau^{-1}$ and reduces to it when $\beta_i = \gamma_i = g_{ij} = 0$. It can be checked [1] that (26) is a generator of $SL(6, Z)$. Formula (27) shows that under this $SL(6, \mathcal{Z})$ generator the zero mode piece of the partition function does not remain invariant, but generates an anomaly $\det A = |\tau|^6$. We will see that such anomaly will be cancelled by the sum over the oscillators, as it happens in the case of the non-chiral boson in two dimensions.

In analogy with the modular group $SL(2, \mathcal{Z})$ which can be generated by two transformations such as $\tau \to \tau + 1$ and $\tau \to -\frac{1}{\tau}$, the mapping class groups of the n-torus, *i.e.* the modular groups $SL(n, \mathcal{Z})$, can be generated by just two transformations as well [15]. It turns out that the other generator of $SL(6, Z)$, the analogue of $\tau \to \tau + 1$, leaves invariant both the zero modes and the oscillators [1] and we will not discuss it further.

To sum over the oscillators, the key point to realize is that the chiral two-form in six dimensions has only three independent degrees of freedom. Although B_{MN} has 15 components, by using gauge invariance and the self-duality condition, one can check that the number of independent components reduce to only three, corresponding to the physical degrees of freedom of the six-dimensional chiral two-form with Spin(4) content (3,1). Since oscillators with different polarizations commute, we can treat each polarization separately and cube the end result.

Starting from the normal mode expansion for B_{MN}, the full expression for the partition function turns out to be

$$Z = Z_{\text{zero modes}} \cdot \operatorname{tr} e^{-2i\pi \sum_{\vec{p} \neq 0} p_6 B^{\kappa\dagger}_{\vec{p}} B^{\kappa}_{\vec{p}} - \pi R_6 \sum_{\vec{p}} \sqrt{G_5^{mn} p_m p_n}\, \delta^{\kappa\kappa}} \tag{28}$$

where $Z_{\text{zero modes}}$ is given in (21) . The index κ labels the three independent polarizations and

$$[B^{\kappa\dagger}_{\vec{p}}, B^{\lambda}_{\vec{p}'}] = \delta^{\kappa\lambda}\, \delta_{\vec{p},\vec{p}'} \tag{29}$$

In (28) p_6 is given by

$$\begin{aligned} p_6 &= -\frac{G^{6m}}{G^{66}} p_m - i\sqrt{\frac{G_5^{mn}}{G^{66}} p_m p_n} \\ &= -\alpha p_1 - (\alpha\beta^i + \gamma^i) p_i - iR_6 \sqrt{G_5^{mn} p_m p_n} \end{aligned} \tag{30}$$

where $2 \leq i \leq 5; 1 \leq m, n \leq 5$. This expression for p_6 can be derived by solving the equation of motion for B_{MN} [1]. The ordering chosen in normal mode expansion of B_{MN} gives rise to the vacuum energy $\sqrt{G_5^{mn} p_m p_n}$ which is essential for modular invariance. It is a divergent sum and needs to be regularized.

THE ANOMALY CANCELLATION

We can now use the standard Fock space argument (7) to do the trace on the oscillators in (28). The answer is

$$Z = Z_{\text{zero modes}} \cdot \left(e^{-\pi R_6 \sum_{\vec{n}} \sqrt{G_5^{lm} n_l n_m}} \prod_{\vec{n} \neq \vec{0}} \frac{1}{1 - e^{-i2\pi p_6}}\right)^3. \tag{31}$$

To understand how the $SL(6, \mathcal{Z})$ invariance of Z works, we separate the product on $\vec{n} = (n, n_\perp) \neq \vec{0}$ into a product on (all n, but $n_\perp \neq (0,0,0,0)$) and on ($n \neq 0, n_\perp =$

$(0,0,0,0))$, where $n_\perp \equiv n_i$ and $n \equiv n_1$. Then (31) separates into the contribution of the '2d massless' scalars and the contribution of the '2d massive' scalars. The former are the modes with zero momentum $n_\perp = 0$ in the transverse direction $i = 2...5$, which appear as massless bosons on the two-torus in the directions 1 and 6. Instead, the modes associated with $n_\perp \neq 0$ correspond to massive bosons on the two-torus. Their partition function is $SL(6, \mathcal{Z})$ symmetric by itself, since there is no anomaly for massive states [1].

The only piece of (31) that has an $SL(6, \mathcal{Z})$ anomaly is the one associated with the '2d massless' modes

$$e^{\frac{R_6}{\pi R_1}\zeta(2)} \prod_{n_1 \neq 0} \frac{1}{1 - e^{2\pi i(\alpha n_1 + i\frac{R_6}{R_1}|n_1|)}} = \left(\eta(\tau)\bar{\eta}(\bar{\tau})\right)^{-1}$$

where η is the Dedekind eta function $\eta(\tau) \equiv e^{\frac{\pi i \tau}{12}} \prod_{n=1}^{\infty}(1 - e^{2\pi i \tau n})$, and ζ is the Riemann zeta function. The term $\zeta(2) = \frac{\pi^2}{6}$ in the exponent comes from using ζ function regularization to regularize the infinite sum $\sum_n |n|$.

Under the $SL(6, \mathcal{Z})$ transformation (26) of sect. 3, $\tau \to -\frac{1}{\tau}$ and

$$\left(\eta(\tau)\bar{\eta}(\bar{\tau})\right)^{-3} \to |\tau|^{-3}\left(\eta(\tau)\bar{\eta}(\bar{\tau})\right)^{-3}. \tag{32}$$

Hence the combination $Z_{\text{zero modes}} \cdot \left(\eta(\tau)\bar{\eta}(\bar{\tau})\right)^{-3}$ is $SL(6, \mathcal{Z})$ invariant. This is how the oscillator anomaly cancels the zero mode anomaly in (27), leaving a modular invariant partition function for the fivebrane.

In conclusion, in spite of the lack of a manifestly covariant Lagrangian, the M-theory fivebrane chiral two-form can be consistently quantized on a six-torus. This result depends heavily on the fact that we compactify on T^6, which has allows us to use normal mode expansion techniques, and would not hold automatically on other spaces.

The expectation [2] indeed is that in the case of compactification on more complicated manifolds (for instance on $\Sigma \times \mathbf{CP}^2$, with Σ a Riemann surface), the partition function will not be modular invariant and will depend instead on the spin structure. The general theory is under investigation [16].

Our approach might instead be useful to compute the partition function on the torus for other chiral p-forms that occur in string theory, such as, for instance, the chiral four-form of type IIB strings in ten-dimensions.

REFERENCES

1. Louise Dolan and Chiara R. Nappi, "A Modular Invariant Partition Function for the Fivebrane", Nucl. Phys. B **530** (1998) 683; hep-th 9806016.
2. E. Witten, "Five-brane Effective Action in M-theory," J. Geom. Phys. **22** (1997) 103; hep-th 9610234.
3. P.S. Howe, E. Sezgin and P.C. West, "Covariant Field Equations of the M-theory Five-brane, Nucl.Phys. **B496** (1997) 191.

4. M. Cederwall, B. Nilsson, P. Sundell, "An Action for the super-5-brane in $D = 11$ Supergravity", J. High Energy Physics 04 (1998) 7; hep-th/9712059.

5. E. Bergshoeff, D. Sorokin and P.K. Townsend, "The M-5brane Hamiltonian," hep-th/9805065.

6. M. Perry and J. H. Schwarz, "Interacting Chiral Gauge Fields in Six Dimensions and Born-Infeld Theory'," Nucl. Phys. **B489** (1997) 47; hep-th/9611065.

7. M. Aganagic, J. Park, C. Popescu and J. H. Schwarz, "Worldvolume Action of the M-theory Fivebrane," Nucl. Phys. **B496** (1997) 191; hep-th/9701166.

8. P. Pasti, D. Sorokin and M. Tonin, "Covariant Action for a D=11 Five-Brane with Chiral Field," Phys. Lett. **B398** (1997) 41; hep-th 9701037; "On Lorentz Invariant Actions for Chiral P-Forms," Phys. Rev. **D52** (1995) 4277; hep-th/9711100.

9. I. Bandos, K. Lechner, A. Nurmagambetov, P. Pasti, D. Sorokin and M. Tonin, "On the Equivalence of Different Formulations of M-theory Fivebrane," Phys. Lett. **408B** (1997) 135; R. Manvelyan, R. Mkrtchyan, H.J.W. Muller-Kirsten, "On Different Formulations of Chiral Bosons", hep-th 9901084.

10. R. Floreanini and R. Jackiw, "Self-Dual Fields as Charge-Density Solitons, Phys. Rev. Lett. **59** (1987) 1873; D. Baleanu and Y. Guler, "Quantization of Floreanini-Jackiw chiral harmonic oscillator", hep-th 9901082.

11. M.B. Green, J. H. Schwarz and E. Witten, *Superstring Theory*, vol. I and II, Cambridge University Press: Cambridge, U.K. 1987.

12. J. Polchinski, *String Theory*, Cambridge University Press, 1999.

13. John H. Schwarz, "Coupling a Self-dual Tensor to Gravity in Six Dimensions,"Phys. Lett. **B395** (1997) 191; hep-th/ 9701008.

14. D. Mumford, *Tata Lectures on Theta* vols. I and II, Boston: Birkhauser 1983; L. Alvarez-Gaume, G. Moore, C. Vafa, "Theta Functions, Modular Invariance, and Strings", Comm. Math. Phys. **106** (1986) 1.

15. H.S.M. Coxeter and W.O.J. Moser, *Generators and Relations for Discrete Groups*, Springer-Verlag: New York 1980.

16. M. Hopkins and I. M. Singer, to appear.

SUPERSYMMETRIC WILSON LOOPS AND SUPER NON-ABELIAN STOKES THEOREM

Robert L. Karp and Freydoon Mansouri

Physics Department
University of Cincinnati
Cincinnati, OH 45221

ABSTRACT

We generalize the standard product integral formalism to supersymmetric product integrals. Using these, we provide an unambiguous mathematical representation of supersymmetric Wilson line and Wilson loop operators and study their properties. We also prove the supersymmetric version of non-abelian Stokes theorem.

1 INTRODUCTION

The notions of Wilson line and Wilson loop [1, 2] have found useful applications in the study of non-abelian gauge theories. In the last few years, there have been significant developments in supersymmetric gauge theories [3] and in superstring theories [4]. In view of this, it is natural to expect that the supersymmetric extensions of Wilson line and Wilson loop will play equally important roles in supersymmetric gauge theories. One of the objectives of the present work is to show that such an extension is indeed possible. In the process of demonstrating this, we will also generalize the non- abelian Stokes theorem to the corresponding supersymmetric case. Using a product integral representation for the Wilson line, it has recently been shown [5] that one can give a mathematically rigorous proof of the non-abelian Stokes theorem. By a suitable generalization of the underlying concepts, we will prove the supersymmetric extension of this theorem.

The supersymmetrized non-abelian Wilson loop also generalizes to the non-abelian case the previous work for abelian supersymmetric gauge theories [6]. One of the main difficulties in dealing with the supersymmetric gauge theories is that the components of the supersymmetric connection and curvature must be constrained so that they can describe the correct number of physical degrees of freedom [7]. In the non-abelian version, the surface representation involves the components of both the field strength and the gauge field. As a result, the unconstrained fields involve not only chiral superfields but also the vector superfield. Although the main focus in this work is the supersymmetric generalization of the concepts described in reference [5], we begin with a summary of the results of that work because most of those results remain valid in this generalization.

2 SOME PROPERTIES OF PRODUCT INTEGRALS

One of the initial motivations for the introduction of product integrals was [8] to solve differential equations of the type

$$Y'(s) = A(s)Y(s). \tag{1}$$

In this expression, $Y(s)$ is an n-dimensional vector, $A(s)$ is a matrix valued function, and prime indicates differentiation. So, for two real numbers a and b, the problem is to obtain $Y(b)$ given $Y(a)$. To deal with this problem, we make a partition $P = \{s_0, s_1, \ldots, s_n\}$ of the interval $[a, b]$, let $\Delta s_k = s_k - s_{k-1}$ for $k = 1, \ldots, n$, and set $a = s_0$, $b = s_n$. Then, solving the differential equation in each subinterval, we can write approximately [8]

$$Y(b) \approx= \prod_{k=1}^{n} e^{A(s_k)\Delta s_k} Y(a) \equiv \Pi_p(A). \tag{2}$$

Since $A(s)$ is matrix valued, the order in this product is important. Let $\mu(P)$ be the length of the longest Δs_k in the partition P. Then, as $\mu(P) \to 0$, we get

$$Y(b) = \lim_{\mu(P)\to 0} \Pi_P(A)Y(a). \tag{3}$$

The limit is clearly independent of $Y(a)$.

The limit of the ordered product on the right hand side of Eq.(3) is the fundamental expression in the definition of a product integral [8]. It is formally defined as

$$\prod_a^x e^{A(s)ds} = \lim_{\mu(P)\to 0} \Pi_P(A) \equiv F(x, a). \tag{4}$$

It is easy to see that $F(x, a)$ satisfies the differential equation

$$\frac{d}{dx} F(x, a) = A(x)F(x, a) \tag{5}$$

with $F(a, a) = 1$. The corresponding integral equation is

$$F(x, a) = 1 + \int_a^x ds\, A(s)F(s, a). \tag{6}$$

Clearly, $F(a, a) = 1$, and $F(x, a)$ is unique. Consider now some of the properties of the product integral matrices. If for each $x \epsilon [a, b]$ the product integral is non-singular, then its determinant is given by

$$\det\left(\prod_a^x e^{A(s)ds}\right) = e^{\int_a^x trA(s)ds}. \tag{7}$$

In analogy with the additive property of ordinary integrals, product integrals have the multiplicative property

$$\prod_z^x e^{A(s)ds} = \prod_y^x e^{A(s)ds} \prod_z^y e^{A(s)ds}. \tag{8}$$

Where $x, y, z \epsilon [a, b]$ and $z \leq y \leq x$. Derivatives with respect to the end points are given by

$$\frac{\partial}{\partial x} \prod_y^x e^{A(s)ds} = A(x) \prod_y^x e^{A(s)ds}, \tag{9}$$

and

$$\frac{\partial}{\partial y} \prod_y^x e^{A(s)ds} = - \prod_y^x e^{A(s)ds} A(y). \tag{10}$$

One of the fundamental features of gauge theories is the notion of parallel transport. To see how it can be formulated in product integral formalism, consider a map $P : [a, b] \to \mathbf{C}_{n \times n}$, which is continuously differentiable. Then $P(x)$ is an indefinite product integral if for a given $A(s)$

$$P(x) = \prod_a^x e^{A(s)ds} P(a). \tag{11}$$

Next, we define an operation known as L operation which is like the logarithmic derivative operation on non-singular functions. Let

$$LP(x) = P'(x) P^{-1}(x), \tag{12}$$

where prime indicates differentiation. Then, from Eq. (11) it follows that: $(LP)(x) = A(x)$. This operation implies that $L(PQ)(x) = LP(x) + P(x)(LQ(x))P^{-1}(x)$.

The L operation is a crucial ingredient in establishing the analog of the fundamental theorem of calculus for product integrals. With the map P as defined above, this theorem states that

$$\prod_a^x e^{(LP)(s)ds} = P(x) P^{-1}(a). \tag{13}$$

From the results given above, it follows that P is a solution of the initial value problem: $P'(x) = (LP)(x)P(x)$. With the unique solution given by Eq. (11), this establishes the fundamental theorem of product integration. Just as in ordinary integration, the knowledge of simple product integrals can be used to evaluate more complicated product integrals. For example, one can prove the *sum rule* for product integrals:

$$\prod_a^x e^{[A(s)+B(s)]ds} = P(x) \prod_a^x e^{P^{-1}(s)B(s)P(s)ds}. \tag{14}$$

Finally, we state two other important properties of product integrals which will be used in the sequel. One is the *similarity theorem* which states that

$$P(x) \prod_a^x e^{B(s)ds} P^{-1}(a) = \prod_a^x e^{[LP(s)+P(s)B(s)P^{-1}(s)]ds}. \tag{15}$$

The other property is differentiation with respect to a parameter. Let

$$P(x, y; \lambda) = \prod_y^x e^{A(s;\lambda)ds}, \tag{16}$$

where λ is a parameter. Then the differentiation with respect to this parameter is given by

$$\frac{\partial}{\partial \lambda} P(x, y; \lambda) = \int_y^x ds P(x, s; \lambda) \frac{\partial}{\partial \lambda} A(s; \lambda) P(s, y; \lambda). \tag{17}$$

3 NON-ALBELIAN STOKES THEOREM

To provide the background for using the product integral formalism of Section 2 to explore the physical properties of gauge theories, we begin with the statement of the problem as it arises in the physics context. Let M be an n-dimensional manifold representing the space-time (target space). Let A be a (connection) 1-form on M. When M is a differentiable manifold, we can choose a local basis dx^μ, $\mu = 1, ..., n$, and express A in terms of its components: $A(x) = A_\mu(x)\, dx^\mu$. We take A to have values in the Lie-algebra, or a representation thereof, of a Lie group. Then, with T_k, $k = 1, .., m$, representing the generators of the Lie group, the components of A can be written as $A_\mu(x) = A^k_\mu(x)\, T_k$. With these preliminaries, we can express the Wilson line of the non-abelian gauge theories in the form [9]

$$W_{ab}(C) = \mathcal{P}e^{\int_a^b A}, \tag{18}$$

where $\mathcal{P}$ indicates path ordering, and C is a path in M. When the path C is closed, the corresponding Wilson line becomes a Wilson loop [9]:

$$W(C) = \mathcal{P}e^{\oint A}. \tag{19}$$

The path C in M can be described in terms of an intrinsic parameter σ, so that for points of M which lie on the path C, $x^\mu = x^\mu(\sigma)$. One can then write $A_\mu(x(\sigma))dx^\mu = A(\sigma)d\sigma$, where

$$A(\sigma) \equiv A_\mu(x(\sigma))\frac{dx^\mu(\sigma)}{d\sigma}. \tag{20}$$

It is the quantity $A(\sigma)$, and the variations thereof, which we will identify with the matrix valued functions of the product integral formalism.

Let us next consider the Wilson loop. We take the 2-surface bounded by the loop to be an orientable submanifold of M. It will be convenient to describe the properties of the 2-surface in terms of its intrinsic parameters σ and τ or σ^a, $a = 0, 1$. So, for the points of the manifold M, which lie on Σ, we have $x = x(\sigma, \tau)$. The components of the 1-form A on Σ can be obtained by means of the vielbeins $v^\mu_a = \partial_a\, x^\mu(\sigma)$. (by standard pull-back construction). Thus, we get $A_a = v^\mu_a\, A_\mu$. The curvature 2-form F of the connection A is given by

$$F = dA + A \wedge A = \frac{1}{2}F_{\mu\nu}\, dx^\mu \wedge dx^\nu. \tag{21}$$

The components of F on Σ can again be obtained by means of the vielbeins:

$$F_{ab} = v^\mu_a\, v^\nu_b\, F_{\mu\nu}. \tag{22}$$

Let us now turn to the definition of Wilson line in terms of product integrals. Consider the continuous map $A : [s_0, s_1] \to \mathbf{C}_{n\times n}$ where $[s_0, s_1]$ is a real interval. Then, we define the Wilson line given above in terms of a product integral as follows:

$$\mathcal{P}e^{\int_{s_0}^{s_1} A(s)ds} \equiv \prod_{s_0}^{s_1} e^{A(s)ds}. \tag{23}$$

Anticipating that we will identify the closed path C over which the Wilson loop is defined with the boundary of a 2-surface, it is convenient to work from the beginning with the matrix valued functions $A(\sigma, \tau)$. This means that our expression for the Wilson line will depend on a parameter. That is, let $A : [\sigma_0, \sigma_1] \times [\tau_0, \tau_1] \to \mathbf{C}_{n\times n}$, where $[\sigma_0, \sigma_1]$ and $[\tau_0, \tau_1]$ are real intervals on the two surface Σ and hence in M. Then, we define a Wilson line

$$P(\sigma, \sigma_0; \tau) = \prod_{\sigma_0}^{\sigma} e^{A_1(\sigma';\tau)d\sigma'} \equiv \mathcal{P}e^{\int_{\sigma_0}^{\sigma} A_1(\sigma';\tau)d\sigma'}. \tag{24}$$

In this expression, $\mathcal{P}$ indicates path ordering with respect to σ, while τ is a parameter. To be able to describe a Wilson loop, we similarly define the Wilson line

$$Q(\sigma; \tau, \tau_0) = \prod_{\tau_0}^{\tau} e^{A_0(\sigma;\tau')d\tau'} \equiv \mathcal{P}e^{\int_{\tau_0}^{\tau} A_0(\sigma;\tau')d\tau'}. \tag{25}$$

In this case, the path ordering is with respect to τ, and σ is a parameter.

In terms of the intrinsic coordinates of such a surface, we can write the Wilson loop operator in the form

$$W(C) = \mathcal{P}e^{\oint A_a d\sigma^a}, \tag{26}$$

where, as mentioned above, $\sigma^a = (\tau, \sigma)$; $a = (0, 1)$. The expression for the Wilson loop depends on the homotopy class of paths in M to which the closed path C belongs. We can, therefore, parameterize the path C in any convenient manner consistent with its homotopy class. In particular, we can break up the path into segments along which either σ or τ remains constant. More explicitly, we write

$$W = W_4 W_3 W_2 W_1. \tag{27}$$

In this expression, W_k, $k = 1, .., 4$, are Wilson lines such that $\tau = const.$ along W_1 and W_3, and $\sigma = const.$ along W_2 and W_4.

To see the advantage of parametrizing the closed path in this manner, consider the exponent of Eq. (26):

$$A_a d\sigma^a = A_0 d\tau + A_1 d\sigma. \tag{28}$$

Along each segment, one or the other of the terms on the right hand side vanishes. For example, along the segment $[\sigma_0, \sigma]$, we have $\tau' = \tau_0 = const.$. As a result, we get for the Wilson lines W_1 and W_2, respectively,

$$W_1 = \prod_{\sigma_0}^{\sigma} e^{A_1(\sigma';\tau_0)d\sigma'} \equiv \mathcal{P}e^{\int_{\sigma_0}^{\sigma} A_1(\sigma';\tau_0)d\sigma'} = P(\sigma, \sigma_0; \tau_0), \tag{29}$$

and

$$W_2 = \prod_{\tau_0}^{\tau} e^{A_0(\sigma;\tau')d\tau'} \equiv \mathcal{P}e^{\int_{\tau_0}^{\tau} A_0(\sigma;\tau')d\tau'} = Q(\sigma; \tau, \tau_0). \tag{30}$$

When the 2-surface Σ requires more than one coordinate patch to cover it, the connections in different coordinate patches must be related to each other in their overlap region by transition functions [10]. Then, the description of Wilson loop in terms of Wilson lines given in Eq. (27) must be suitably augmented to take this complication into account. The product integral representation of the Wilson line and the composition rule for product integrals given by Eq. (8) will still make it possible to describe the corresponding Wilson loop as a composite product integral. For definiteness, we will confine ourselves to the representation given by Eq. (27).

It is convenient for later purposes to define two composite Wilson line operators U and T according to

$$U(\sigma, \tau) = Q(\sigma; \tau, \tau_0)\, P(\sigma, \sigma_0; \tau); \quad T(\sigma; \tau) = P(\sigma, \sigma_0; \tau)\, Q(\sigma_0; \tau, \tau_0). \tag{31}$$

Using the first of these, we have: $W_2 W_1 = U(\sigma, \tau)$. Similarly, we have for the two remaining Wilson lines $W_3 = P^{-1}(\sigma, \sigma_0; \tau)$; $W_4 = Q^{-1}(\sigma_0; \tau, \tau_0)$. From the Eq. (31), it follows that $W_4 W_3 = T^{-1}(\sigma, \tau)$. In terms of the quantities T and U, the Wilson loop operator will take the compact form

$$W = T^{-1}(\sigma; \tau) U(\sigma; \tau). \tag{32}$$

As a first step in the proof of the non-abelian Stokes theorem, we obtain the action of the L operator on W. It is easy to show that

$$L_\tau W = T^{-1}(\sigma, \tau)[A_0(\sigma, \tau) - L_\tau T(\sigma, \tau)]T(\sigma, \tau). \tag{33}$$

Next, we prove the analog of Eq. (13), which applies to an elementary Wilson line, for the composite Wilson loop operator defined in Eq. (27). We have shown [5] that the Wilson loop operator can be expressed in the form

$$W = \prod_{\tau_0}^{\tau} e^{T^{-1}(\sigma, \tau')[A_0(\sigma, \tau') - L_\tau T(\sigma, \tau')]T(\sigma, \tau') d\tau'}. \tag{34}$$

Finally, we can prove [5] that the above Wilson loop operator can be expressed as a surface integral of the field strength:

$$W = \prod_{\tau_0}^{\tau} e^{\int_{\sigma_0}^{\sigma} T^{-1}(\sigma'; \tau') F_{01}(\sigma'; \tau') T(\sigma'; \tau') d\tau'}, \tag{35}$$

where F_{01} is the 0-1 component of the non-abelian field strength. We note that in this expression the ordering of the operators is defined with respect to τ whereas σ is a parameter. Recalling the antisymmetry of the components of the field strength, we can rewrite this expression in terms of path ordered exponentials familiar from the physics literature:

$$W = \mathcal{P}_\tau e^{\frac{1}{2} \int_\Sigma d\sigma^{ab} T^{-1}(\sigma; \tau) F_{ab}(\sigma; \tau) T(\sigma; \tau)}, \tag{36}$$

where $d\sigma^{ab}$ is the area element of the 2-surface. Despite appearances, it must be remembered that σ and τ play very different roles in this expression.

4 A SECOND PROOF

To illustrate the power and the flexibility of the product integral formalism, we give here a variant of the previous proof for the non-abelian Stokes theorem. This time the proof makes essential use of the differentiation with respect to a parameter given by Eq. (17). We start by evaluating the analog of the logarithmic derivative of W. It is given by [5]

$$\begin{aligned} L_\tau W = \frac{\partial W}{\partial \tau} W^{-1} = \; & T^{-1}(\sigma; \tau) \left[A_0(\sigma; \tau) - P(\sigma, \sigma_0; \tau) A_0(\sigma_0; \tau) P^{-1}(\sigma, \sigma_0; \tau) \right. \\ & \left. - \partial_\tau P(\sigma, \sigma_0; \tau) P^{-1}(\sigma, \sigma_0; \tau) \right] T(\sigma; \tau). \end{aligned} \tag{37}$$

Now we can use Eq. (17) to evaluate the derivative of the product integral with respect to the parameter τ:

$$\partial_\tau P(\sigma, \sigma_0; \tau) = \int_{\sigma_0}^{\sigma} d\sigma' P(\sigma, \sigma'; \tau) \partial_\tau A_1(\sigma'; \tau) P(\sigma', \sigma_0; \tau). \tag{38}$$

Then, after some simple manipulations using the defining equations for the various terms in Eq. (37), we get:

$$T^{-1}(\sigma;\tau)\partial_\tau P(\tau)P^{-1}(\tau)T(\sigma;\tau) = \int_{\sigma_0}^{\sigma} d\sigma' T^{-1}(\sigma';\tau)\partial_\tau A_1(\sigma';\tau)T(\sigma';\tau). \quad (39)$$

Using Eq. (8) and the fact that $P(\sigma_0,\sigma_0;\tau)=1$, we can write the rest of Eq. (37) as an integral too:

$$T^{-1}(\sigma;\tau)[A_0(\sigma;\tau) - P(\sigma,\sigma_0;\tau)A_0(\sigma_0;\tau)P^{-1}(\sigma,\sigma_0;\tau)]T(\sigma;\tau) =$$
$$= \int_{\sigma_0}^{\sigma} d\sigma'\, P^{-1}(\sigma',\sigma_0;\tau)(\partial_\tau A_0(\sigma',\tau) + [A_0(\sigma',\tau), A_1(\sigma',\tau)])P(\sigma',\sigma_0;\tau). \quad (40)$$

Combining, we obtain:

$$L_\tau W = \frac{\partial W}{\partial \tau} W^{-1} = \int_{\sigma_0}^{\sigma} d\sigma' T^{-1}(\sigma',\tau)F_{01}(\sigma',\tau)T(\sigma',\tau). \quad (41)$$

Using Eq. (13), we are immediately led to Eq. (35) which was obtained by the previous method of proof.

5 CONVERGENCE ISSUES

The definitions of Wilson lines and Wilson loops as currently conceived in the physics literature involve exponentials of operators. The standard method of making sense out of such exponential operators in the physics literature is through their power series expansion:

$$\mathcal{P}e^{\int_a^b A(x)dx} = \sum_{n=0}^{\infty} \frac{1}{n!}\mathcal{P}\left(\int_a^b A(x)dx\right)^n, \quad (42)$$

where a typical path ordered term in the sum has the form

$$\frac{1}{n!}\mathcal{P}(\int_a^b A(x)dx)^n = \int_a^b dx_1 \int_a^{x_1} dx_2 \ldots \int_a^{x_{n-1}} dx_n\, A(x_1)A(x_2)\ldots A(x_n). \quad (43)$$

Such a power series expansion is purely formal, and it is not clear *á priori* that the series (42) is well defined and convergent. Indeed, in previous attempts at proving the non-abelian Stokes theorem by other methods, the convergence of such series has been assumed. One important advantage of our product integral approach is that we can make the definition of the path ordered exponentials precise. This will enable us to prove that the series of partial sums converges uniformly to the product integral. The proof is contained, as a special case, in the following two theorems valid for all product integrals [8]: Given the continuous function $A : [a,b] \to \mathbf{C}_{n\times n}$, and given $x, y \in [a,b]$, let $L(x,y) = \int_x^y ||A(s)||ds$. Also let $J_0(x,y) = I$, and for $n \geq 1$ define iteratively $J_n(x,y) := \int_x^y A(s)J_{n-1}(s,y)ds$. Then for any $n \geq 0$ the following holds:

$$||\prod_x^y e^{A(s)ds} - \sum_{k=0}^{n} J_k(x,y)|| \leq \frac{1}{(n+1)!}|L(x,y)|^{n+1}e^{L(a,b)}. \quad (44)$$

This estimate is uniform for all x, y in the interval $[a,b]$. It then follows that, with A and $J_k(x,y)$ given above, we have, in the same notation,

$$\prod_x^y e^{A(s)ds} = \sum_{k=0}^{\infty} J_k(x,y). \quad (45)$$

The series on the right hand side of this expression converges uniformly for any $x, y \in [a,b]$.

6 THE GAUGE TRANSFORMS OF WILSON LINES AND WILSON LOOPS

Under a gauge transformation, the components of the connection, i.e. the gauge potentials, transform according to [9]: $A_\mu(x) \longrightarrow g(x)A_\mu(x)g^{-1}(x) - g(x)\partial_\mu g(x)^{-1}$. The components of the field strength (curvature) transform covariantly: $F_{\mu\nu}(x) \longrightarrow g(x)F_{\mu\nu}(x)g^{-1}(x)$. Using the product integral formalism, we want to derive the effect of these gauge transformations on Wilson lines and Wilson loops.

Let us start with the Wilson line defined by Eq. (24). Under the gauge transformation this quantity transforms as

$$P(\sigma,\sigma_0;\tau) = \prod_{\sigma_0}^{\sigma} e^{A_1(\sigma';\tau)d\sigma'} \longrightarrow \prod_{\sigma_0}^{\sigma} e^{[g(\sigma';\tau)A_1(\sigma';\tau)g^{-1}(\sigma';\tau) - g(\sigma';\tau)\partial_\sigma g^{-1}(\sigma';\tau)]d\sigma'}. \tag{46}$$

By Eq. (14), the gauge transformed Wilson line is

$$\prod_{\sigma_0}^{\sigma} e^{[g(\sigma';\tau)A_1(\sigma';\tau)g^{-1}(\sigma';\tau) + L_\sigma g(\sigma';\tau)]d\sigma'}. \tag{47}$$

Moreover, it follows that $\prod_{\sigma_0}^{\sigma} e^{L_\sigma g(\sigma';\tau)d\sigma'} = g(\sigma;\tau)g^{-1}(\sigma_0;\tau)$. Then, the gauge transform of $P(\sigma,\sigma_0;\tau)$ will take the form

$$g(\sigma;\tau)g^{-1}(\sigma_0;\tau)\prod_{\sigma_0}^{\sigma} e^{g(\sigma_0;\tau)A_1(\sigma';\tau)g^{-1}(\sigma_0;\tau)}. \tag{48}$$

Finally, we can readily see that the constant terms in the exponents can be factored from the product integral so that we get $P(\sigma,\sigma_0;\tau) \longrightarrow g(\sigma;\tau)P(\sigma,\sigma_0;\tau)g^{-1}(\sigma_0;\tau)$. In the physicist's notation, the result can be stated as

$$\mathcal{P}e^{\int_a^b A_\mu(x)dx^\mu} \longrightarrow g(b)\left(\mathcal{P}e^{\int_a^b A_\mu(x)dx^\mu}\right)g(a). \tag{49}$$

For a closed path, the points a and b coincide. As a result, the corresponding Wilson loop operator transforms gauge covariantly.

For consistency, we expect that the surface integral representation of the Wilson loop also transforms covariantly under gauge transformations. To show this explicitly, we note from Eq. (32) that in this case we need to know how the operator $T(\sigma,\tau)$ transforms under gauge transformations. It is easy to see that the transform of this composite Wilson line is given by

$$T(\sigma;\tau) = P(\sigma,\sigma_0;\tau)\,Q(\sigma_0;\tau,\tau_0) \longrightarrow g(\sigma;\tau)T(\sigma;\tau)g^{-1}(\sigma_0;\tau_0). \tag{50}$$

Then, it is straight forward to show that the surface integral representation of Wilson loop transforms gauge covariantly:

$$W \longrightarrow g(\sigma_0;\tau_0)\prod_{\tau_0}^{\tau} e^{\int_{\sigma_0}^{\sigma} T^{-1}(\sigma';\tau')F_{10}(\sigma';\tau')T(\sigma';\tau')dt'} g^{-1}(\sigma_0;\tau_0). \tag{51}$$

In the physics notation, this transformation law takes the form

$$\mathcal{P}e^{\oint_C A_\mu(x)dx^\mu} \longrightarrow g(a)\left(\mathcal{P}e^{\oint_C A_\mu(x)dx^\mu}\right)g^{-1}(a), \tag{52}$$

where a is a point on the loop C.

An important consequence of the gauge covariance of the Wilson loop operator is that the trace of Wilson loop is gauge invariant.

7 SUPERSYMMETRIC PRODUCT INTEGRALS

To provide a supersymmetric generalization of the notions of Wilson line and Wilson loop in terms of product integrals, we must address two questions. The first question has to do with the fact that in a supersymmetric gauge theory quantities such as connection and curvature have values in a Grassmann algebra. To be able to explore the properties of these theories in terms of product integrals, we must first ensure that Grassmann valued product integrals exist and are consistent. The second question has to do with the peculiarities of supersymmetric gauge theories. We will address the first question in this section and leave the second one for the next two sections.

To generalize the product integral formalism to the supersymmetric case, we must reexamine every one of the properties of ordinary product integrals. In so far as product integrals are products of exponentials of matrix valued functions, we expect that almost all of their properties can be extended to the supersymmetric case. This is because in a supersymmetric product integral, the analog of the quantity $A(s)$ in the exponent must belong to the even element of the Grassmann algebra. So must the quantity ds. To establish the existence of supersymmetric product integrals, one of the crucial steps is to obtain the estimates of the norms of the supermatrices and their inverses. In other words, we must specify a norm on the Grassmann algebra of bosonic and fermionic variables. A normed algebra that is complete with respect to the topology induced by the norm is known as a Banach algebra.

The Banach algebra structure of the Grassmann algebra has been considered in the literature [11]. Consider for definiteness the finite dimensional complex vector space $\mathbf{C}^n$, and the Grassmann algebra generated by the anticommuting quantities $\theta^1, \theta^2, \ldots, \theta^m$. In this case, a generic element of the algebra can be written as a linear combination of the generators $\theta^{i_1}\theta^{i_2}\ldots\theta^{i_k}$, $k = 0, \ldots, m$, with complex coefficients $a_{i_1 i_2 \ldots i_k}$. In other words, as a vector space the Grassmann algebra is 2^m dimensional.

The norm on the above vector space can be defined as the sum of the moduli of the coefficients. For example the Grassmann algebra generated by a single θ has general element $x = a + b\theta$. Then, with $a, b \in \mathbf{C}$, we define the norm as $||x|| = |a| + |b|$, where $|a|, |b|$ are the complex moduli. Here we sidestep such questions as the uniqueness of the norm, etc. They have been addressed elsewhere [12]. From this definition of the norm, one can show that the norm of the product of any two elements x and y of the Grassmann algebra satisfies the following inequality: $||x \cdot y|| \leq ||x|| \cdot ||y||$. This result is true not only for the above simple example but for the general Grassmann algebra generated by $\theta^1, \theta^2, \ldots, \theta^m$. In other words, with respect to this norm, the Grassmann algebra becomes a Banach algebra. This identification allows us to extend to supersymmetric product integrals almost all the theorems [8] which establish the properties of ordinary product integrals.

Another crucial issue is the inversion of Grassmann valued matrices and their determinants. Since such matrices of interest to us form a super group, they are necessarily invertible. However, in the expression for the determinant the operation of trace must, for consistency, be replaced with supertrace operation. We also have a theorem for the inversion of an element of the Grassmann algebra (see [11]), namely it is invertible if its bosonic part is invertible (in other words non-zero). This is always true for our case of exponentiated objects.

Let us next consider the impact of the above generalization on the proof of the (super) non-abelian Stokes theorem [8, 5]. All the constructions such as the step function and the point value approximant that were given in previous sections and were discussed in more detail in [5] go through in the context of Grassmann valued matrices.

We are also able to demonstrate the uniform convergence of the norms. The theorem leading to the definition of the product integral is altered somewhat to accommodate the new definition of norms but nevertheless remains valid. Other theorems such as those concerning the differentiability of the product integral, determinant, invertability, the L-operation, sum rule, similarity rule, and the differentiation with respect to a parameter are valid with respect to the new definition of the norm. Since in the non-supersymmetric case our proofs of the non-abelian Stokes theorem rely on these theorems it is natural to expect that the proofs remain valid in the supersymmetric case also. We validate this expectation by a step by step analysis of all aspects of the proof [12].

8 SUPERSYMMETRIC WILSON LINES AND LOOPS

The general conclusion which we wish to draw from the arguments presented in the previous section is that we can express supersymmetric Wilson lines and Wilson loops in terms of supersymmetric product integrals. What remains is to consider in detail how supersymmetry affects the structure of the matrix valued functions which enter the expressions for Wilson line and Wilson loop. In other words, we seek the supersymmetric generalizations of the developments in section 3. We will follow closely the notation of reference [6].

Let, in standard two component spinor notation, the local coordinates of a superspace be given by $Z^A = (x^{\alpha\dot{\alpha}}, \theta^\alpha, \theta^{\dot{\alpha}})$. Also let the components of a supersymmetric connection Γ be given by $\Gamma_A = \Gamma_A(x^{\alpha\dot{\alpha}}, \theta^\alpha, \theta^{\dot{\alpha}})$. In terms of local coordinates, the connection Γ with values in a super Lie algebra can be expressed as $\Gamma = \Gamma_A dZ^A$. Then, in the notation of Section 3, the supersymmetric Wilson line can be written as

$$W_{ab}(C) = \mathcal{P}e^{\int_a^b \Gamma}, \tag{53}$$

where $\mathcal{P}$ indicates path ordering, and C is a path not in superspace but in M. When the path C is closed, the corresponding Wilson line becomes a Wilson loop:

$$W(C) = \mathcal{P}e^{\oint \Gamma}. \tag{54}$$

Just as in the non-supersymmetric case, the path C in M can be described in terms of an intrinsic parameter σ, so that for points of Z which lie on the path C, $Z^A = Z^A(\sigma)$. One can then write $\Gamma_A(Z(\sigma))\, dZ^A = \Gamma(\sigma)d\sigma$, where

$$\Gamma(\sigma) \equiv \Gamma_\mu \frac{dx^\mu}{d\sigma} + \Gamma_\alpha \frac{d\theta^\alpha}{d\sigma} + \Gamma_{\dot{\alpha}} \frac{d\theta^{\dot{\alpha}}}{d\sigma}. \tag{55}$$

It is this supersymmetrized version of $A(\sigma)$, and the variations thereof defined on the 2-surface bounded by the Wilson loop, which we will identify with the matrix valued functions of the supersymmetric product integral formalism. To obtain the pull-back of the supersymmetric connection Γ, on the 2-surface, we must replace the bosonic vielbeins by their supersymmetric counter parts [6]. Then, the components of the supersymmetric field strength F_{AB} on the 2-surface will take the form

$$F_{ab} = v_a^A v_b^B F_{AB}. \tag{56}$$

With these generalizations, the rest of the developments of Section 3 from Eq. (23) on will go through for the supersymmetric theories if we simply replace $A(s)$ with $\Gamma(s)$. This will amount to a proof of super non-abelian Stokes theorem in terms of super product integrals. In other words, with the appropriate interpretation of the symbols involved, Eq. (36) is equally applicable to supersymmetric gauge theories.

9 THE USE OF UNCONSTRAINED SUPERFIELDS

Although the mathematical proof of the super non-abelian Stokes theorem was sketched above, we must still clarify the physical content of the theory. It is well known that in supersymmetric gauge theories the field strength F_{AB} contains more degrees of freedom than is required by supersymmetry and gauge invariance. As a result, it is necessary to impose constraints on the components of the field strength to eliminate the unphysical degrees of freedom [7]. They are taken to be

$$F_{\alpha\beta} = F_{\alpha\dot{\beta}} = F_{\dot{\alpha}\dot{\beta}} = 0. \tag{57}$$

It is then straight forward to show [7] that the non-vanishing components of the field strength can be expressed in terms of an unconstrained chiral superfield W_α. Moreover, the field strength is derivable from the connection Γ according to

$$F_{AB} = D_{[A}\Gamma_{B\}} - T^{C}_{AB}\Gamma_C. \tag{58}$$

In this expression, D is covariant derivative, T is torsion tensor, and the quantity $[A, B\}$ represents a commutator except when both A and B are fermionic indices. In that case, it is an anticommutator. The above constraints on the components of the field strength also impose constraints on the components Γ_A of the connection. It follows that we must also express the components of the connection in terms of unconstrained superfields. To this end, we first define a gauge and supersymmetric covariant derivative ∇_A, with $A = (\alpha\dot{\alpha}, \alpha, \dot{\alpha})$, such that, acting on a superfield Φ, it transforms as

$$(\nabla_A\Phi)' = e^{-i\Lambda}(\nabla_A\Phi). \tag{59}$$

Here Λ is also a chiral superfield satisfying $\bar{D}_{\dot{\alpha}}\Lambda = D_\alpha\Lambda^\dagger = 0$. This implies that

$$\nabla'_A = e^{-i\Lambda}\nabla_A e^{i\Lambda}. \tag{60}$$

Keeping in mind the structure of the gauge covariant derivative in non-supersymmetric gauge theories, we must determine the form of ∇_A so that it would be consistent with it in the absence of supersymmetry. Let us choose (chiral representation) $\nabla_{\dot{\alpha}} = \bar{D}_{\dot{\alpha}}$, so that $\nabla'_{\dot{\alpha}} = \nabla_{\dot{\alpha}}$. Moreover, we define

$$\nabla_\alpha = e^{-V}D_\alpha e^V, \tag{61}$$

where V is a vector superfield. It is easy to show that this ∇_α has the correct transformation properties. Then, the vector component of ∇ can be expressed as $\nabla_{\alpha\dot{\alpha}} = -\frac{i}{2}\{\nabla_\alpha, \nabla_{\dot{\alpha}}\}$. Writing

$$\nabla_A = D_A + \Gamma_A, \tag{62}$$

we can see that we must identify the supersymmetric connection Γ_α according to

$$\Gamma_\alpha = e^{-V}(D_\alpha e^V). \tag{63}$$

It transforms correctly as a gauge field provided $e^{V'} = e^{-i\Lambda^\dagger}e^V e^\Lambda$. In Wess-Zumino gauge, the expression for Γ_α simplifies to $-i\Gamma_\alpha = D_\alpha V + \frac{1}{2}[D_\alpha V, V]$. Just as in the abelian case, we define the non-abelian chiral superfield as

$$W_\alpha = ig^{-1}\bar{D}^2\Gamma_\alpha = g^{-1}\bar{D}^2e^{-V}D_\alpha e^V. \tag{64}$$

With the components of the connection Γ and the chiral superfield W constructed according to the above recipe, it is straight forward to show that all the constraint equations are satisfied. Therefore, the correct expressions for the components of the connection Γ which enter the definitions of the supersymmetric Wilson line and Wilson loop are the ones given above.

This work was supported in part by the Department of Energy under the contract number DOE-FGO2-84ER40153. The hospitality of Aspen Center for Physics in the Summer of 1998 is also gratefully acknowledged.

REFERENCES

[1] K.G. Wilson, Phys. Rev. **D 10**, 2445 (1974).

[2] A.M. Polyakov, Phys. Lett. **59B**, 82 (1975).

[3] N. Seiberg and E. Witten, Nucl. Phys. **B 426**, 19 (1994).

[4] J.G. Polchinski, *String Theory*, vol I,II
Cambridge Monographs on Mathematical Physics, 1998.

[5] R.L. Karp, F. Mansouri, J.S. Rno, Cincinnati preprint UCTP101.99.

[6] M. Awada and F. Mansouri, Phys. Lett. **B 384**, 111 (1996) and **B 387**, 75 (1996).

[7] J. Wess and J. Bagger, *Introduction to Supersymmetry*, second edition,
Princeton University Press, 1992.

[8] J.D. Dollard, C.N. Friedman, *Product Integration*, Addison Wesley, 1979.

[9] Our conventions follow closely M.E. Peskin, D.V. Schroeder, *An Introduction to Quantum Field Theory*, Addison-Wesley, 1995.

[10] T.T. Wu and C.N. Yang, Phys. Rev. **D12**, 3845 (1975).

[11] R. Cianci, *Introduction to supermanifolds*,
Monographs and textbooks in physical science, 1990.

[12] R.L. Karp, F. Mansouri, in preparation.

TYPE IIB STRING THEORY ON $ADS_3 \times S^3 \times T^4$

L. Dolan

Department of Physics
University of North Carolina
Chapel Hill, North Carolina 27599-3255

INTRODUCTION

String theory in a background of anti-de Sitter(AdS) space times a compact manifold has recently proved important in probing the large N limit of a conformal field theory that lives on the AdS boundary [1−3]. In order to investigate this duality further, it is useful to formulate a string perturbation theory for the string on AdS, in terms of vertex operators, so that string tree amplitudes can be calculated for non-zero values of α', thus extending the low energy supergravity field theory calculation. This enables one to calculate in the boundary conformal field theory at small as well as large 't Hooft coupling.

As an example of this holographic principle, we discuss the perturbative formulation of type IIB superstring theory on $AdS_3 \times S^3 \times T^4$, which ultimately could describe states of a $D = 2$ superconformal field theory (SCFT) on the boundary of AdS_3. See [1-11].

In order to satisfy the equations of motion, where the metric is that of $AdS_3 \times S^3 \times T^4$, both the dilaton and a two-form field must take on non-vanishing vacuum expectation values (vevs). We will primarily discuss the case for a Neveu-Schwarz (NS) two-form field with non-zero vev, where the vertex operators factor into holomorphic and anti-holomorphic pieces. For Ramond-Ramond (RR) two-forms with non-zero vevs, the vertex operators are more complicated[10].

In this proceedings, we describe the worldsheet conformal field theory of the type IIB string on $AdS_3 \times S^3 \times T^4$ in terms of its current algebra. Also we give the Kaluza-Klein spectrum derived by first compactifying type IIB supergravity in ten dimensions on T^4 to six dimensions, and then compactifying the resulting $D = 6$ massless theory on $AdS_3 \times S^3$ to three dimensions, keeping all Kaluza-Klein levels at this last stage. In this tower, we identify the massless three-dimensional supermultiplets, their representations under the gauge group $SU(2)^2$, and their vertex operators.

Confluence of Cosmology, Massive Neutrinos, Elementary Particles, and Gravitation
Edited by Kursunoglu *et al.*, Kluwer Academic / Plenum Publishers, New York, 1999.

WORLDSHEET CONFORMAL FIELD THEORY

As one of the simpler examples of string theory on anti-de Sitter space, we describe type IIB string theory on $AdS_3 \times S^3 \times T^4$. For a Minkowsi metric on AdS_3, this space is the group manifold of $SU(1,1)$ and has isometries $SU(1,1)^2$. (Note that for AdS_3 with a Euclidean metric, the space is a coset $SL(2,C)/SU(2)$ which has isometries $SL(2,C)$). Since S^3 is the group manifold of $SU(2)$, we can describe the worldsheet conformal field theory by the following commuting current algebras:

The $SU(1,1)$ Kac-Moody current is

$$J^A(z) = Q^A(z) + \check{J}^A(z) \tag{1}$$

where

$$\check{J}^A(z) \equiv -\frac{i}{2k} f^A{}_{BC} \psi^B(z) \psi^C(z) \tag{2a}$$

$$\psi^A(z)\psi^B(\zeta) = k\eta^{AB}(z-\zeta)^{-1} \tag{2b}$$

$$\check{J}^A(z)\check{J}^B(\zeta) = \frac{\check{k}\eta_{AB}}{(z-\zeta)^2} + \frac{if^{AB}{}_C \check{J}^C(\zeta)}{(z-\zeta)} \tag{3a}$$

and $\check{k} = -1$. Here $\eta^{AB} = \eta_{AB} = \mathrm{diag}(1,1,-1)$ is the metric on the group manifold of $SU(1,1) \sim SL(2,R)$, with $f^{ABC} = f^{AB}{}_D \eta^{DC} = \epsilon^{ABC}$ and $\epsilon^{123} = 1$, and $f^{AB}{}_C f_{ABE} = C_\psi \eta_{CE}$ with $C_\psi = -2$. Also,

$$Q^A(z)\psi^B(\zeta) = 0 \qquad \text{so} \qquad Q^A(z)\check{J}^B(\zeta) = 0\,,$$

$$Q^A(z)Q^B(\zeta) = \frac{k_Q\,\eta_{AB}}{(z-\zeta)^2} + \frac{if^{AB}{}_C Q^C(\zeta)}{(z-\zeta)} \tag{3b}$$

$$J^A(z)J^B(\zeta) = \frac{k\,\eta_{AB}}{(z-\zeta)^2} + \frac{if^{AB}{}_C J^C(\zeta)}{(z-\zeta)} \tag{3c}$$

where $k = k_Q - 1$, and the level of the Kac-Moody algebra (KMA) in (3c) is $x = \frac{2k}{\psi^2} = \frac{2k}{C_\psi}\tilde{h} = 2k$, since $\tilde{h} = -2$ is the dual Coxeter number of the Lie algebra $SU(1,1)$, and $\psi^2 = 1$ in our normalization. The level of the KMA in (3a) is $\check{x} = 2\check{k} = -2$, and in (3b) the level is $x_Q = 2k_Q$. The SU(1,1) super Kac-Moody algebra is (2b), (3c) together with

$$J^A(z)\psi^B(\zeta) = if^{AB}{}_C \psi^C(\zeta)(z-\zeta)^{-1}\,. \tag{4}$$

The $SU(2)$ current is

$$T^a(z) = q^a(z) + \check{T}^a(z) \tag{5}$$

where

$$\check{T}^a(z) \equiv -\frac{i}{2k}\epsilon_{abc}\chi^b(z)\chi^c(z) \tag{6a}$$

$$\chi^a(z)\chi^b(\zeta) = k\eta^{ab}(z-\zeta)^{-1} \tag{6b}$$

$$\check{T}^a(z)\check{T}^b(\zeta) = \frac{\check{k}\,\delta^{ab}}{(z-\zeta)^2} + \frac{i\epsilon_{abc}\check{T}^c(\zeta)}{(z-\zeta)} \tag{7a}$$

where $\check{k} = 1$ and $\epsilon_{abc}\epsilon_{abe} = c_\psi \delta_{ce}$, with $c_\psi = 2$ and $\epsilon_{123} = 1$. Also,

$$q^a(z)\chi^b(\zeta) = 0 \quad \text{so} \quad q^a(z)\check{T}^b(\zeta) = 0\,,$$

$$q^a(z)q^b(\zeta) = \frac{k_q\,\delta_{ab}}{(z-\zeta)^2} + \frac{if_{abc}q^c(\zeta)}{(z-\zeta)} \tag{7b}$$

$$T^a(z)T^b(\zeta) = \frac{k\,\delta_{ab}}{(z-\zeta)^2} + \frac{if_{abc}T^c(\zeta)}{(z-\zeta)} \tag{7c}$$

where $k = k_q + 1$, and the level of the Kac-Moody algebra (KMA) in (7c) is $x = \frac{2k}{\psi^2} = \frac{2k}{c_\psi}\tilde{h} = 2k$, since $\tilde{h} = 2$ is the dual Coxeter number of the Lie algebra $SU(2)$, and $\psi^2 = 1$. The level of the KMA in (7a) is $\check{x} = 2$, and in (7b) the level is $x_q = 2k_q$. The SU(2) super KMA is (6b), (7c) with

$$T^a(z)\chi^b(\zeta) = i\epsilon_{abc}\chi^c(\zeta)(z-\zeta)^{-1}\,. \tag{8}$$

The T^4 compactification has a $U(1)^4$ SKMA and a worldsheet superVirasoro algebra with $c = 6$:

$$a^i(z)a^j(\zeta) = (z-\zeta)^{-2}\delta^{ij} \tag{9a}$$

$$\lambda^i(z)\lambda^j(\zeta) = (z-\zeta)^{-1}\delta^{ij} \tag{9b}$$

$$a^i(z)\lambda^j(\zeta) = 0\,. \tag{9c}$$

The Virasoro algebra for the above current algebras is

$$\begin{aligned} L(z) = & \frac{1}{2(k_Q-1)}\eta_{AB} : Q^A(z)Q^B(z) : + \frac{1}{2k}\eta_{AB} : \partial\psi^A(z)\,\psi^B(z) : \\ & + \frac{1}{2(k_q+1)} : q^a(z)q^a(z) : + \frac{1}{2k} : \partial\chi^a(z)\,\chi^a(z) : \\ & + \frac{1}{2} : a^i(z)a^i(z) : + \frac{1}{2} : \partial\lambda^i(z)\lambda^i(z) : \end{aligned} \tag{10}$$

with central charge $c = c_{SU(1,1)} + c_{SU(2)} + c_{T^4} = 15$ since

$$\begin{aligned} c_{SU(1,1)} &= \frac{2k_Q\cdot 3}{2k_Q - 2} + \frac{3}{2} = \frac{9}{2} + \frac{3}{k} \\ c_{SU(2)} &= \frac{2k_q \cdot 3}{2k_q + 2} + \frac{3}{2} = \frac{9}{2} - \frac{3}{k} \\ c_{T^4} &= 6 \end{aligned} \tag{11}$$

where the levels of the both the $SU(1,1)$ and $SU(2)$ total KMA's have been chosen to be k, i.e. $k_Q = k_q + 2$, so that this critical string theory has total central charge of 15. The worldsheet supercurrent is

$$\begin{aligned} G(z) = & -\frac{i}{6k}f_{ABC}\psi^A(z)\psi^B(z)\psi^C(z) + \frac{1}{k}\eta_{AB}\psi^A(z)Q^A(z) \\ & - \frac{i}{6k}f_{abc}\chi^a(z)\chi^b(z)\chi^c(z) + \frac{1}{k}\chi^a(z)q^a(z) \\ & + \lambda^i(z)a^i(z)\,. \end{aligned} \tag{12}$$

There are similar antiholographic current algebras for the left movers. For the Minkowski metric on AdS_3, the right and left movers $J^A(z)$ and $\bar{J}^A(\bar{z})$ are real currents. For the Euclidean metric, they satisfy $(J^A(z))^* = \bar{J}^A(\bar{z})$, *i.e.* $J^A(z) = j^A(z) + ik^A(z)$, $\bar{J}^A(\bar{z}) = j^A(\bar{z}) - ik^A(\bar{z})$ for $j^A(z)$ and $k^A(z)$ real currents.[4]

The zero modes of the affine $SU(1,1)$ algebra J_0^A form $SU(1,1)$, where $J^A(z) = \sum_{n\in\mathbf{Z}} J_n^A z^{-n-1}$. States which carry $SU(1,1)$ representations are labelled by the eigenvalues of J_0^3 and the quadratic Casimir $Q = \frac{1}{2}(J_0^+ J_0^- + J_0^- J_0^+) - J_0^3 J_0^3$:

$$
\begin{aligned}
J_0^3|j,m\rangle &= m|j,m\rangle \\
Q|j,m\rangle &= -j(j+1)|j,m\rangle .
\end{aligned}
\tag{13}
$$

Here $J^{\pm}(z) = J^1(z) \pm iJ^2(z)$.

Unitary representations of $SU(1,1)$ are infinite-dimensional (apart from the singlet $|j=0, m=0\rangle$) and listed as follows:

The discrete unitary series D_j^- consists of a highest weight state $|j,j\rangle$ which satisfies $J_0^+|j,j\rangle = 0$, with a tower $|j,m\rangle$, with decreasing $m = j, j-1, j-2, \ldots$. Here $j = -\frac{1}{2}, -1, -\frac{3}{2}, \ldots$. A no-ghost analysis of the free spectrum which requires physical states ψ (i.e. those satisfying $L_n\psi = 0$, $G_r\psi = 0$ for $n, r \geq 0$) to have non-negative norm further truncates j to $-k-1 < j < 0$. (k is half the KMA level).[5]

The discrete unitary series D_j^+ consists of a lowest weight state $|j,-j\rangle$ which satisfies $J_0^-|j,-j\rangle = 0$, with a tower $|j,m\rangle$, with increasing $m = -j, -j+1, -j+2, \ldots$. Here $j = -\frac{1}{2}, -1, -\frac{3}{2}, \ldots$. The no-ghost analysis further restricts j to $-k-1 \leq j < 0$.

The continuous unitary series C_j^0 are the states $|j,m\rangle$, with $j = -\frac{1}{2} + i\kappa$, $\kappa \in \mathbf{R}$ and $m \in \mathbf{Z}$.

The continuous unitary series $C_j^{\frac{1}{2}}$ are the states $|j,m\rangle$, with $j = -\frac{1}{2} + i\kappa$, $\kappa \in \mathbf{R}$ and $m \in \mathbf{Z} + \frac{1}{2}$.

The exceptional unitary series are the states $|j,m\rangle$, with continuous real j where $-\frac{1}{2} \leq j < 0$ and $m \in \mathbf{Z}$.

KALUZA-KLEIN SPECTRUM

We consider the Kaluza-Klein spectrum found by compactifying the fields of the $D = 6, N = (2,0)$ supergravity theory with n tensor multiplets on $AdS_3 \times S^3$. This is a useful set of fields[8], since $n = 21$ gives the massless fields obtained by considering type IIB superstring on $R^6 \times K3$ at a generic point in the $K3$ moduli space, and $n = 5$ gives most of the massless fields[6] obtained by compactifying the type IIB superstring on $R^6 \times T^4$. We label the Kaluza-Klein states by

$$
D^{(\ell_1,\ell_2)}(E_0, s_0)(R \times S)
\tag{14}
$$

where (ℓ_1, ℓ_2) is the highest weight of the gauge group $SO(4) \sim SU(2)^2$ (coming from the isometries of S^3) whose isospins $(j, \bar{j})$ are $j = \frac{1}{2}(\ell_1 + \ell_2)$, $\bar{j} = \frac{1}{2}(\ell_1 - \ell_2)$. Allowed values have $j = 0, \frac{1}{2}, 1 \ldots$; $\bar{j} = 0, \frac{1}{2}, 1 \ldots$, so $\ell_1 \geq |\ell_2|$. Similarly, the $(m, \bar{m})$ eigenvalues (also called $(h, \bar{h})$) of the $J_0^3, \bar{J}_0^3$ generators of $SU(1,1)^2$, the isometry group of Minkowski AdS_3, are related to the AdS energy E_0 and helicity s_0 by $h = \frac{1}{2}(E_0 + s_0)$ and $\bar{h} = \frac{1}{2}(E_0 - s_0)$. R denotes the representation of the $SO(4)_R$

subgroup (unbroken by the compactification on $AdS_3 \times S^3$) of the $SO(5)$ R symmetry group of this $D = 6$ theory, and S denotes the representation of of the SO(n) group in which the supergravity multiplet is a singlet, and the n tensor multiplets are a vector.

The supergroup of this theory is $SU(1,1/2)^2$, which has 16 fermionic generators and 12 even generators in $SU(1,1)^2 \times SU(2)^2$.

There is a tower of spin 2 supermultiplets in a singlet of $SO(n)$, a tower of spin 1 supermultiplets in a singlet of $SO(n)$, a tower of spin 1 supermultiplets in the n (vector) representation of $SO(n)$, and one spin $\frac{1}{2}$ supermultiplet in the vector representation of $SO(n)$. Here spin refers to AdS helicity, in particular the largest s_0 value found in the supermultiplet.

The states in these towers describe massless and massive fields of $D = 3$, $N = 8$ gauged supergravity, where the supergravity multiplet has no propagating degrees of freedom. The spin $\frac{1}{2}$ supermultiplet describes n copies of a matter multiplet with eight AdS fermions ($s_0 = \frac{1}{2}$), and eight AdS scalars ($s_0 = 0$).

The spin 2 tower is

$$\begin{array}{ll}
s_0 = 2 & D^{(\ell+1,0)}(\ell+3,2)(0,0) \\
s_0 = \frac{3}{2} & D^{(\ell+\frac{3}{2},\frac{1}{2})}(\ell+\frac{5}{2},\frac{3}{2})(2_+,0) \\
 & D^{(\ell+\frac{1}{2},\frac{1}{2})}(\ell+\frac{7}{2},\frac{3}{2})(2_-,0) \\
s_0 = 1 & D^{(\ell+2,1)}(\ell+2,1)(0,0) \\
 & D^{(\ell+1,1)}(\ell+3,1)(4,0) \\
 & D^{(\ell,1)}(\ell+4,1)(0,0) \\
s_0 = \frac{1}{2} & D^{(\ell+\frac{3}{2},\frac{3}{2})}(\ell+\frac{5}{2},\frac{1}{2})(2_-,0) \\
 & D^{(\ell+\frac{1}{2},\frac{3}{2})}(\ell+\frac{7}{2},\frac{1}{2})(2_+,0) \\
s_0 = 0 & D^{(\ell+1,2)}(\ell+3,0)(0,0) \\
s_0 = 0 & D^{(\ell+1,-2)}(\ell+3,0)(0,0) \\
s_0 = -\frac{1}{2} & D^{(\ell+\frac{1}{2},-\frac{3}{2})}(\ell+\frac{7}{2},-\frac{1}{2})(2_-,0) \\
 & D^{(\ell+\frac{3}{2},-\frac{3}{2})}(\ell+\frac{5}{2},-\frac{1}{2})(2_+,0) \\
s_0 = -1 & D^{(\ell,-1)}(\ell+4,-1)(0,0) \\
 & D^{(\ell+1,-1)}(\ell+3,-1)(4,0) \\
 & D^{(\ell+2,-1)}(\ell+2,-1)(0,0) \\
s_0 = -\frac{3}{2} & D^{(\ell+\frac{1}{2},-\frac{1}{2})}(\ell+\frac{7}{2},-\frac{3}{2})(2_+,0) \\
 & D^{(\ell+\frac{3}{2},\frac{1}{2})}(\ell+\frac{5}{2},-\frac{3}{2})(2_-,0) \\
s_0 = -2 & D^{(\ell+1,0)}(\ell+3,-2)(0,0)\,.
\end{array} \tag{15}$$

There is one supermultiplet for each ℓ, where $\ell = -1, 0, 1, \ldots$ with the restriction that $\ell_1 \geq |\ell_2|$ in the notation of (14). The spin 2 supermultiplet at level ℓ has 16 $(\ell+1)(\ell+3)$ states for $\ell \geq 0$.

The spin 2 tower for level $\ell = -1$ describes the $AdS_3, N = 8$ gauged supergravity multiplet with a gravitino, eight gravitini, and six vector fields, none of which are physical degrees of freedom:

$$
\begin{array}{llll}
 & (j,\bar{j}) & (h,\bar{h}) & \\
s_0 = 2 & (0,0) & (2,0) & D^{(0,0)}(2,2)(0,0) \\
s_0 = \frac{3}{2} & (\frac{1}{2},0) & (\frac{3}{2},0) & D^{(\frac{1}{2},\frac{1}{2})}(\frac{3}{2},\frac{3}{2})(2_+,0) \\
s_0 = 1 & (1,0) & (1,0) & D^{(1,1)}(1,1)(0,0) \\
s_0 = -1 & (0,1) & (0,1) & D^{(1,-1)}(1,-1)(0,0) \\
s_0 = -\frac{3}{2} & (0,\frac{1}{2}) & (0,\frac{3}{2}) & D^{(\frac{1}{2},-\frac{1}{2})}(\frac{3}{2},-\frac{3}{2})(2_-,0) \\
s_0 = -2 & (0,0) & (0,2) & D^{(0,0)}(2,-2)(0,0)
\end{array}
\qquad (16)
$$

Here we have listed the $(j,\bar{j})$ and $(h,\bar{h})$ representations explicitly.

The tower of spin 1 supermultiplets in a singlet of $SO(n)$ is

$$
\begin{array}{ll}
s_0 = 1 & D^{(\ell+1,-1)}(\ell+3,1)(0,0) \\
s_0 = \frac{1}{2} & D^{(\ell+\frac{3}{2},-\frac{1}{2})}(\ell+\frac{5}{2},\frac{1}{2})(2_+,0) \\
 & D^{(\ell+\frac{1}{2},-\frac{1}{2})}(\ell+\frac{7}{2},\frac{1}{2})(2_-,0) \\
s_0 = 0 & D^{(\ell+2,0)}(\ell+2,0)(0,0) \\
s_0 = 0 & D^{(\ell+1,0)}(\ell+3,0)(4,0) \\
s_0 = 0 & D^{(\ell,0)}(\ell+4,0)(0,0) \\
s_0 = -\frac{1}{2} & D^{(\ell+\frac{1}{2},\frac{1}{2})}(\ell+\frac{7}{2},-\frac{1}{2})(2_+,0) \\
 & D^{(\ell+\frac{3}{2},\frac{1}{2})}(\ell+\frac{5}{2},-\frac{1}{2})(2_-,0) \\
s_0 = -1 & D^{(\ell+1,1)}(\ell+3,-1)(0,0)\,.
\end{array}
\qquad (17)
$$

Supermultiplets in this tower with physical degrees of freedom start at level $\ell = 0$.

The tower of spin 1 supermultiplets in the n vector representation of $SO(n)$ is similar to (16), but in this case there is also a physical multiplet at level $\ell = -1$:

$$
\begin{array}{ll}
s_0 = 1 & D^{(\ell+1,-1)}(\ell+3,1)(0,n) \\
s_0 = \frac{1}{2} & D^{(\ell+\frac{3}{2},-\frac{1}{2})}(\ell+\frac{5}{2},\frac{1}{2})(2_+,n) \\
 & D^{(\ell+\frac{1}{2},-\frac{1}{2})}(\ell+\frac{7}{2},\frac{1}{2})(2_-,n) \\
s_0 = 0 & D^{(\ell+2,0)}(\ell+2,0)(0,n) \\
s_0 = 0 & D^{(\ell+1,0)}(\ell+3,0)(4,n) \\
s_0 = 0 & D^{(\ell,0)}(\ell+4,0)(0,n) \\
s_0 = -\frac{1}{2} & D^{(\ell+\frac{1}{2},\frac{1}{2})}(\ell+\frac{7}{2},-\frac{1}{2})(2_+,n) \\
 & D^{(\ell+\frac{3}{2},\frac{1}{2})}(\ell+\frac{5}{2},-\frac{1}{2})(2_-,n) \\
s_0 = -1 & D^{(\ell+1,1)}(\ell+3,-1)(0,n)\,.
\end{array}
\qquad (18)
$$

The is supermultiplet is self-conjugate and has $8n(\ell+2)^2$ Bose states and $8n(\ell+2)^2$ Fermi states, for $\ell \geq -1$. The spin 1 supermultiplet in the n of $SO(n)$ for level $\ell = -1$

is actually a spin $\frac{1}{2}$ supermultiplet describing n copies of a matter multiplet with 8 bosonic and 8 fermionic degrees of freedom:

$$\begin{array}{llll} & (j,\bar{j}) & (h,\bar{h}) & \\ s_0=\frac{1}{2} & 2(0,\frac{1}{2}) & (1,\frac{1}{2}) & (D^{(\frac{1}{2},-\frac{1}{2})}(\frac{3}{2},\frac{1}{2})(2_+,n) \\ s_0=0 & (\frac{1}{2},\frac{1}{2}) & (\frac{1}{2},\frac{1}{2}) & D^{(1,0)}(1,0)(0,n) \\ s_0=0 & 4(0,0) & (1,1) & D^{(0,0)}(2,0)(4,n) \\ s_0=-\frac{1}{2} & 2(\frac{1}{2},0) & (\frac{1}{2},1) & D^{(\frac{1}{2},\frac{1}{2})}(\frac{3}{2},-\frac{1}{2})(2_-,n)\,. \end{array} \tag{19}$$

All the states in (17) come from the n tensor multiplets of the $D=6$ theory.

For the compactification of the type IIB superstring on $AdS_3\times S^3\times T^4$, there is also another spin 1 supermultiplet[6], and $n=5$ for the supermultiplets listed above. For this case, the states in (19) labelled by

$$D^{(0,0)}(2,0)(4,5) \tag{20}$$

describe 20 massless scalars in a singlet of the gauge group $SU(2)^2$ with conformal weights $(h,\bar{h})=(1,1)$. 16 of these come from the NS-NS sector, and 4 come from the R-R sector. We review their vertex operators below. These scalars can have vevs which are the 20 moduli of the $D=6$, $N=(2,2)$ supergravity theory compactified on $AdS_3\times S^3$.

VERTEX OPERATORS

16 of the moduli listed in (20) have NS-NS vertex operators given by

$$V^{i\bar{i}}(z,\bar{z})=e^{-\phi(z)-\phi(\bar{z})}\lambda^i(z)\bar{\lambda}^{\bar{i}}(\bar{z})V'_{j_{SU(2)},m',\bar{m}'}(z,\bar{z})V_{j_{SU(1,1)},m,\bar{m}}(z,\bar{z}) \tag{21}$$

where $j=\bar{j}=j_{SU(2)}$, $V'_{j_{SU(2)},m',m'}(z,\bar{z})$ are the vertex operators for the basis of states corresponding to $q^a(z)$. Also $j_{SU(1,1)}$ for the left and right non-compact algebras are the same, and $V_{j_{SU(1,1)},m,m}(z,\bar{z})$ are the vertex operators for the basis of states corresponding to $Q^A(z)$. In (21), to describe the 16 massless scalars, we set $j_{SU(2)}=0$, so $m'=\bar{m}'=0$, and $j_{SU(1,1)}=-1$, with $(m,\bar{m})\equiv(h\bar{h})=(1,1)$, *i.e.* the lowest weight of a discrete unitary series $D^+_{j=-1}$, with $J_0^-|j_{SU(1,1)},m\equiv h=-j_{SU(1,1)}=1\rangle$. In (21), the fields $\phi(z)$ and $\bar{\phi}(\bar{z})$ are the $N=1$ ghosts with worldsheet conformal dimension $\frac{1}{2}$. The 16 massless scalars which are singlets under $SU(2)^2$ have RR superpartner scalars labelled as $4(j,\bar{j})=4(\frac{1}{2},\frac{1}{2})$ with conformal weights $(h,\bar{h})=(\frac{1}{2},\frac{1}{2})$.

The remaining 4 moduli listed by (20) have RR vertex operators again with $(j,\bar{j})=(0,0)$ and $(h,\bar{h})=(1,1)$. These are the superpartners of NS-NS vertex operators with $(j,\bar{j})=(\frac{1}{2},\frac{1}{2})=(h,\bar{h})$.

CONCLUSIONS

In this proceedings we have discussed the current algebras which define a worldsheet conformal field theory operator algebra for the type IIB superstring compactified on $AdS_3 \times S^3 \times T^4$, in the case where the background supports a NS-NS 2-form vev. Furthermore we displayed the Kaluza-Klein spectrum for the supergravity field theory associated with this compactification, and identified certain massless scalars in this spectrum as moduli. The vertex operators for these scalars were described in terms of their $SU(1,1)^2$ and $SU(2)^2$ quantum numbers of the supergroup $SU(1,1/2)^2$.

REFERENCES

1. J. Maldacena, "The Large N Limit of Superconformal Field Theories and Supergravity," hep-th/9711200.
2. S. Gubser, I. Klebanov, and A. Polyakov, "Gauge Theory Correlators from Noncritical String Theory," hep-th/9802109.
3. E. Witten, "Anti-de Sitter Space and Holography," hep-th/9802150.
4. A. Giveon, D. Kutasov, and N. Seiberg, "Comments on String Theory on ADS(3)," hep-th/9806194.
5. J.M. Evans, M.R. Gaberdiel and M.I. Perry, "The No-ghost Theorem for AdS_3 and the Stringy Exclusion Principle," hep-th/9806024.
6. F. Larsen, "The Perturbative Spectrum of Black Holes in $N = 8$ Supergravity," Nucl. Phys. **B536** (1998) 258-278, hep-th/9805208.
7. J. de Boer, "Six-dimensional Supergravity on $S^3 \times AdS_3$ a 2d Conformal Field Theory," hep-th/9806104.
8. S. Deger, A. Kaya, E. Sezgin and P. Sundell, "Spectrum of $D = 6, N = 4b$ Supergravity on $AdS_3 \times S^3$," Nucl. Phys. **B536** (1998) 110-140, hep-th/9804166.
9. D. Kutasov, F. Larsen and R.G. Leigh, "String Theory in Magnetic Monopole Backgrounds," hep-th/9812027.
10. N. Berkovits, C. Vafa, and E. Witten, *Conformal Field Theory of AdS Background with Ramond-Ramond Flux*, hep-thg/9802098.
11. L. Dolan and M. Langham, *Symmetric Subgroups of Gauged Supergravities and AdS String Theory Vertex Operators*, hep-th/9901030.

FROM THREEBRANES TO LARGE N GAUGE THEORIES

Igor R. Klebanov

Joseph Henry Laboratories
Princeton University
Princeton, New Jersey 08544

ABSTRACT

This is a brief introductory review of the AdS/CFT correspondence and of the ideas that led to its formulation. Emphasis is placed on dualities between conformal large N gauge theories in 4 dimensions and string backgrounds of the form $AdS_5 \times X_5$. Attempts to generalize this correspondence to asymptotically free theories are also included.

1 INTRODUCTION

It is well-known that string theory originated from attempts to understand the strong interactions [1]. However, after the emergence of QCD as the theory of hadrons, the dominant theme of string research shifted to the Planck scale domain of quantum gravity [2]. Although in hadron physics one routinely hears about flux tubes and the string tension, the majority of particle theorists gave up hope that string theory might lead to an exact description of the strong interactions. Now, however, for the first time we can say with confidence that at least some strongly coupled gauge theories have a dual description in terms of strings. Let me emphasize that one is not talking here about effective strings that give an approximate qualitative description, but rather about an exact duality. At weak coupling a convenient description of the theory involves conventional perturbative methods; at strong coupling, where such methods are intractable, the dual string description simplifies and gives exact information about the theory. The best established examples of this duality are conformal gauge theories where the

Confluence of Cosmology, Massive Neutrinos, Elementary Particles, and Gravitation
Edited by Kursunoglu *et al.*, Kluwer Academic / Plenum Publishers, New York, 1999.

so-called AdS/CFT correspondence [3, 4, 5] has allowed for many calculations at strong coupling to be performed with ease. In these notes I describe, from my own personal perspective, some of the ideas that led to the formulation of the AdS/CFT correspondence. I will also speculate on the future directions. For the sake of brevity I will mainly discuss the AdS_5/CFT_4 case which is most directly related to 4-dimensional gauge theories.

It has long been believed that the best hope for a string description of non-Abelian gauge theories lies in the 't Hooft large N limit. A quarter of a century ago 't Hooft proposed to generalize the $SU(3)$ gauge group of QCD to $SU(N)$, and to take the large N limit while keeping $g_{YM}^2 N$ fixed [6]. In this limit each Feynman graph carries a topological factor N^χ, where χ is the Euler characteristic of the graph. Thus, the sum over graphs of a given topology can perhaps be thought of as a sum over world sheets of a hypothetical "QCD string." Since the spheres (string tree diagrams) are weighted by N^2, the tori (string one-loop diagrams) – by N^0, etc., we find that the closed string coupling constant is of order N^{-1}. Thus, the advantage of taking N to be large is that we find a weakly coupled string theory. It is not clear, however, how to describe this string theory in elementary terms (by a 2-dimensional world sheet action, for example). This is clearly an important problem: the free closed string spectrum is just the large N spectrum of glueballs. If the quarks are included, then we also find open strings describing the mesons. Thus, if methods are developed for calculating these spectra, and it is found that they are discrete, then this provides an elegant explanation of confinement. Furthermore, the $1/N$ corrections correspond to perturbative string corrections.

Many years of effort, and many good ideas, were invested into the search for an exact gauge field/string duality [7]. One class of ideas, exploiting the similarity of the large N loop equation with the string Schroedinger equation, eventually led to the following fascinating speculation [8]: one should not look for the QCD string in four dimensions, but rather in five, with the fifth dimension akin to the Liouville dimension of non-critical string theory [9]. This leads to a picture where the QCD string is described by a two-dimensional world sheet sigma model with a curved 5-dimensional target space. At that stage it was not clear, however, precisely what target spaces are relevant to gauge theories. Luckily, we now do have answers to this question for a variety of conformal large N gauge models. The route that leads to this answer, and confirms the idea of the fifth dimension, involves an unexpected detour via black holes and Dirichlet branes. We turn to these subjects next.

2 D-BRANES vs. BLACK HOLES AND p-BRANES

A few years ago it became clear that, in addition to strings, superstring theory contains soliton-like "membranes" of various internal dimensionalities called Dirichlet branes (or D-branes) [10]. A Dirichlet p-brane (or Dp-brane) is a $p+1$ dimensional hyperplane in $9+1$ dimensional space-time where strings are allowed to end, even in theories where all strings are closed in the bulk of space-time. In some ways a D-brane is like a topological defect: when a closed string touches it, it can open open up and turn into an open string whose ends are free to move along the D-brane. For the end-points of such a string the $p+1$ longitudinal coordinates satisfy the conventional free (Neumann) boundary conditions, while the $9-p$ coordinates transverse to the Dp-brane have the fixed (Dirichlet) boundary conditions; hence the origin of the term "Dirichlet brane." In a seminal paper [11] Polchinski showed that the Dp-brane is a BPS saturated object which preserves 1/2 of the bulk supersymmetries and carries an elementary unit of charge with respect to the $p+1$ form gauge potential from the Ramond-Ramond sector of type II superstring. The existence of BPS objects carrying such charges is required by non-perturbative string dualities [12]. A striking feature of the D-brane formalism is that it provides a concrete (and very simple) embedding of such objects into perturbative string theory.

Another fascinating feature of the D-branes is that they naturally realize gauge theories on their world volume. The massless spectrum of open strings living on a Dp-brane is that of a maximally supersymmetric $U(1)$ gauge theory in $p+1$ dimensions. The $9-p$ massless scalar fields present in this supermultiplet are the expected Goldstone modes associated with the transverse oscillations of the Dp-brane, while the photons and fermions may be thought of as providing the unique supersymmetric completion. If we consider N parallel D-branes, then there are N^2 different species of open strings because they can begin and end on any of the D-branes. N^2 is the dimension of the adjoint representation of $U(N)$, and indeed we find the maximally supersymmetric $U(N)$ gauge theory in this setting [13]. The relative separations of the Dp-branes in the $9-p$ transverse dimensions are determined by the expectation values of the scalar fields. We will be primarily interested in the case where all scalar expectation values vanish, so that the N Dp-branes are stacked on top of each other. If N is large, then this stack is a heavy object embedded into a theory of closed strings which contains gravity. Naturally, this macroscopic object will curve space: it may be described by some classical metric and other background fields, such as the Ramond-Ramond $p+1$ form potential. Thus, we have two very different descriptions of the stack of Dp-branes: one in terms of the $U(N)$ supersymmetric gauge theory on its world volume, and the other in terms of the classical Ramond-Ramond charged p-brane background of the type II closed superstring theory. It is the relation between these two descriptions that is at the heart of the recent progress in understanding the connections between gauge fields and strings.[1] Of course, more work is needed to make this relation precise.

2.1 COUNTING THE ENTROPY

The first success in building this kind of correspondence between black hole metrics and D-branes was achieved by Strominger and Vafa [15]. They considered 5-dimensional supergravity obtained by compactifying 10-dimensional type IIB theory on a 5-dimensional compact manifold (for example, the 5-torus), and constructed a class of black holes carrying 2 separate $U(1)$ charges. These solutions may be viewed as generalizations of the well-known 4-dimensional charged (Reissner-Nordstrom) black hole. For the Reissner-Nordstrom black hole the mass is bounded from below by a

[1]There are other similar relations between large N SYM theories and gravity stemming from the BFSS matrix theory conjecture [14].

quantity proportional to the charge. In general, when the mass saturates the lower (BPS) bound for a given choice of charges, then the black hole is called extremal. The extremal Strominger-Vafa black hole preserves 1/8 of the supersymmetries present in vacuum. Also, the black hole is constructed in such a way that, just as for the Reissner-Nordstrom solution, the area of the horizon is non-vanishing at extremality [15]. In general, an important quantity characterizing black holes is the Bekenstein-Hawking entropy which is proportional to the horizon area:

$$S_{BH} = \frac{A_h}{4G} , \tag{1}$$

where G is the Newton constant. Strominger and Vafa calculated the Bekenstein-Hawking entropy of their extremal solution as a function of the charges and succeeded in reproducing this result with D-brane methods. To build a D-brane system carrying the same set of charges as the black hole, they had to consider intersecting D-branes wrapped over the compact 5-dimensional manifold. For example, one may consider D3-branes intersecting over a line or D1-branes embedded inside D5-branes. The $1+1$ dimensional gauge theory describing such an intersection is quite complicated, but the degeneracy of the supersymmetric BPS states can nevertheless be calculated in the D-brane description valid at weak coupling. For reasons that will become clear shortly, the description in terms of black hole metrics is valid only at very strong coupling. Luckily, due to the supersymmetry, the number of states does not change as the coupling is increased. This ability to extrapolate the D-brane counting to strong coupling makes a comparison with the Bekenstein-Hawking entropy possible, and exact agreement is found in the limit of large charges [15]. In this sense the collection of D-branes provides a "microscopic" explanation of the black hole entropy.

This correspondence was quickly generalized to black holes slightly excited above the extremality [16, 17]. Further, the Hawking radiation rates and the absorption cross-sections were calculated and successfully reproduced by D-brane models [16, 18, 19]. Since then this system has been receiving a great deal of attention. However, some detailed comparisons are hampered by the complexities of the dynamics of intersecting D-branes: to date there is no first principles approach to the lagrangian of the $1+1$ dimensional conformal field theory on the intersection.

For this and other reasons it has turned out very fruitful to study a similar correspondence for simpler systems which involve parallel D-branes only [20, 21, 22, 23, 24]. Our primary motivation is that, as explained above, parallel Dp-branes realize $p+1$ dimensional $U(N)$ SYM theories, and we may learn something new about them from comparisons with Ramond-Ramond charged black p-brane classical solutions. These solutions in type II supergravity have been known since the early 90's [25, 26]. The metric and dilaton backgrounds may be expressed in the following simple and universal form:

$$ds^2 = H^{-1/2}(r)\left[-f(r)dt^2 + \sum_{i=1}^{p}(dx^i)^2\right] + H^{1/2}(r)\left[f^{-1}(r)dr^2 + r^2 d\Omega_{8-p}^2\right] , \tag{2}$$

$$e^{\Phi} = H^{(3-p)/4}(r) ,$$

where

$$H(r) = 1 + \frac{L^{7-p}}{r^{7-p}} , \qquad f(r) = 1 - \frac{r_0^{7-p}}{r^{7-p}} ,$$

and $d\Omega_{8-p}^2$ is the metric of a unit $8-p$ dimensional sphere. The horizon is located at $r = r_0$ and the extremality is achieved in the limit $r_0 \to 0$. A solution with $r_0 \ll L$ is called near-extremal. In contrast to the situation encountered for the Strominger-Vafa black hole, the Bekenstein-Hawking entropy vanishes in the extremal limit. Just like the stacks of parallel D-branes, the extremal solutions are BPS saturated: they preserve 16 of the 32 supersymmetries present in the type II theory. For $r_0 > 0$ the p-brane carries some excess energy E above its extremal value, and the Bekenstein-Hawking entropy is also non-vanishing. The Hawking temperature is then defined by $T^{-1} = \partial S_{BH}/\partial E$.

The correspondence between the entropies of the p-brane solutions (2) and those of the $p+1$ dimensional SYM theories was first considered in [20, 21]. Among these solutions $p = 3$ has a special status: in the extremal limit $r_0 \to 0$ the 3-brane solution is perfectly non-singular [27]. This is evidenced by the fact that the dilaton Φ is constant for $p = 3$, but blows up at $r = 0$ for all other extremal solutions. In [20] the Bekenstein-Hawking entropy of a near-extremal 3-brane of Hawking temperature T was compared with the entropy of the $\mathcal{N} = 4$ supersymmetric $U(N)$ gauge theory (which lives on N coincident D3-branes) heated up to the same temperature. The results turned out to be quite interesting. The Bekenstein-Hawking entropy expressed in terms of the Hawking temperature T and the number N of elementary units of charge was found to be

$$S_{BH} = \frac{\pi^2}{2} N^2 V_3 T^3 \ , \tag{3}$$

where V_3 is the spatial volume of the 3-brane. This was compared with the entropy of a free $U(N)$ $\mathcal{N} = 4$ supermultiplet, which consists of the gauge field, $6N^2$ massless scalars and $4N^2$ Weyl fermions. This entropy was calculated using the standard statistical mechanics of a massless gas (the black body problem), and the answer turned out to be

$$S_0 = \frac{2\pi^2}{3} N^2 V_3 T^3 \ . \tag{4}$$

It is remarkable that the 3-brane geometry captures the T^3 scaling characteristic of a conformal field theory (in a CFT this scaling is guaranteed by the extensivity of the entropy and the absence of dimensionful parameters).[2] Also, the N^2 scaling indicates the presence of $O(N^2)$ unconfined degrees of freedom, which is exactly what we expect in the $\mathcal{N} = 4$ supersymmetric $U(N)$ gauge theory. On the other hand, the relative factor of 3/4 between S_{BH} and S_0 at first appeared mysterious and was interpreted by many as a subtle failure of the D3-brane approach to black 3-branes. As we will see shortly, however, the relative factor of 3/4 is not a contradiction but rather a prediction about strongly coupled $\mathcal{N} = 4$ SYM theory at finite temperature.

[2]Other examples of the "conformal" behavior of the Bekenstein-Hawking entropy include the 11-dimensional 5-brane and membrane solutions [21]. For the 5-brane, $S_{BH} \sim N^3 T^5 V_5$, while for the membrane $S_{BH} \sim N^{3/2} T^2 V_2$. The microscopic description of the 5-brane solution is in terms of a large number N of coincident singly charged 5-branes of M-theory, whose chiral world volume theory has $(0, 2)$ supersymmetry. Similarly, the membrane solution describes the large N behavior of the CFT on N coincident elementary membranes. The entropy formulae suggest that these theories have $O(N^3)$ and $O(N^{3/2})$ massless degrees of freedom respectively. These predictions of supergravity [21] are non-trivial and still mysterious. Since the geometry of the 5-brane throat is $AdS_7 \times S^4$, and that of the membrane throat is $AdS_4 \times S^7$, these systems lead to other interesting examples of the AdS/CFT correspondence.

2.2 FROM ABSORPTION CROSS-SECTIONS TO TWO-POINT CORRELATORS

Almost a year after the entropy comparisons [20, 21] I came back to the 3-branes (and also to the 11-dimensional membranes and 5-branes) and tried to interpret absorption cross-sections for massless particles in terms of the world volume theories [22]. This was a natural step beyond the comparison of entropies, and for the Strominger-Vafa black holes the D-brane approach to absorption was initiated earlier in [16, 18]. For the system of N coincident D3-branes it was interesting to inquire to what extent the supergravity and the weakly coupled D-brane calculations agreed. For example, they might scale differently with N or with the incident energy. Even if the scaling exponents agreed, the overall normalizations could differ by a subtle numerical factor similar to the 3/4 found for the 3-brane entropy. Surprisingly, the low-energy absorption cross-sections turned out to agree exactly!

To calculate the absorption cross-sections in the D-brane formalism one needs the low-energy world volume action for coincident D-branes coupled to the massless bulk fields. Luckily, these couplings may be deduced from the D-brane Born-Infeld action. For example, the coupling of 3-branes to the dilaton Φ, the Ramond-Ramond scalar C, and the graviton $h_{\alpha\beta}$ is given by [22, 23]

$$S_{\rm int} = \frac{\sqrt{\pi}}{\kappa} \int d^4x \left[\operatorname{tr} \left(\tfrac{1}{4}\Phi F_{\alpha\beta}^2 - \tfrac{1}{4} C F_{\alpha\beta} \tilde{F}^{\alpha\beta} \right) + \tfrac{1}{2} h^{\alpha\beta} T_{\alpha\beta} \right] , \tag{5}$$

where $T_{\alpha\beta}$ is the stress-energy tensor of the $\mathcal{N}=4$ SYM theory. Consider, for instance, absorption of a dilaton incident on the 3-brane at right angles with a low energy ω. Since the dilaton couples to $\operatorname{tr} F_{\alpha\beta}^2$ it can be converted into a pair of back-to-back gluons on the world volume. The leading order calculation of the cross-section for weak coupling gives [22]

$$\sigma = \frac{\kappa^2 \omega^3 N^2}{32\pi} , \tag{6}$$

where $\kappa = \sqrt{8\pi G}$ is the 10-dimensional gravitational constant (note that the factor N^2 comes from the degeneracy of the final states which is the number of different gluon species). This result was compared with the absorption cross-section by the extremal 3-brane geometry,

$$ds^2 = \left(1 + \frac{L^4}{r^4}\right)^{-1/2} \left(-dt^2 + dx_1^2 + dx_2^2 + dx_3^2\right) + \left(1 + \frac{L^4}{r^4}\right)^{1/2} \left(dr^2 + r^2 d\Omega_5^2\right) . \tag{7}$$

This geometry may be viewed as a semi-infinite throat which for $r \gg L$ opens up into flat $9+1$ dimensional space. Waves incident from the $r \gg L$ region partly reflect back and partly penetrate into the the throat region $r \ll L$. The relevant s-wave radial equation turns out to be [22]

$$\left[\frac{d^2}{d\rho^2} - \frac{15}{4\rho^2} + 1 + \frac{(\omega L)^4}{\rho^4} \right] \psi(\rho) = 0 , \tag{8}$$

where $\rho = \omega r$. For a low energy $\omega \ll 1/L$ we find a high barrier separating the two asymptotic regions. The low-energy behavior of the tunneling probability may be calculated by the so-called matching method, and the resulting absorption cross-section is [22]

$$\sigma_{SUGRA} = \frac{\pi^4}{8}\omega^3 L^8 \ . \tag{9}$$

In order to compare (6) and (9) we need a relation between the radius of the throat, L, and the number of D3-branes, N. Such a relation follows from equating the ADM tension of the extremal 3-brane solution to N times the tension of a single D3-brane, and one finds [20]

$$L^4 = \frac{\kappa}{2\pi^{5/2}} N \ . \tag{10}$$

Substituting this into (9), we find that the supergravity absorption cross-section agrees exactly with the D-brane one, without any relative factor like 3/4.

This result was a major surprise to me, and I started searching for its explanation. The most important question is: what is the range of validity of the two calculations? Since $\kappa \sim g_{st}\alpha'^2$, (10) gives $L^4 \sim N g_{st}\alpha'^2$. Supergravity can only be trusted if the length scale of the 3-brane solution is much larger than the string scale $\sqrt{\alpha'}$, i.e. for $N g_{st} \gg 1$.[3] Of course, the incident energy also has to be small compared to $1/\sqrt{\alpha'}$. Thus, the supergravity calculation should be valid in the "double-scaling limit" [22]

$$\frac{L^4}{\alpha'^2} \sim g_{st} N \to \infty \ , \qquad \omega^2\alpha' \to 0 \ . \tag{11}$$

If the description of the black 3-brane by a stack of many coincident D3-branes is correct, and we presume that it is, then it *must* agree with the supergravity results in this limit. Since $g_{st} \sim g^2_{\rm YM}$, this corresponds to the limit of *infinite* 't Hooft coupling in the $\mathcal{N} = 4$ $U(N)$ SYM theory. Since we also want to send $g_{st} \to 0$ in order to suppress the string loop corrections, we necessarily have to take the large N limit. To summarize, the supergravity calculations are expected to give exact information about the $\mathcal{N} = 4$ SYM theory in the limit of large N and large 't Hooft coupling [22].

Coming back to the entropy problem, we now see that the Bekenstein-Hawking entropy calculation applies to the $g^2_{\rm YM}N \to \infty$ limit of the theory, while the free field calculation applies to the $g^2_{\rm YM}N \to 0$ limit. Thus, the relative factor of 3/4 is not a discrepancy: it relates two different limits of the theory. Indeed, on general grounds we expect that in the 't Hooft large N limit the entropy is given by [28]

$$S = \frac{2\pi^2}{3} N^2 f(g^2_{\rm YM}N) V_3 T^3 \ . \tag{12}$$

The function f is certainly not constant: for example, a recent two-loop calculation [29] shows that its perturbative expansion is

$$f(g^2_{\rm YM}N) = 1 - \frac{3}{2\pi^2} g^2_{\rm YM}N + \ldots \tag{13}$$

Thus, the Bekenstein-Hawking entropy in supergravity, (3), is translated into the prediction that $f(g^2_{\rm YM}N \to \infty) = 3/4$. In fact, a recent string theory calculation of the leading strong coupling correction gives [28]

$$f(g^2_{\rm YM}N) = \frac{3}{4} + \frac{45}{32}\zeta(3)(2g^2_{\rm YM}N)^{-3/2} + \ldots \ . \tag{14}$$

[3] A similar conclusion applies to the Strominger-Vafa black hole [15].

This is consistent with $f(g_{\rm YM}^2 N)$ being a monotonic function which interpolates between 1 at $g_{\rm YM}^2 N = 0$ and 3/4 at $g_{\rm YM}^2 N = \infty$.

Although we have sharpened the region of applicability of the supergravity calculation (9), we have not yet explained why it agrees with the leading order perturbative result (6) on the D3-brane world volume. After including the higher-order SYM corrections, the general structure of the absorption cross-section in the large N limit is expected to be [24]

$$\sigma = \frac{\kappa^2 \omega^3 N^2}{32\pi} a(g_{\rm YM}^2 N) \ , \tag{15}$$

where

$$a(g_{\rm YM}^2 N) = 1 + b_1 g_{\rm YM}^2 N + b_2 (g_{\rm YM}^2 N)^2 + \ldots$$

For agreement with supergravity, the strong 't Hooft coupling limit of $a(g_{\rm YM}^2 N)$ should be equal to 1 [24]. In fact, a stronger result is true: all perturbative corrections vanish and $a = 1$ independent of the coupling. This was first shown explicitly in [24] for the graviton absorption. The absorption cross-section is related to the imaginary part of the two-point function $\langle T_{\alpha\beta}(p) T_{\gamma\delta}(-p)\rangle$ in the SYM theory. In turn, this is determined by a conformal "central charge" which satisfies a non-renormalization theorem: it is completely independent of the 't Hooft coupling.

In general, the two-point function of a gauge invariant operator in the strongly coupled SYM theory may be read off from the absorption cross-section for the supergravity field which couples to this operator in the world volume action [24]. Some examples of this field operator correspondence may be read off from (5). Thus, we learn that the dilaton absorption cross-section measures the imaginary part of $\langle \mathrm{tr}\, F_{\alpha\beta}^2(p)\, \mathrm{tr}\, F_{\gamma\delta}^2(-p)\rangle$, the Ramond-Ramond scalar absorption cross-section measures the imaginary part of $\langle \mathrm{tr}\, F_{\alpha\beta}\tilde{F}^{\alpha\beta}(p)\, \mathrm{tr}\, F_{\gamma\delta}\tilde{F}^{\gamma\delta}(-p)\rangle$, etc. The agreement of these two-point functions with the weak-coupling calculations performed in [22, 23] is explained by supersymmetric non-renormalization theorems. Thus, the proposition that the $g_{\rm YM}^2 N \to \infty$ limit of the large N $\mathcal{N} = 4$ SYM theory can be extracted from the 3-brane of type IIB supergravity has passed its first consistency checks.

3 THE AdS/CFT CORRESPONDENCE

The circle of ideas reviewed in the previous section received a seminal development by Maldacena [3] who also connected it for the first time with the QCD string idea. Maldacena made a simple and powerful observation that the "universal" region of the 3-brane geometry, which should be directly identified with the $\mathcal{N} = 4$ SYM theory, is the throat, i.e. the region $r \ll L$.[4] The limiting form of the metric (7) is

$$ds^2 = \frac{L^2}{z^2}\left(-dt^2 + d\vec{x}^2 + dz^2\right) + L^2 d\Omega_5^2 \ , \tag{16}$$

where $z = \frac{L^2}{r} \gg L$. This metric describes the space $AdS_5 \times S^5$ with equal radii of curvature L. One also finds that the self-dual 5-form Ramond-Ramond field strength has constant flux through this space (the field strength term in the Einstein equation

[4]Related ideas were also pursued in [30].

effectively gives a positive cosmological constant on S^5 and a negative one on AdS_5). Thus, Maldacena conjectured that type IIB string theory on $AdS_5 \times S^5$ should be somehow dual to the large N $\mathcal{N} = 4$ SYM theory.

Maldacena's argument was based on the fact that the low-energy ($\alpha' \to 0$) limit may be taken directly in the 3-brane geometry and is equivalent to the throat ($r \to 0$) limit. Another way to motivate the identification of the gauge theory with the throat is to think about the absorption of massless particles considered in the previous section. In the D-brane description, a particle incident from the asymptotic infinity is converted into an excitation of the stack of D-branes, i.e. into an excitation of the gauge theory on the world volume. In the supergravity description, a particle incident from the asymptotic (large r) region tunnels into the $r \ll L$ region and produces an excitation of the throat. The fact that the two different descriptions of the absorption process give identical cross-sections supports the identification of excitations of $AdS_5 \times S^5$ with the excited states of the $\mathcal{N} = 4$ SYM theory.

Another strong piece of support for this identification comes from symmetry considerations [3]. The isometry group of AdS_5 is $SO(2,4)$, and this is also the conformal group in $3+1$ dimensions. In addition we have the isometries of S^5 which form $SU(4) \sim SO(6)$. This group is identical to the R-symmetry of the $\mathcal{N} = 4$ SYM theory. After including the fermionic generators required by supersymmetry, the full isometry supergroup of the $AdS_5 \times S^5$ background is $SU(2,2|4)$, which is identical to the $\mathcal{N} = 4$ superconformal symmetry. We will see that in theories with reduced supersymmetry the compact S^5 factor becomes replaced by other compact spaces X_5, but AdS_5 is the "universal" factor present in the dual description of any large N CFT and realizing the $SO(2,4)$ conformal symmetry. One may think of these backgrounds as type IIB theory compactified on X_5 down to 5 dimensions. Such Kaluza-Klein compactifications of type IIB supergravity were extensively studied in the mid-eighties [31, 32, 33], and special attention was devoted to the $AdS_5 \times S^5$ solution because it is a maximally supersymmetric background [34, 35]. It is remarkable that these early works on compactification of type IIB theory were actually solving large N gauge theories without knowing it.

As Maldacena has emphasized, however, it is important to go beyond the supergravity limit and think of the $AdS_5 \times X_5$ space as a background of string theory [3]. Indeed, type IIB strings are dual to the electric flux lines in the gauge theory, and this provides a natural set-up for calculating correlation functions of the Wilson loops. Furthermore, if N is sent to infinity while $g^2_{\rm YM}N$ is held fixed and finite, then there are finite string scale corrections to the supergravity limit [3, 4, 5] which proceed in powers of

$$\frac{\alpha'}{L^2} \sim \left(g^2_{\rm YM}N\right)^{-1/2} . \tag{17}$$

If we wish to study finite N, then there are also string loop corrections in powers of

$$\frac{\kappa^2}{L^8} \sim N^{-2} . \tag{18}$$

As expected, taking N to infinity enables us to take the classical limit of the string theory on $AdS_5 \times X_5$. However, in order to understand the large N gauge theory with finite 't Hooft coupling, we should think of the $AdS_5 \times X_5$ as the target space of a

2-dimensional sigma model describing the classical string physics [4]. The fact that after the compactification on X_5 the string theory is 5-dimensional supports Polyakov's idea [8]. In AdS_5 the fifth dimension is related to the radial coordinate and, after a change of variables $z = Le^{-\varphi/L}$, the sigma model action turns into a special case of the general ansatz proposed in [8],

$$I = \frac{1}{2}\int d^2\sigma[(\partial_i\varphi)^2 + a^2(\varphi)(\partial_i X^\mu)^2 + \ldots] , \tag{19}$$

where $a(\varphi) = e^{\varphi/L}$. It is clear, however, that the string sigma models dual to the gauge theories are of rather peculiar nature. The new feature revealed by the D-brane approach, which is also a major stumbling block, is the presence of the Ramond-Ramond background fields. Little is known to date about such 2-dimensional field theories and, in spite of recent new insights [36], an explicit solution is not yet available.

3.1 CORRELATION FUNCTIONS AND THE BULK/BOUNDARY CORRESPONDENCE

Maldacena's work provided a crucial insight that the $AdS_5 \times S^5$ throat is the part of the 3-brane geometry that is most directly related to the $\mathcal{N} = 4$ SYM theory. It is important to go further, however, and explain precisely in what sense the two should be identified and how physical information can be extracted from this duality. Major strides towards answering these questions were made in two subsequent papers [4, 5] where essentially identical methods for calculating correlation functions of various operators in the gauge theory were proposed. As we mentioned in section 2.2, even prior to [3] some information about the field/operator correspondence and about the two-point functions had been extracted from the absorption cross-sections. The reasoning of [4] was a natural extension of these ideas.

One may motivate the general method as follows. When a wave is absorbed, it tunnels from the asymptotic infinity into the throat region, and then continues to propagate toward smaller r. Let us separate the 3-brane geometry into two regions: $r \gtrsim L$ and $r \lesssim L$. For $r \lesssim L$ the metric is approximately that of $AdS_5 \times S^5$, while for $r \gtrsim L$ it becomes very different and eventually approaches the flat metric. Signals coming in from large r may be thought of as disturbing the "boundary" of AdS_5 at $r \sim L$, and then propagating into the bulk. This suggests that, if we discard the $r \gtrsim L$ part of the 3-brane metric, then we have to cut off the radial coordinate of AdS_5 at $r \sim L$, and the gauge theory correlation functions are related to the response of the string theory to boundary conditions. Guided by this idea, [4] proposed to identify the generating functional of connected correlation functions in the gauge theory with the extremum of the classical string action subject to the boundary conditions that $\phi(x^\lambda, z) = \phi_b(x^\lambda)$ at $z = L$ (at $z = \infty$ all fluctuations are required to vanish):[5]

$$W[\phi_b(x^\lambda)] = S_{\phi_b(x^\lambda)} \ . \tag{20}$$

W generates the connected Green's functions of the gauge theory operator that corresponds to the field ϕ in the sense explained in section 2.2, while $S_{\phi_b(x^\lambda)}$ is the extremum

[5]As usual, in calculating the correlation functions in a CFT it is convenient to carry out the Euclidean continuation. On the string theory side we then have to use the Euclidean version of AdS_5.

of the classical string action subject to the boundary conditions. An essentially identical prescription was also proposed in [5] with a somewhat different motivation. If we are interested in the correlation functions at infinite 't Hooft coupling, then the problem of extremizing the classical string action reduces to solving the equations of motion in type IIB supergravity whose form is known explicitly [34]. Note that from the point of view of the metric (16) the boundary conditions are imposed not at $z = 0$ (which would be a true boundary of AdS_5) but at some finite value $z = z_{cutoff}$. It does not matter which value it is since it can be changed by an overall rescaling of the coordinates (z, x^λ) which leaves the metric unchanged. The physical meaning of this cut-off is that it acts as a UV cut-off in the gauge theory [4, 37]. Indeed, the radial coordinate of AdS_5 is to be thought of as the effective energy scale of the gauge theory [3], and decreasing z corresponds to increasing energy. In some calculations one may remove the cut-off from the beginning and specify the boundary conditions at $z = 0$, but in others the cut-off is needed at intermediate stages and may be removed only at the end [38].

There is a growing literature on explicit calculations of correlation functions following the proposal of [4, 5]. In these notes we will limit ourselves to a brief discussion of the 2-point functions. Their calculations show that the dimensions of gauge invariant operators are determined by the masses of the corresponding fields in AdS_5 [4, 5]. For scalar operators this relation is

$$\Delta = 2 + \sqrt{4 + (mL)^2}\ . \tag{21}$$

Therefore, the operators in the $\mathcal{N} = 4$ large N SYM theory naturally break up into two classes: those that correspond to the Kaluza-Klein states of supergravity and those that correspond to massive string states. Since the radius of the S^5 is L, the masses of the Kaluza-Klein states are proportional to $1/L$. Thus, the dimensions of the corresponding operators are independent of L and therefore independent of $g_{\rm YM}^2 N$. On the gauge theory side this is explained by the fact that the supersymmetry protects the dimensions of certain operators from being renormalized: they are completely determined by the representation under the superconformal symmetry [39, 40]. All families of the Kaluza-Klein states, which correspond to such BPS protected operators, were classified long ago [35].

On the other hand, the masses of string excitations are $m^2 = \frac{4n}{\alpha'}$ where n is an integer. For the corresponding operators the formula (21) predicts that the dimensions do depend on the 't Hooft coupling and, in fact, blow up for large $g_{\rm YM}^2 N$ as $2\left(n g_{\rm YM}\sqrt{2N}\right)^{1/2}$. This is a highly non-trivial prediction of the AdS/CFT duality which has not yet been verified on the gauge theory side.

It is often stated that the gauge theory lives on the boundary of AdS_5. A more precise statement is that the gauge theory corresponds to the entire AdS_5, with the effective energy scale measured by the radial coordinate. In this correspondence the bare (UV) quantities in the gauge theory are indeed specified at the boundary of AdS_5. In calculating the correlation functions it is crucial that the boundary values of various fields in AdS_5 act as the sources in the gauge theory action which couple to gauge invariant operators as in (5). A similar connection arises in the calculation of Wilson loop expectation values [41]. A Wilson loop is specified by a contour in x^λ space placed at $z = z_{cutoff}$. One then looks for a minimal area surface in AdS_5 bounded by this contour and evaluates the Nambu action I_0 which is proportional to the area. The semiclassical value of the Wilson loop is then e^{-I_0}. This prescription, which is motivated by the duality between fundamental strings and electric flux lines, gives interesting results which are consistent with the conformal invariance [41]. For example, the quark-antiquark potential scales as $\sqrt{g^2_{\rm YM}N}/|\vec{x}|$. Note that this strong coupling result is different from the weak coupling limit where we have $V \sim g^2_{\rm YM}N/|\vec{x}|$.

3.2 CONFORMAL FIELD THEORIES AND EINSTEIN MANIFOLDS

As we mentioned above, the duality between strings on $AdS_5 \times S^5$ and the $\mathcal{N} = 4$ SYM is naturally generalized to dualities between strings on $AdS_5 \times X_5$ and other conformal gauge theories. The 5-dimensional compact space X_5 is required to be a postively curved Einstein manifold, i.e. one for which $R_{\mu\nu} = \Lambda g_{\mu\nu}$ with $\Lambda > 0$. The number of supersymmetries in the dual gauge theory is determined by the number of Killing spinors on X_5.

The simplest examples of X_5 are the orbifolds S^5/Γ where Γ is a discrete subgroup of $SO(6)$ [42, 43]. In these cases X_5 has the local geometry of a 5-sphere. The dual gauge theory is the IR limit of the world volume theory on a stack of N D3-branes placed at the orbifold singularity of R^6/Γ. Such theories typically involve product gauge groups $SU(N)^k$ coupled to matter in bifundamental representations [44].

Constructions of the dual gauge theories for Einstein manifolds X_5 which are not locally equivalent to S^5 are also possible. The simplest example is the Romans compactification on $X_5 = T^{1,1} = (SU(2) \times SU(2))/U(1)$ [32, 45]. It turns out that the dual gauge theory is the conformal limit of the world volume theory on a stack of N D3-branes placed at the singularity of a certain Calabi-Yau manifold known as the conifold. This turns out to be the $\mathcal{N} = 1$ superconformal field theory with gauge group $SU(N) \times SU(N)$ coupled to two chiral superfields in the $(\mathbf{N}, \overline{\mathbf{N}})$ representation and two chiral superfields in the $(\overline{\mathbf{N}}, \mathbf{N})$ representation [45]. This theory has an exactly marginal quartic superpotential which produces a critical line related to the radius of $AdS_5 \times T^{1,1}$.

4 TOWARDS NON-CONFORMAL GAUGE THEORIES IN FOUR DIMENSIONS

In the preceding sections I hope to have convinced the reader that type IIB strings on $AdS_5 \times X_5$ shed genuinely new light on four-dimensional conformal gauge theories. While many insights have already been achieved, I am convinced that a great deal remains to be learned in this domain. We should not forget, however, that the prize question is whether this duality can be extended to QCD or at least to other gauge theories which exhibit the asymptotic freedom and confinement. It is immediately clear that this will not be easy because, as we remarked in section 3, a string approach to weakly coupled gauge theory has not yet been fully developed (the well-understood supergravity limit describes gauge theory with very large 't Hooft coupling). On the other hand, the asymptotic freedom makes the coupling approach zero in the UV region [46]. Nevertheless, there may be some at least qualitative approaches to non-conformal gauge theories that shed light on the essential physical phenomena.

One such approach, proposed by Witten [47], builds on the observation that thermal gauge theories are described by near-extremal p-brane solutions [20, 21]. It is also known that the high temperature limit of a supersymmetric gauge theory in $p+1$ dimensions is described by non-supersymmetric gauge theory in p dimensions. Thus, 3-dimensional non-supersymmetric gauge theory is dual to the throat region of the near-extremal 3-brane solution which turns out to have the geometry of a black hole in AdS_5 [47] (similar black holes were studied long ago by Hawking and Page [48]). Similarly, 4-dimensional non-supersymmetric gauge theory is dual to the near-horizon region of the near-extremal 4-brane solution [47]. Witten calculated the Wilson loop expectation values in these geometries and showed that they satisfy the area law. Furthermore, calculations of the glueball masses produce discrete spectra with strong resemblance to the lattice simulations [49]. Unfortunately, this supergravity model has some undesirable features as well: for example, the presence in the geometry of a large $8-p$ dimensional sphere introduces into the spectrum families of light "Kaluza-Klein glueballs" which are certainly absent from the lattice results. Presumably, the root of the problems is that the bare 't Hooft coupling is taken to be large, while in order to achieve the conventional continuum limit it has to be sent to zero along a renormalization group trajectory.

A pessimistic conclusion would be that little more can be done at present because the supergravity approximation is supposed to be poor at weak 't Hooft coupling. Nevertheless, I feel that one should not give up attempts to understand the asymptotic freedom on the string side of the duality. In fact, some progress in this direction was recently achieved in [50, 51, 52] following Polyakov's suggestion [53] on how to break supersymmetry. Polyakov argued that a string dual of non-supersymmetric gauge theory should have world sheet supersymmetry without space-time supersymmetry. Examples of such theories include the type 0 strings, which are NSR strings with a non-chiral GSO projection which breaks the space-time supersymmetry [54].

There are two type 0 theories, A and B, and both have no space-time fermions in their spectra but produce modular invariant partition functions [54]. The massless bosonic fields are as in the corresponding type II theory (A or B), but with the doubled set of the Ramond-Ramond (R-R) fields. The type 0 theory also contains a tachyon, which is why it has not received much attention thus far. In [53, 50] it was suggested, however, that the presence of the tachyon does not spoil its application to large N gauge theories. A well-established route towards gauge theory is via the D-branes which were first considered in the type 0 context in [55]. Large N gauge theories, which are constructed on N coincident D-branes of type 0 theory, may be shown to contain no open string tachyons [50, 53].

In [50] the presence of a bulk tachyon was turned into an advantage because it gives rise to the renormalization group flow. There the $3+1$ dimensional $SU(N)$ theory coupled to 6 adjoint massless scalars was constructed as the low-energy description of N coincident electric D3-branes of type 0B theory.[6] The conjectured dual type 0 background thus carries N units of electric 5-form flux. The dilaton decouples from the $(F_5)^2$ terms in the effective action, and the only source for it originates from the tachyon mass term,

$$\nabla^2\Phi = \tfrac{1}{8}m^2 e^{\frac{1}{2}\Phi}T^2 , \qquad m^2 = -\tfrac{2}{\alpha'} . \tag{22}$$

Thus, the tachyon background induces a radial variation of Φ. Since the radial coordinate is related to the energy scale of the gauge theory, the effective coupling decreases toward the ultraviolet. In [51, 52] the UV limit of the type 0B background dual to the gauge theory was studied in more detail and a solution was found where the geometry is $AdS_5 \times S^5$ while the 't Hooft coupling flows logarithmically. A calculation of the quark-antiquark potential showed qualitative agreement with what is expected in an asymptotically free theory.

These results raise the hope that the AdS/CFT duality can indeed be generalized to asymptotically free gauge theories. While we are still far from constructing reliable string duals of such theories, the availability of new ideas on this old and difficult problem makes me hopeful that more surprises lie ahead.

ACKNOWLEDGMENTS

I am grateful to S. Gubser, A. Peet, A. Polyakov, A. Tseytlin and E. Witten, my collaborators on parts of the material reviewed in these notes. I also thank B. Kursunoglu and other organizers of Orbis Scientiae '98, especially L. Dolan (the convener of the string session), for sponsoring a very interesting conference. This work was supported in part by the NSF grant PHY-9802484 and by the James S. McDonnell Foundation Grant No. 91-48.

REFERENCES

[1] Y. Nambu, "Quark model and the factorization of the Veneziano amplitude," in *Symmetries and Quark Models*, ed. R. Chand, Gordon and Breach (1970); H. B. Nielsen, "An almost physical interpretation of the integrand of the n-point Veneziano amplitude," submitted to the 15th International Conference on High Energy Physics, Kiev (1970); L. Susskind, "Dual-symmetric theory of hadrons," *Nuovo Cim.* **69A** (1970) 457.

[2] J. Scherk and J. Schwarz, "Dual models for non-hadrons," *Nucl. Phys.* **B81** (1974) 118.

[3] J. Maldacena, "The Large N limit of superconformal field theories and supergravity," *Adv. Theor. Math. Phys.* **2** (1998) 231, hep-th/9711200.

[6] In the type 0B theory the 5-form R-R field strength F_5 is not constrained to be selfdual. Therefore, it is possible to have electrically or magnetically charged 3-branes. This should be contrasted with the situation in the type IIB theory where the 5-form strength is constrained to be selfdual and, thus, only the selfdual 3-branes are allowed.

[4] S.S. Gubser, I.R. Klebanov, and A.M. Polyakov, "Gauge theory correlators from noncritical string theory," *Phys. Lett.* **B428** (1998) 105, hep-th/9802109.

[5] E. Witten, "Anti-de Sitter space and holography," *Adv. Theor. Math. Phys.* **2** (1998) 253, hep-th/9802150.

[6] G. 't Hooft, "A planar diagram theory for strong interactions," *Nucl. Phys.* **B72** (1974) 461.

[7] See, for example, A.M. Polyakov, "Gauge Fields and Strings," Harwood Academic Publishers (1987).

[8] A.M. Polyakov, "String theory and quark confinement," *Nucl. Phys. B (Proc. Suppl.)* **68** (1998) 1, hep-th/9711002.

[9] A.M. Polyakov, "Quantum geometry of bosonic strings," *Phys. Lett.* **B103** (1981) 207.

[10] For a review, see J. Polchinski, "TASI lectures on D-branes," hep-th/9611050.

[11] J. Polchinski, "Dirichlet branes and Ramond-Ramond charges," *Phys. Rev. Lett.* **75** (1995) 4724, hep-th/9510017.

[12] C.M. Hull and P.K. Townsend, "Unity of superstring dualities," *Nucl. Phys.* **B438** (1995) 109; P.K. Townsend, "The eleven-dimensional supermembrane revisited," *Phys. Lett.* **B350** (1995) 184; E. Witten, "String theory dynamics in various dimensions," *Nucl. Phys.* **B443** (1995) 85.

[13] E. Witten, "Bound states of strings and p-branes," *Nucl. Phys.* **B460** (1996) 335, hep-th/9510135.

[14] T. Banks, W. Fischler, S. Shenker and L. Susskind, "M Theory as a matrix model: a conjecture," *Phys. Rev.* **D55** (1997) 5112, hep-th/9610043.

[15] A. Strominger and C. Vafa, "Microscopic origin of the Bekenstein-Hawking entropy," *Phys. Lett.* **B379** (1996) 99 , hep-th/9601029.

[16] C.G. Callan and J.M. Maldacena, "D-brane approach to black hole quantum mechanics," *Nucl. Phys.* **B472** (1996) 591, hep-th/9602043.

[17] G. Horowitz and A. Strominger, "Counting states of near-extremal black holes," *Phys. Rev. Lett.* **77** (1996) 2368, hep-th/9602051.

[18] A. Dhar, G. Mandal and S. Wadia, "Absorption vs. decay of black holes in string theory and T symmetry," *Phys. Lett.* **B388** (1996) 51; S. Das and S. Mathur, "Comparing decay rates for black holes and D-branes," *Nucl. Phys.* **B478** (1996) 561.

[19] J.M. Maldacena and A. Strominger, "Black hole grey body factors and D-brane spectroscopy," *Phys. Rev.* **D55** (1997) 861.

[20] S.S. Gubser, I.R. Klebanov, and A.W. Peet, "Entropy and temperature of black 3-branes," *Phys. Rev.* **D54** (1996) 3915, hep-th/9602135; A. Strominger, unpublished.

[21] I.R. Klebanov and A.A. Tseytlin, "Entropy of near-extremal black p-branes," *Nucl. Phys.* **B475** (1996) 164, hep-th/9604089.

[22] I.R. Klebanov, "World volume approach to absorption by nondilatonic branes," *Nucl. Phys.* **B496** (1997) 231, hep-th/9702076.

[23] S.S. Gubser, I.R. Klebanov, and A.A. Tseytlin, "String theory and classical absorption by three-branes," *Nucl. Phys.* **B499** (1997) 217, hep-th/9703040.

[24] S.S. Gubser and I.R. Klebanov, "Absorption by branes and Schwinger terms in the world volume theory," *Phys. Lett.* **B413** (1997) 41, hep-th/9708005.

[25] G.T. Horowitz and A. Strominger, "Black strings and p-branes," *Nucl. Phys.* **B360** (1991) 197.

[26] M.J. Duff and J.X. Lu, "The selfdual type IIB superthreebrane," *Phys. Lett.* **B273** (1991) 409; M.J. Duff, R. Khuri, J.X. Lu, "String solitons," *Phys. Rept.* **259** (1995) 213.

[27] G.W. Gibbons and P.K. Townsend, "Vacuum interpolation in supergravity via super p-branes," *Phys. Rev. Lett.* **71** (1993) 3754, hep-th/9307049.

[28] S.S. Gubser, I.R. Klebanov, and A.A. Tseytlin, "Coupling constant dependence in the thermodynamics of N = 4 supersymmetric Yang-Mills theory," *Nucl. Phys.* **B534** (1998) 202, hep-th/9805156.

[29] A. Fotopoulos, T.R. Taylor, "Comment on two loop free energy in $N = 4$ supersymmetric Yang-Mills theory at finite temperature," hep-th/9811224.

[30] S. Hyun, "U-duality between Three and Higher Dimensional Black Holes," hep-th/9704005;
K. Sfetsos and K. Skenderis, "Microscopic derivation of the Bekenstein-Hawking entropy formula for non-extremal black holes," hep-th/9711138.

[31] M. Günaydin, L.J. Romans and N. Warner, "Compact and non-compact gauged supergravity in five dimensions," *Nucl. Phys.* **B272** (1986) 598.

[32] L. Romans, "New compactifications of chiral $N = 2$, $d = 10$ supergravity," *Phys. Lett.* **B153** (1985) 392.

[33] M.J. Duff, B.E.W. Nilsson and C.N. Pope, "Kaluza-Klein supergravity," *Phys. Rep.* **130** (1986) 1.

[34] J.H. Schwarz, "Covariant field equations of chiral $N = 2, D = 10$ supergravity," *Nucl. Phys.* **B226** (1983) 269.

[35] H.J. Kim, L.J. Romans and P. van Nieuwenhuizen, "Mass spectrum of chiral ten dimensional $\mathcal{N} = 2$ supergravity on $\mathbf{S}^5$," *Phys. Rev.* **D32** (1985) 389; M. Günaydin and N. Marcus, "The spectrum of the S^5 compactification of the chiral $N = 2, D = 10$ supergravity and the unitary supermultiplets of $U(2,2|4)$," *Class. Quant. Grav.* **2** (1985) L11.

[36] R.R. Metsaev and A.A. Tseytlin, "Type IIB superstring action in $AdS_5 \times S^5$ background," *Nucl. Phys.* **B533** (1998) 109, hep-th/9805028; R. Kallosh and A.A. Tseytlin, "Simplifying superstring action on $AdS_5 \times S^5$," *J. High Energy Phys.* **9810** (1988) 016, hep-th/9808088.

[37] L. Susskind and E. Witten, "The holographic bound in Anti-de Sitter space," hep-th/9805114.

[38] D.Z. Freedman, S. Mathur, A. Matusis and L. Rastelli, "Correlation functions in the $\mathrm{CFT}_D/\mathrm{AdS}_{D+1}$ correspondence," hep-th/9804058.

[39] P.S. Howe and P.C. West, "Superconformal invariants and extended supersymmetry," *Phys. Lett.* **B400** (1997) 307.

[40] S. Ferrara, C. Fronsdal and A. Zaffaroni, "On $N = 8$ supergravity on AdS_5 and $N = 4$ superconformal Yang-Mills theory," *Nucl. Phys.* **B532** (1998) 153, hep-th/9802203.

[41] J. Maldacena, "Wilson loops in large N field theories," *Phys. Rev. Lett.* **80** (1998) 4859, hep-th/9803002; S.-J. Rey and J. Yee, "Macroscopic strings as heavy quarks in large N gauge theory and anti-de Sitter supergravity", hep-th/9803001.

[42] S. Kachru and E. Silverstein, "4d conformal field theories and strings on orbifolds," hep-th/9802183.

[43] A. Lawrence, N. Nekrasov and C. Vafa, "On conformal field theories in four dimensions," hep-th/9803015.

[44] M.R. Douglas and G. Moore, "D-branes, quivers, and ALE instantons," hep-th/9603167.

[45] I.R. Klebanov and E. Witten, "Superconformal field theory on threebranes at a Calabi-Yau singularity," *Nucl. Phys.* **B536** (1998) 199, hep-th/9807080.

[46] D.J. Gross and F. Wilczek, "Ultraviolet behavior of nonabelian gauge theories," *Phys. Rev. Lett.* **30** (1973) 1343; H.D. Politzer, "Reliable perturbative results for strong interactions?" *Phys. Rev. Lett.* **30** (1973) 1346.

[47] E. Witten, "Anti-de Sitter space, thermal phase transition, and confinement in gauge theories," *Adv. Theor. Math. Phys.* **2** (1998) 505, hep-th/9803131.

[48] S.W. Hawking and D. Page, "Thermodynamics of black holes in anti de Sitter space," *Commun. Math. Phys.* **87** (1983) 577.

[49] C. Csaki, H. Ooguri, Y. Oz and J. Terning, "Glueball mass spectrum from supergravity," hep-th/9806021; R. De Mello Koch, A. Jevicki, M. Mihailescu and J. Nunes, "Evaluation of glueball masses from supergravity," hep-th/9806125; A. Hashimoto and Y. Oz, "Aspects of QCD dynamics from string theory," hep-th/9809106.

[50] I. R. Klebanov and A. A. Tseytlin, "D-Branes and dual gauge theories in type 0 strings," hep-th/9811035.

[51] J. Minahan, "Glueball mass spectra and other issues for supergravity duals of QCD models," hep-th/9811156.

[52] I. R. Klebanov and A. A. Tseytlin, "Asymptotic freedom and infrared behavior in the type 0 string approach to gauge theory," hep-th/9812089.

[53] A.M. Polyakov, "The wall of the cave," hep-th/9809057.

[54] L. Dixon and J. Harvey, "String theories in ten dimensions without space-time supersymmetry", *Nucl. Phys.* **B274** (1986) 93; N. Seiberg and E. Witten, "Spin structures in string theory", *Nucl. Phys.* **B276** (1986) 272; C. Thorn, unpublished.

[55] O. Bergman and M. Gaberdiel, "A non-supersymmetric open string theory and S-Duality," *Nucl. Phys.* **B499** (1997) 183, hep-th/9701137.

INDEX